Metal Toxicity in Plants: Perception, Signaling and Remediation

Dharmendra K. Gupta • Luisa M. Sandalio
Editors

Metal Toxicity in Plants: Perception, Signaling and Remediation

Springer

Editors
Dharmendra K. Gupta
Bioquímica, Biología Celular y Molecular de Plantas
Estación Experimental Del Zaidín
CSIC
Apartado 419
E-18008 Granada
Spain
guptadk1971@gmail.com

Luisa M. Sandalio
Bioquímica, Biología Celular y Molecular de Plantas
Estación Experimental Del Zaidín
CSIC
Apartado 419
E-18008 Granada
Spain
luisamaria.sandalio@eez.csic.es

ISBN 978-3-642-22080-7 e-ISBN 978-3-642-22081-4
DOI 10.1007/978-3-642-22081-4
Springer Heidelberg Dordrecht London New York

Library of Congress Control Number: 2011937548

Printed on acid-free paper

Springer is part of Springer Science+Business Media (www.springer.com)

Preface

The extensive increase of world population and industrial management has produced numerous environmental problems such as pollution (e.g. water, air, soil, noise and radiation), accumulation of heavy metals in soil and reduction in water quality. These facts can produce severe deterioration of natural resources, disturbance of ecosystems and affect human health. The term "heavy metal" refers to metallic elements with a high specific gravity (more than 5) or density which are very toxic even at very low concentrations. Some of these elements are referred as the trace elements, including iron (Fe), copper (Cu), manganese (Mn), molybdenum (Mo), cobalt (Co) and zinc (Zn), which are essential for biological systems in small quantities by participating in redox reactions and acting as enzyme cofactors (Sanitá di Toppi and Gabbrielli 1999). However, these metals can be toxic at high concentrations. Other heavy metals, such as cadmium (Cd), mercury (Hg), lead (Pb), aluminum (Al) or arsenic (As), have no function as nutrients and are very toxic to plants, animals and humans. The toxicity of these metals is based on their chemical properties which allow them to promote the production of reactive oxygen species (ROS), inactivation of enzymes, basically by reaction with SH-groups, and displacement of other cations or metals from proteins (Sanitá di Toppi and Gabbrielli 1999).

Heavy metals appear in the environment through natural sources or by anthropogenic activities such as mining, fossil fuel combustion, phosphate fertilizers used in agriculture and metal-working industries (Clemens 2006). These human activities have produced a severe environmental concern in some parts of the world because of the contamination by metals in day-to-day life, which can even compromise the health of future generations, due to the persistence of the metals in the environment by their bioaccumulation through the food chain (Clemens 2006). Tolerance to heavy metals in plants may be defined as the ability to survive in a soil that is toxic to other plantss and is manifested by an interaction between the genotype and its environment (McNair et al. 2000). Some plants have developed resistance to high metal concentrations, basically by two mechanisms, avoidance and tolerance. The first mechanism involved exclusion of metals outside the roots, and the second mechanism consists basically in complexing the metals to avoid protein and enzyme inactivation. Some plants can also accumulate metals in their tissues at concentrations higher than those found in the soil, and these plants as referred as hyperaccumulator. Most hyperaccumulator plant species belongs to Brassicaceae family. Heavy metal hyperaccumulation in plants is due to a combination of metal transporters and chelator molecules. Chelation of metals in cytosols by high affinity ligands is potentially a very important mechanism of heavy metal detoxification and tolerance. Potential ligands include amino acids, nicotianamine, phytochelatins and metallothioneins (Clemens 2001). Phytochelatins have been the most widely studied in plants with a general structure $(\gamma\text{-Glu Cys})_n\text{-Gly}$ where $n = 2\text{–}11$, and are rapidly induced in plants by heavy metal treatments (Rauser 1995). Hyperaccumulation can be exploited as a very useful tool to clean contaminated soils, water and sediments by the process called phytoremediation which essentially uses green plants to clean-up contaminants.

During the last two decades, ROS has gain importance in different aspects of heavy metal stress. Under physiological conditions, there is a balance between production and scavenging of ROS in all cell compartments. However, this balance could be perturbed by a number of adverse environmental factors. One of the major consequences of heavy metal action is enhanced production of ROS giving rise to damage to membranes, nucleic acids, and proteins (Halliwell and Gutteridge 2000). However, ROS are double-faced molecules acting as signal molecules regulating a large gene network in response against biotic and abiotic stress. On the other hand, nitric oxide (NO) also gained much importance in the last decade, as basically NO is a gaseous reactive molecule with a pivotal signaling role in many developmental and cell response processes (Besson-Bard et al. 2008). Recently, an increasing number of studies have been reported on the effects of NO alleviating toxicity of heavy metal including Cd and As (Xiong et al. 2010). Changes in the levels of both molecules are associated in the perception of stress and can trigger the defence cellular responses against adverse environmental conditions. In plants, hormones also play a critical role in the regulation of growth/development and modulation in plant responses against stresses. ROS and plant hormones interplay in the regulation of those processes, although the mechanisms involved are not well known in most cases.

The number of publications focused on heavy metal toxicity in plants has been growing exponentially in the last decade. The purpose of this book is to present the most recent advances in this field, mainly on the uptake and transport of heavy metals in plants, mechanisms of toxicity, perception of metals and the regulation of cell responses under metal stress. Another key feature of this book is related to the studies in recent years on signaling and remediation processes taking advantage of recent technological advances including "omic" approaches. Transcriptomic, proteomic and metabolomic studies have become very important tools to analyze the dynamics of changes in gene expression, and the profiles of protein and metabolites under heavy metal stress. This information is also very useful to draw the complex signaling and metabolic network induced by heavy metals in which hormones and reactive oxygen species also have an important role. Understanding the mechanism involved in sequestration and hyperaccumulation is very important in order to develop new strategies of phytoremediation are reviewed in several chapters of this book. The information included in this book will bring very stimulating insights into the mechanism involved in the regulation of plant response to heavy metals, which in turn will contribute to improving our knowledge of cell regulation under metal stress and the use of plants for phytoremediation.

The editors are grateful to the authors for contributing their time, knowledge and enthusiasm to bring this book into being.

Granada, Spain

Dr. Dharmendra Kumar Gupta
Dr. Luisa Maria Sandalio

Reference

Besson-Bard A, Pugin A, Wendehenne D (2008) New insights into nitric oxide signalling in plants. Annu Rev Plant Biol 59: 21–39

Clemens S (2001) Molecular mechanisms of plant metal tolerance and homeostasis. Planta 212: 475–486

Clemens S (2006) Toxic metal accumulation, responses to exposure and mechanisms of tolerance in plants. Biochimie 88: 1707–1719

Halliwell B, Gutteridge JMC (1989) Free radicals in biology and medicine, 3rd edn. Clarendon Press, Oxford

McNair MR, Tilstone GH, Smith SS (2000) The genetics of metal tolerance and accumulation in higher plants. In Terry N and Banuelos G (ed) Phytoremediation of contaminated soil and water. Luis, Boca Raton, pp 235–250

Nriagu JO, Pacyna JM (1988) Quantitative assessment of worldwide contamination of air, water and soils with trace metals. Nature 333:134–139

Rauser WE (1995) Phytochelatins and related peptides: structure, biosynthesis, and function. Plant Physiol 109: 1141–1149

Sanitá di Toppi L, Gabbrielli R (1999) Response to cadmium in higher plants. Environ Exp Bot 41:105–130

Xiong J, Fu G, Tao L, Zhu C (2010) Roles of nitric oxide in alleviating heavy metal toxicity in plants. Arch Biochem Biophy 497: 13–20

Contents

Heavy Metal Bindings and Their Interactions with Thiol Peptides and Other Biological Ligands in Plant Cells

Mashiro Inouhe, Huagang Huang, Sanjay Kumar Chaudhary, and Dharmendra Kumar Gupta

Abstract Plants have developed their potentials for uptake, transport and accumulation of terrestrial elements in order to coordinate their developmental and life-cycle performance. The utilization and toxicity of the metallic elements in plants are principally based on their own chemical properties in water and the interaction with their counterpart anions and cooperative molecules. Biochemical partners of the metals are various organic ligands composed of C, H, O, N, P, or S. Their roles are shared by two cell sites – the outside *apoplast* and the inside *symplast*. The apoplast equips the polymeric ligands of polysaccharides, phenolics, and proteins with carboxylic and some other functional groups capable of conjugating metals in the cell surfaces, but excess heavy metals in the primary cell wall are toxic to plants. Mobile organics in the apoplast have another function in xylem transport or biological interactions in the rhizosphere underground. The symplast (and vacuole) contains a variety of organic ligands such as organic acids, amino acids, polyamines, nicotianamine, phytates, soluble phenolics, and thiol-peptides called *cadystins* or *phytochelatins* (PCs). These can bind most heavy metals to make the

M. Inouhe (✉)
Department of Biology and Environmental Sciences, Graduate School of Science and Engineering, Ehime University, 790-8577 Matsuyama, Ehime, Japan
e-mail: inouhe@sci.ehime-u.ac.jp

H. Huang
Ministry of Education key laboratory of Environmental Remediation and Ecosystem Health, Zhejiang University, Hangzhou 310029, China
e-mail: huanghg82@gmail.com

S.K. Chaudhary
Department of Botany, University of Lucknow, Lucknow 226007, India
e-mail: sanjay_env@rediffmail.com

D.K. Gupta
Departamento de Bioquimica, Biologia Cellular y Molecular de Plantas, Estacion Experimental Del Zaidin, CSIC, Granada 18008, Spain
e-mail: guptadk1971@gmail.com

D.K. Gupta and L.M. Sandalio (eds.), *Metal Toxicity in Plants: Perception, Signaling and Remediation*, DOI 10.1007/978-3-642-22081-4_1,

lesser toxic binding forms and hence affecting their movements, transports, accumulations and their final fates in vivo in plants. PCs have the general structure of (γ-glutamyl-cysteinyl)$_n$-glycine ($n = 2$–11) and they are synthesized from glutathione ($n = 1$). The PC–metal conjugates are formed in the cytoplasm and transported to vacuole to make more stable complex mixtures with inorganic sulfur (S^{2-}). By contrast, little evidence supports the idea that PCs have a central role in xylem transport or the immobilization in shoots of heavy metals. Hyperaccumulators of Cd, Zn, Ni or As have a feature to carry out massive transport of them from root to shoot using other prevailing O- or N-bond ligands, besides the ability to form PCs. These suggest that the distinctive mechanisms for metal transports through the xylem sap system may be established independently of the PC-detoxification mechanism in the roots. Intentional and practical readjustment of the PC-dependent versus PC-independent systems in situ can improve the relative efficiency of the heavy metal mobility to shoot sites and the total accumulation capacity in the vascular plants.

1 Introduction

As a consequence of the industrial revolution there is an enormous and increasing demand for heavy metals that leads to highly anthropogenic emission into the biosphere (Ayres 1992). Apart from some emissions into the atmosphere in the form of dust particles or gases, these heavy metals stay largely in the aquatic and soil phases of this planet. Contamination also occurs extensively or locally even under natural environmental conditions where there are no directly connected human activities. Heavy metal pollution of environment is one of major ecological concern because of its impact on human health through the food chain and its high persistence in the environment (Piechalak et al. 2002). Meanwhile, various species of plants are very useful for cleaning up the metal-contaminated soil or water as a very eco-friendly technique called phytoremediation. This technology based on the potential and capacity of plants capable of accumulating heavy metals to shoot sites via root with no remarkable metabolic impediment or growth retardation of the organs. Here, what is required is an understanding of the plant mechanisms: how the plant neutralizes the toxic metals in roots (detoxification mechanism), how it transports them from roots to shoot (transport mechanism), and how it stores or fixes them stably in a special shoot sites (accumulation/immobilization mechanism), otherwise discharge or elimination will occur. All these mechanisms are closely connected to the problem of which biological ligands are bio-synthesized, co-transported, and further utilized for the respective metals in plants (Fig. 1).

The tolerance characteristics of plants to heavy metal ions are diverse among the metal ions involved (Foy et al. 1978; Woolhouse 1983; Verkleij and Schat 1990). Especially a group of metals called "Borderline class" metals including Mn, Zn, Fe, Ni, Cd, Pb and Cu etc. are capable of binding to multiple types of naturally occurring chemicals or components in plants, although "Class A" metals, such as K, Ca, Na, Mg, Al, and Cs prefer the O-donor ligands, all of which bind through

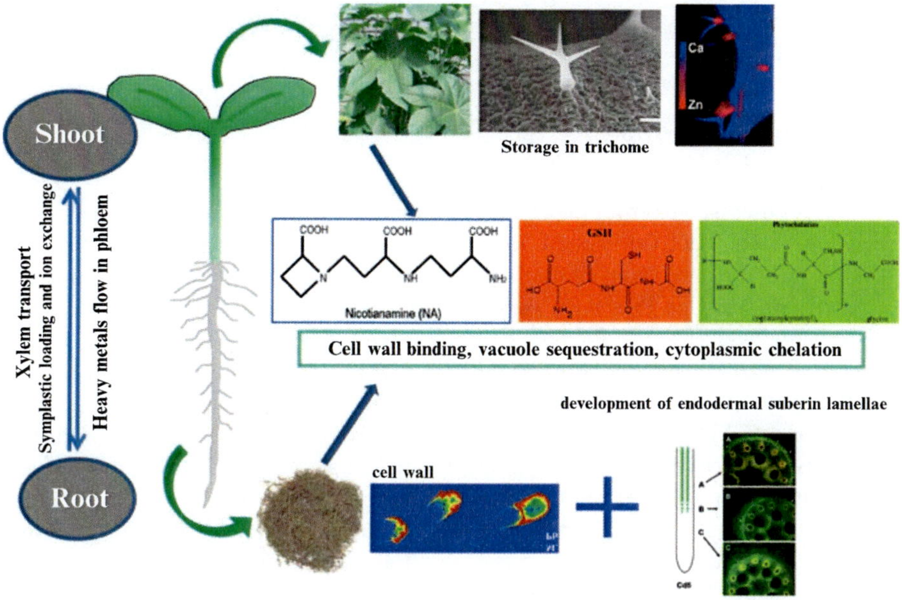

Fig. 1 Simplified scheme involved in heavy metal accumulation and homeostasis in plants

oxygen (–COOH, –H_2PO_4, –OH, –CHO etc.), rather than the S- or N-bond ligands (–SH, –SS–, –NH_2, =NH etc.) preferred by "Class B" metals (Woolhouse 1983). Nevertheless, the tolerance against those toxic ions can be expressed in a highly specific manner for each metal in general in plants, and co-tolerance appears relatively rare (Hall 2002; Inouhe 2005). One of the fundamental bases of the mechanisms can be addressed to either the alteration of the metal-sensitive metabolism and structure or the development of new metal-sequestering principles within some cellular compartments (Mehra and Winge 1991). As for the latter detoxification mechanism, various types of metal-binding complexes have been identified from plants. Among them the best characterized are phytochelatins (PCs) and the related thiol-peptides. Details of the structures, biosynthesis, analytical methods, genetics and the other many aspects of them are available in many publications (Rauser 1995, 1999; Zenk 1996; Cobbett and Goldsbrough 2002; Inouhe 2005). Furthermore, a variety of other organic ligands capable of conjugating to various metals in vivo have been reported with their possible roles similar to or distinct from those of PCs in plants (Callahan et al. 2006; Sharma and Dietz 2006; Haydon and Cobbett 2007). Based on recent information, we here survey their biochemical characteristics and the possible functions in bindings, detoxification, transport and accumulation of representative heavy metals such as Cd, Zn, Cu, and Ni in plant cells. Next, their localization and distribution in different sites of the plant body including their consolidate bindings to polymeric ligands in the structures are compared to facilitate our understanding on the possible roles of PCs and non-PC ligands contained in them.

2 Biological Ligands for Heavy Metal Conjugation and Detoxification in Plant Cells

2.1 Phytochelatins

To protect themselves from the toxicity of metal ions, plant cells have developed a mechanism to inactivate metal ions thus preventing enzymatic and structural proteins (Kneer and Zenk 1992). This mechanism consists of the biosynthesis of a set of iso-peptides PCs with varying chain lengths such as (γ-Glu-Cys)$_n$-Gly; where $n = 2$–11 (Fig. 1). PCs (or *cadystins*) were first discovered in fission yeast *Schizosaccharomyces pombe* exposed to Cd (Murasugi et al. 1981) and then in many plants (Grill et al. 1989; Rauser 1995). PCs are formed directly from glutathione (GSH, a reduced form) by the activity of PC synthase (γ-Glu-Cys dipeptidyl transpeptidase: EC 2.3.2.15), in the last step of the following metabolic sequence: Glu + Cys $\rightarrow$ γ-Glu-Cys (γEC peptide) $\rightarrow$ γ-Glu-Cys-Gly (GSH) $\rightarrow$ PCs. The first and second steps of this sequence are mediated by γEC synthetase (EC 6.3.2.2) and GSH synthetase (EC 6.3.2.3), respectively. PC synthase (PCS) consists of 95,000 Mr tetramers of protein subunits and has a Km of 6.7 mM for GSH, and its activities to produce PCs are post-translationally regulated by a range of heavy metals and metalloids (Grill et al. 1989). This enzyme continues the reaction until the activating metal ions are chelated by the PCs formed, providing an auto-regulated mechanism of the PC biosynthesis in which the reaction products chelate the activating metals thereby terminating the reaction (Loeffler et al. 1989).

Since the first isolation of PC synthase gene (*PCS1*, *CAD2*) in 1999 (Clemens et al. 1999; Ha et al. 1999; Vatamaniuk et al. 1999), various PCS genes have been isolated from different species of plants and other organisms such as yeast, nematode, slime molds and cyanobacteria (Vatamaniuk et al. 2002; Tsuji et al. 2004; Pal and Rai 2010). The PCS activities have been detected in plants such as *Silene cucubalis* (Grill et al. 1989), *Arabidopsis* (Howden et al. 1995), *Pisum sativum* (Klapheck et al. 1995), *Cicer arietinum* (Gupta et al. 2002), and tomato (Chen et al. 1997), but not in azuki bean (Inouhe et al. 2000). In tomato, PCS activity was detected mainly in the roots and stems and not leaves or fruit (Chen et al. 1997), but the tissue-specific PCS expression or PC biosynthesis are not well understood in the other plants.

PCs play an important role in detoxification of various heavy metal ions in plants (Rauser 1995; Zenk 1996; Cobbett 2000). Chelation of heavy metals with PCs produced in cytoplasm and compartmentalization of the PC-metal complexes in vacuoles are generally considered as the "first line" of defence mechanisms by plants (Clemens 2006). PC synthesis can be stimulated in cells exposed to Cd and various other metals such as Cu, Zn, Pb and Ag, or metalloid As, and thePCs formed are capable of binding to all these ions via the sulfhydryl (−SH) and carboxyl (−COOH) residues (Grill et al. 1987). *Arabidopsis* mutants lacking enzymes involved in GSH synthesis (Howden and Cobbett 1992) or deficient in PCS activity

(Howden et al. 1995) were hypersensitive to Cd. Inhibition studies of PC biosynthesis via GSH using either mutants or inhibitor further demonstrated fundamental roles of PCs in the metal detoxification in yeast, fungi, green algae, aquatic plants, and many higher plants and their cell cultures (Inouhe 2005). In addition, overexpression of PCS genes efficiently increases the Cd-tolerance in plants as well as in yeast and bacteria. For example, transgenic plants of *Brassica juncea*, overexpressing GSH synthetase, γ-glutamylcysteine synthetase or PCS, are more tolerant to Cd stress (Zhu et al. 1999a, b; Wawrzyn'ski et al. 2006; Gasic and Korban 2007). However, there are exceptions to such a relationship. Firstly, some transgenic *Arabidopsis* lines overexpressing PCS are hypersensitive to Cd since these are probably depleted in GSH pools and thus more susceptible to Cd-induced oxidative stress (Li et al. 2004). The discrepancy suggests that the tolerance levels of plants to heavy metal toxicity may be correlated to the total levels or balance of "thiol" compounds in the cells (Cobbett and Goldsbrough 2002; Gupta et al. 2002). In yeast *Saccharomyces cerevisiae*, exposure of cells to Cd led to a global drop in sulfur-containing protein synthesis and in a redirection of sulfur metabolite fluxes towards the GSH pathway (Lafaye et al. 2005). More recently, simultaneous overexpression of GSH synthetase and PCS in *Arabidopsis* was found to increase the tolerance and accumulation of Cd and As (Guo et al. 2008), which also supports the need to maintain a proper balance of thiol metabolism under stress conditions. Secondly, besides the metabolic balance, transports of PC-metal conjugates from cytoplasm to vacuole are required for metal tolerance and accumulation in plant cells (Clemens 2006). In *B. juncea*, a change of expression of a GSH transporter BjGT1 in response to Cd exposure has been reported (Bogs et al. 2003) also indicating that GSH plays a prominent role in Cd accumulation and detoxification. ABC transporters have been identified in yeast and fission yeast that directly mediate the vacuolar transport of Cd complexes and thus are involved in the final step of Cd detoxification (Ortiz et al. 1995; Li et al. 1997). Recent analyses of AtMRPs, a subfamily of *Arabidopsis* ABC transporters, showed that AtMRP3 was induced by Cd and not by oxidative stress (Bovet et al. 2003), suggesting that ABC transporters in plants, as in yeast, are involved in heavy metal fluxes.

Massive PC production is accompanied by a coordinated transcriptional induction of biosynthesis of enzymes involved in sulfate uptake (Nocito et al. 2002; Herbette et al. 2006) and assimilation into Cys (Harada et al. 2001; Gupta et al. 2002; Weber et al. 2006) and GSH (Xiang and Oliver 1998; Wawrzyn'ski et al. 2006). This suggests the requirement for the reduced sulfur in the PC biosynthesis and heavy-metal responses of plants. Sulfur is taken up by roots and translocated to different organs through specific transporters on membranes and mainly in the apoplastic route. Sulfate transporters of Group 1 (e.g. SULTR1;1 and SULTR1;2) are the high-affinity transporters expressed primarily in roots of sulfur-starved plants and they function to overcome sulfur limiting conditions (Leustek 2002). Expression of Group 1 sulfate transporters is negatively regulated by cytokinins through their receptor gene *CRE1* (Maruyama-Nakashita et al. 2004). Thus, a decline in the cytokinin content (Veselov et al. 2003) may indirectly indicate increased expression of Group 1 sulfate transporters. Sulfate transporters from

Group 2 (e.g. SULTR2;1) are involved in xylem loading, while those of Group 4 (SULTR4;1 and SULTR4;2) are localized in vacuoles and chloroplasts (Leustek 2002) and thus may play an important role in transport of sulfate from roots to shoots and finally to chloroplasts, an organelle where major fraction of sulfate is assimilated to Cys after a series of reactions: sulfate + ATP → APS (adenosine 5′–phosphosulfate) → sulfite → sulfide → Cys. These four steps are mediated by ATP sulfurylase, APS reductase, ferredoxin-dependent sulfite reductase, and O-acetylserine (thiol) lyase, respectively. Then the synthesised Cys and GSH in the source organs are transported to roots and other sink organs by translocation and further used for PC formation.

The long-distance transports between source and sink organs are essential for the nutritional correlations in vascular plants. As a typical example, PCs might play a role in Cd transport from root to shoot demonstrating that a PC-dependent "overflow protection mechanism" would contribute to keeping Cd accumulation low in the root, causing extra Cd transport to the shoot (Gong et al. 2003). However, overexpression of *Arabidopsis* PCS in tobacco plants enhances Cd tolerance and accumulation but not its translocation to the shoot (Pomponi et al. 2006). Some levels of PCs are detected in phloem sap in rice (Kato et al. 2010) but not in xylem sap in *Arabidopsis halleri* (Ueno et al. 2008). Thus the special role of PCs in long-distance transport of heavy metals has not been fully substantiated in plants, especially hyper-accumulating species.

Chickpea roots are capable of forming a substantial level of thiol compounds that are apparently different from GSH and PCs, the major compounds identified are homo-phytochelatins (hPCs), consisting mainly of hPC_2 and hPC_3. These peptides are synthesized from homo-glutathione (hGSH) in response to Cd and As almost to the equivalent levels of PCs, but not to Cu, Zn, Ni and Co, suggesting that hPCs may have an important role in Cd and As-sequestering and signaling in chickpea roots (Gupta et al. 2002, 2004). Some other PC-related peptides were reported in different plant sources (Table 1). Although their physiological roles in the absence or presence of heavy metals are not well understood at present, PCs and PC-related peptides can be thought to have a role in the homeostasis and metabolism of essential metal ions in plants (Rauser 1999; Zenk 1996; Cobbett 2000). In vitro experiments have shown that PC-Cu and PC-Zn complexes could reactivate the apoforms of the copper-dependent enzyme diamino-oxidase and the Zn-dependent enzyme carbonic anhydrase, respectively (Thumann et al. 1991). In addition, roles for PCs in Fe or sulfur metabolism have also been proposed (Zenk 1996;

Table 1 Various PC-like peptides produced by plants and yeast

PC-related γ-EC peptides	Structure	Plant sources
Homophytochelatin	(γ -Glu-Cys)n-Ala	Leguminosae
Hydroxymethyl-PC	(γ -Glu-Cys)n-Ser	Gramineae
iso-Phytochelatin (Glu)	(γ -Glu-Cys)n-Glu	Maize
iso-Phytochelatin (Gln)	(γ -Glu-Cys)n-Gln	Horse radish
Desglycine phytochelatin	(γ -Glu-Cys)n	Maize, yeast

Adapted from Rauser (1995); Zenk (1996); Klapheck et al. (1995); Inouhe (2005)

Toppi and Gabbrielli 1999). PCs and PC-related peptides are thiol compounds functionally equivalent or superior to Cys and GSH. These are therefore biologically active compounds that function to prevent oxidative stress in plant cells (Gupta et al. 2010).

2.2 Organic Acids, Nicotianamine, Amino Acids, and Phytates

Organic acids (OAs) have been associated with metal hyperaccumulation and tolerance in a range of plant species and have been proposed as important cellular ligands for Zn, Cd and Ni (Salt et al. 1999; Kupper et al. 2004). The carboxylic acids known to be present in high concentrations in the cell vacuoles of photosynthetic tissues include citric, isocitric, oxalic, tartaric, malic, malonic and aconitic (Callahan et al. 2007). Many studies have implied that these acids play a role in hyperaccumulation (Rauser 1999; Salt et al. 1999; Romheld and Awad 2000; Chiang et al. 2006). Analysis of tissues from metal hyperaccumulator species using X-ray absorption techniques has identified OAs as the predominant ligands. By X-ray absorption spectrometry (XAS) and extended X-ray absorption fine structure (EXAFS) analysis, citrate was identified as the predominant ligand for Zn in leaves of *Thlaspi caerulescens* (Salt et al. 1999). Similarly, Ni-citrate accounted for one-quarter of the Ni species in leaves of the Ni hyperaccumulator *T. goesingense* and in the related nonaccumulator *T. arvense* (Kramer et al. 2000). The identification of the vacuole as the major subcellular compartment for Zn, Cd and Ni and the favoring of the formation of metal-OA complexes in the acidic environment of the vacuolar lumen suggest that citrate and malate are probably relevant only as ligands for these metals within vacuoles (Kramer et al. 2000; Ma et al. 2005).

Studies have demonstrated that the primary constituents of root exudates are low-molecular weight organic acids (LMWOAs) that play essential roles in making sparingly soluble soil Fe, P, and other metals available to growing plants (Romheld and Awad 2000). Acetic, lactic, glycolic, malic, maleic, and succinic acids were found in rhizosphere soils of tobacco and sunflower (Chiang et al. 2006). Concentrations of these LMWOAs exudates increased with increasing amendment of Cd concentrations in the rhizosphere soils. After the loss of H^+, each acid contains a COO^- group, which binds to the cations. Correlation coefficients between concentrations of Cd amendment versus LMWOAs exudates of tobacco and sunflower were 0.85 and 0.98, respectively (Chiang et al. 2006). Positive correlations have been found between external Zn and organic acid concentrations in the roots of hyperaccumulator plants *A. halleri* (Zhao et al. 2000). These results suggest that the different levels of LMWOAs present in the rhizosphere soil may play an important role in the solubilization of heavy metals that bind with soil particles into soil solution and followed by uptake by plants. However, this mechanism does not draw a sharp line between toxic and essential metals for uptake and further utilization. This role may be covered by other specific biological ligands or transporters in the root and shoot tissues.

Nicotianamine (NA), a non-proteinaceous amino acid synthesized in all plants by the condensation of three S-adenosyl-methionine molecules through the activity of the enzyme nicotianamine synthase (NAS), is ubiquitously present in higher plants (Fig. 1). It is known to be involved in chelation of metals such as Fe, Cu, Zn for their enhanced extraction by roots and/or transport to shoost, especially under mineral-deficient conditions (Takahasi et al. 2003; Mari et al. 2006). However, recent evidence supports their possible functions in heavy metal-tolerance and hyperaccumulation in plants. The hyperaccumulation of Zn and Cd is a constitutive property of the metallophyte *A. halerii*. Recently, Weber et al. (2004) have used *Arabidopsis* gene chips to identify those genes that are more active in roots of *A. halleri* than *A. thaliana* under controlled conditions. Two genes showing highest levels of expression in *A. halleri* roots code for a NAS and a putative Zn^{2+} uptake system. In addition, roots of *A. halleri* also show higher levels of both NA and NAS. *A. halleri* presents a 2-fold increase of its NA root content probably linked to the constitutive expression of the *AhNAS2* gene. Expression of NAS in *S. pombe* cells has demonstrated that formation of NA can confer Zn^{2+} tolerance. Taken together, these observations suggest active roles of NA in plant Zn homeostasis and NAS in hyperaccumulation of Zn in *A. halleri* (Weber et al. 2004). Recently, it was reported that the overexpression of *TcNAS* in *A. thaliana* transgenic plants also confers Ni resistance (Pianelli et al. 2005), strengthening the idea that NA could play a role in metal tolerance and hyperaccumulation.

Plant cells contain many other small organic ligands with variable functional groups, including amino acids, polyamines, nucleotides, phytates and other phosphate sugars. Of these, polyamines appear to act as a messenger or a molecule to stabilize or protect the cell membranes rather than as direct binding ligands to toxic heavy metals (Sharma and Dietz 2006). Nucleotides, phytates and sugar phosphates can conjugate to Ca, Mn, Mg, Al and other metals through their O-bonds. Especially, the importance of phytates in coordination and storage of phosphate and metals such as Zn, Mg, and K in vacuole and cytoplasm and also in the detoxification of Cd has been widely suggested (Van Steveninck et al. 1992; Hayden and Cobbett 2006). Amino acids are the most abundant amphoteric ions with variable forms and residues, existing in 10–100 mM orders of concentrations and serving multiple functions in plant cells. Cysteine (Cys) is a thiol compound that has a S-donor residue equivalent to a GSH molecule. However, its internal level does not usually exceed that of GSH or PCs, probably because of the restricted supply of total S available for it and its quick turnover and utilization for the other thiol ligands and proteins. Acidic amino acids, glutamic acid (Glu) and aspartic acid (Asp), provide an extra carboxyl group (−COOH), and their amides, glutamine (Gln) and asparagine (Asn), provide an acid amide group consisting of both O- and N-donors ($-CO\text{-}NH_2$). All these are generally rich in phloem sap, for example, at near 300 mM in cereals and 50 mM in some dicotyledonous plants (Oshima et al. 1990; Winter et al. 1992), and can be potential ligands for translocational metal cations. Histidine (His) is the most characterized imidazole (=NH)-containing amino acid that plays a central role in binding to and transport of Ni, especially in Ni-hyper-accumulating plants (Kramer et al. 2000; Callahan et al. 2006). Two His molecules can make a stable complex chelating to one Ni (Callahan et al. 2006). Furthermore, proline (Pro) has been most extensively studied for its unique and important function as a compatible solute in many plants affected by water-deficit and salinity stress, but interestingly, heavy metals such as Cu, Cd, Zn or Pb also significantly stimulate the accumulation and/or biosynthesis of Pro in many plants (Sharma and Dietz 2006). Possible roles of Pro as a direct N-donor ligand conjugating to heavy metals are not established as yet, but will be more attractive in combination with its role as osmotic protectant or antioxidant under complex conditions including salinity and drought stress.

As mentioned above, there are possible interactions between different soluble organic ligands and different metals in cytoplasm, vacuole and other apoplastic solutions in shoots and roots. These solutions also contain inorganic anions such as sulfate, phosphate, nitrate, borate, carbonate, chloride and silicate. These inorganic anions and counterpart cations affect the organic ligand's interactions with metals in each site at different but almost constant pH conditions (Callahan et al. 2006). Some bindings between metals and ligands are not specific and not stable, especially under varied pH and ion-strength conditions. Conversely the regulated conditions can promise a unique and established mechanism for metal transport and binding systems in land plants.

2.3 *Soluble Phenolics*

At the end of this section on the soluble form of metal-binding ligands, we introduce a unique but increasingly well-recognized example of phenolics. Phenolic compounds are derived mainly from trans-cinnamic acid, which is formed from L-phenylalanine in a reaction catalyzed by L-phenylalanine ammonia-lyase (PAL). These compounds are constitutively expressed in higher plants and can effectively prevent oxidative stress caused by unfavorable environmental factors. Since the levels of phenolics are affected sensitively by heavy metal accumulations, they are suitable candidates to act as biomarkers (Santiago et al. 2000). Such compounds can be used as early indicators of environmental stress on a target organism before morphological or ultrastructural damage occurs. They are also useful as cytological and biochemical indicators because they are compartmented as secondary metabolites at the different tissue- and sub-cellular levels in response to the environment, and the specified localization reflects their biochemical properties or roles in plants. In general, glycosides of phenolics are localized in hydrophilic regions of the cell such as vacuoles and apoplasts, while aglycones are localized in lipophilic regions (Sakihama et al. 2002). All these are also known as potent bio-ligands capable of binding or precipitating heavy metal ions in different cell sites. Furthermore, known insoluble phenolics such as lignin are localized in the cell wall especially more differentiated secondary cell walls in many plants and can perform as metal-accumulating polymeric ligands. Tong et al. (2004) have reported that compartmentation and the formation of complexes with phenol derivatives in the vacuole may be another example of the mechanisms of resistance to heavy metals. Precipitation of phenolics generally revealed a significant higher electron-opacity over all protoplasm in bilberry leaves collected in a polluted forest in comparison to leaves from an unpolluted locality (Bialonska et al. 2007). These results indicate that the distribution and properties of phenolics depend on the level of heavy metals accumulated in the cell and the phenolics accumulated in vacuoles and apoplasts may play a significant role in scavenging of free radicals produced in plant cells (Bialonska et al. 2007). In a herbaceous plant chamomile (*Matricaria chamomilla*), soluble phenolics in the root and leaf rosettes were elevated by high doses of Cu and Cd, whereby Cu had a more expressive effect in roots and Cd in leaf rosettes, respectively (Kovacik et al. 2008). Low doses of Cd and Cu did not affect soluble phenolics in either the leaf rosettes or the roots. Recently, Janas et al. (2010) suggested that higher phenolics accumulation in vacuoles and cell walls of lentil (*Lens culinaris* Medic.) seedlings treated with Cu ions might be involved in scavenging ROS produced in the Cu-treated plant cells. They also confirmed that the induction of phenolics in Cu-treated seedlings had an important role in the lentil root protection against this metal. The concentration of polyphenolic compounds (particularly isoflavonoids like genistein and genistein-(malonyl)-glucoside) was significantly higher for lupin (*Lupinus albus* L.) roots when grown in a 20-μM Cu solution as compared to the control, and these phenolic compounds can bind Cu ions (Jung et al. 2003). In addition, plants exposed to 20 and 62 μM Cu accumulated high Cu amounts in root cell walls whereas only low amounts reached the symplasm. Therefore, it is proposed that the complexation of Cu^{2+} in the rhizosphere and in the roots apoplasm by phenolic compounds could have restricted Cu toxicity to the plant (Jung et al. 2003). Going back further, Suresh and Subramanyam (1998) had already studied the role of polyphenolic compounds involved in Cu binding onto the cell walls of fungus *Neurospora crassa*. Their ESR (electron spin resonance) and FTIR (Fourier transformation infrared) studies of the Cu-polyphenol complexes indicated Cu to be bound as Cu(I) present in a distorted octahedral geometry and bound through oxygens belonging to phenolic hydroxyls and/or nitrite groups. The authors proposed that both groups might participate in a binding mechanism and supposed that nitrophenols are the responsible ligands located in the cell wall. Similar bindings are likely in plant cells.

3 Heavy Metal Localization and Distribution

3.1 Localization of Heavy Metals in Cells and Tissues of Different Plant Organs

As shown in Fig. 2, general mechanisms for detoxification and accumulation of heavy metals in plants are the distribution of the metals to apoplastic compartments

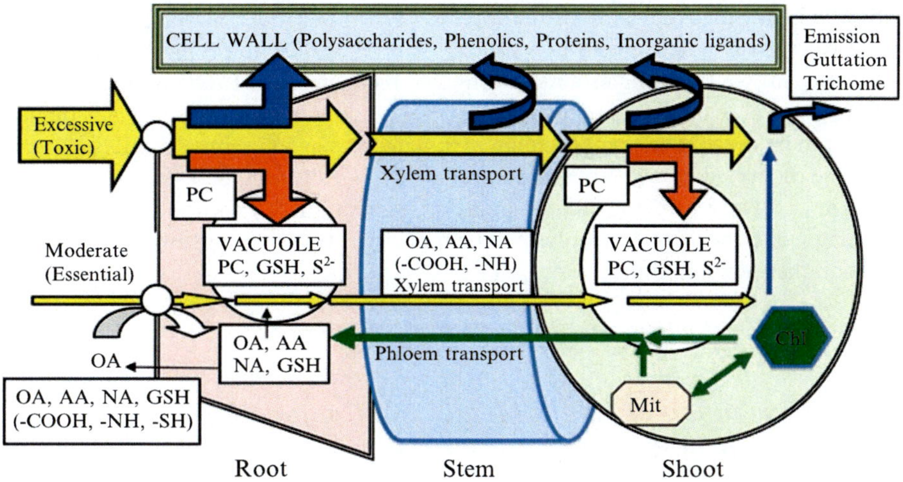

Fig. 2 Possible metal localization and presence of major metal-binding ligands in a model plant with a standard root, stem and shoot system. In each organ, tissues and cells are conventionally divided into apoplasic and symplastic sites. The former including xylem (sap) in the conductive tissues of each organ, and rizosphere connected to or surrounding the root system underground, and also in some cases vacuoles (apoplast but inside the protoplasm). The latter includes phloem (sap) and cytoplasm in each organ. The xylem and phloem systems support large parts of the stem and other tissues, and they play considerable roles in mineral/water transport from root to shoot and vice versa, with assimilatives as long-distance transports. Trichomes in shoot (leaf) also consist of apoplastic and symplastic sites but develop their special structure and functions for metal binding and accumulation. *Mit* mitochondria, *Chl* chloroplast, *PC*, phytochelatin, *GSH* glutathione, *OA* organic acid, *NA* nicotianamine, *AA* amino acid, *-COOH* carboxyl group, *-NH* amino- or imino- group, *-SH* sulfhydryl group

like cell walls or trichome, and the chelation of the metals by a ligand in cytoplasm, followed by the sequestration of the metal–ligand complex into the vacuole, in the different organs such as roots, stems and leaves (Yang et al. 2005). Generally, the heavy metal contents in plant organs decrease in the following sequence; root > leaves > stems > inflorescence > seeds. However, this order sometimes varies with plant species, especially in hyperaccumulators, of which the shoots have the highest heavy metal content. Roots usually manifest the maximum content of heavy metals. Leaves vary with age in their ability to accumulate heavy metals, some heavy metals accumulate preferentially in the youngest leaves of plants, whereas in others, the maximum content is found in senescing leaves. Preventing Cd ions from entering the cytosol by the plant cell walls theoretically represents the best detoxification mechanism (Ma et al. 2005). Cd stress may be alleviated by sequestration of Cd in the cell wall or the vacuole in Cd-tolerant genotypes of barley, especially in short-term Cd-exposed experiments. Cell walls of the root can act as a first barrier against Cd stress in immobilizing excesses of Cd (Wu et al. 2005). Available evidence suggests that Cd binds to the secondary wall and middle lamellae in maize roots (Khan et al. 1984). On the other hand, in bush bean, Cd was mainly bound to pectic sites and hystidyl groups of the cell wall in roots and leaves (Leita et al. 1996). In white lupin, the cell wall was found to retain up to 47% of the absorbed Cd in leaves, 51% in stems, and 42% in the roots, although 20–40% of total Cd was associated with PCs (Vazquez et al. 2006), implying that this plant may use cell wall binding as a more effective mechanism of Cd detoxification than PCs. However, excess and non-specific metal binding to primary cell walls did not appear to be the tolerance mechanism in tomato suspension-cultured cells and roots of some dicotyledonous plants (Inouhe et al. 1991, 1994). In these cases, where the cells are actively growing, the cytoplasmic formation of PCs followed by metal binding and transport to vacuoles can be more effective mechanisms of Cd detoxification than wall bindings.

3.2 Distribution of Heavy Metals and Conjugating Ligands in Root

Besides bioavailability, uptake and translocation efficiencies determine metal accumulation and distribution in plants (Clemens 2006). Roots are the plant organs in closest contact with metal-contaminated soils; therefore, they are the most affected by metals. Resistance to excess metals can be achieved by avoidance when the plant is able to restrict metal uptake into the cells, or tolerance when the plant is able to survive in the presence of excess metals inside. Having been taken up by the root and transported to various cells and tissues within the plant, heavy metals concentrate there to cause injury in a sensitive plant, or as an inactivated form in a tolerant plant.

Cd-tolerant tobacco species (*Nicotiana rustica*) indicated greater labeled cadmium (^{109}Cd) content in the roots than the leaves, the major part of which was stored in the distal part as a tolerance strategy (Bovet et al. 2006). In hyperaccumulator *A. halleri* roots exposed to 100 μM Cd and 500 μM Zn hydroponically, Zn and Cd accumulated in the cell walls of the rhizodermis (root epidermis), mainly due to precipitation of Zn/Cd phosphates (Kupper et al. 2000). In roots, scanning electron microscope combined with energy dispersive spectrometry (SEM-EDS) confirmed that the highest Zn concentration was found in xylem parenchyma cells and epidermal cells, while for Cd, a gradient was observed with the highest Cd concentration in rhizodermal and cortex cells, followed by central cylinder. Light microscope results showed that Zn and Cd distributed mainly along the walls of epidermis, cortex, endodermis and some xylem parenchyma (Hu et al. 2009). Energy-dispersed X-ray (EDX) microanalysis revealed details about the subcellular localization of Cd in *A. thaliana*, ecotype Columbia (Van Belleghem et al. 2007). The results indicated that the localizations of Cd in the root cortex were associated with phosphorus (Cd/P) in the apoplast and sulfur (Cd/S) in the symplast, suggesting phosphate and PC sequestration, respectively. In the endodermis, sequestration of Cd/S was present as fine granular deposits in the vacuole and as large granular deposits in the cytoplasm. In the central cylinder, symplastic accumulation followed a distinct pattern illustrating the importance of passage cells for the uptake of Cd. Furthermore, in the apoplast, a shift of Cd/S granular deposits from the middle lamella towards the plasmalemma was observed. Large amounts of precipitated Cd in the phloem suggest retranslocation from the shoot (Van Belleghem et al. 2007). On the other hand, subcellular localization of Pb and Cd in *Iris pseudacorus* showed that numerous Pb deposits were found on the inner surface of dead cell walls in the cortex treated with 2,070 mg L^{-1} Pb, there were no Pb deposits in the cell walls and cytoplasm of the neighbor cells (Zhou et al. 2010). Cd deposits were found in the cell wall and on the outer surface of the cells in a triangular intercellular space bordering with three cortical cells treated with 1,000 mg L^{-1} Cd for 16 days sand culture. The ultrastructure showed that Cd deposits in some cell walls were not well distributed and not found in the cytoplasma and vacuoles, showing that Cd was mainly transported by the way of apoplasts (Zhou et al. 2010); Han et al. (2007) found similar results that some Cd deposits were located not only in the cell walls but also in the vicinity of the plasma membranes and membrane-bound organelles in the root cells of *Iris lactea* var. *chinensis*. This observation also supports the apoplastic transport of Cd in the plan but cannot exclude the possibility that Cd deposits accumulated in the cell walls might negatively affect the enzymes and other protein functions in this compartment.

The increase of the cell walls (CWs) capacity to bind Pb by formation of cell wall thickenings (CWTs) rich in JIM5 pectins, callose and lipids in *Funaria hygrometrica* plant cells treated with Pb might be regarded as the next step in the development of the plant resistance strategy against this metal based on immobilizing toxic ions within apoplast (Krzeslowska et al. 2009). Binding metal ions within CWs is the important resistance strategy of plant cells in response to Cd (Fig. 1). This has been shown recently for *T. caerulescens* (Wojcik et al. 2005); *Salix viminalis* (Vollenweider et al. 2006) and *Linum usitatissimum* (Douchiche et al. 2007, 2010). In the last named, it was found moreover that exposing plants to Cd resulted in significant increases of both the cell wall thickness and JIM5 pectins formation level in CWs (Douchiche et al. 2007). In *S. viminalis*, the main Cd sink was pectin-rich collenchyma CWs of the veins. Moreover, also in this case, the amount of pectins slightly increased in collenchyma cells in response to Cd. Active storage of Cd in this plant was indicated by homogeneous CWTs containing cellulose and proanthocyanidins (Vollenweider et al. 2006). Thus, similarly to *Funaria* protonemata treated with Pb, both *L. usitatissimum* seedlings tissues and *S. viminalis* collenchyma increased the capacity of cell walls for Cd detoxification by formation of thicker cell wall and increasing the level of polysaccharides, especially that of pectin (Krzeslowska et al. 2009).

3.3 Distribution of Heavy Metals and Conjugating Ligands in Shoots

As already noted, there are well-documented differences across plant species in the partitioning of Cd between organs. Compared to other toxic metals or metalloids (e.g., Pb and As), Cd has a higher propensity to accumulate in shoots other than the roots. Still, there is normally more Cd in roots than in leaves, and even less in fruits and seeds (Wagner 1993). The tendency of tobacco plants to translocate Cd quite efficiently to the leaves contributes to the fact that tobacco smoke is an important Cd source for smokers (Lugon Moulin et al. 2004). But recently, some research showed that tobacco develops an original mechanism of metal detoxification by the exudation of metal/Ca-containing particles through leaf trichomes (Choi et al. 2001; Choi and Harada 2005; Sarret et al. 2006).

An energy-dispersive X-ray (EDX) analysis system equipped to variable pressure scanning electron microscopy (VP-SEM) revealed that the tobacco trichomes exudates contain amounts of heavy metals. Overexpression of cysteine synthase confers Cd tolerance to tobacco, and the endogenous concentration of Cd was 20% less in transgenic plants than in wild-type plants. The numbers of both long and short trichomes in the transgenic plants were 25% higher than in that of wild-type plants, indicating the active excretion of Cd from trichomes in transgenic plants (Harada and Choi 2008). Upon Cd or Zn treatment, the number of trichomes was increased more than 2-fold (Choi et al. 2001; Sarret et al. 2006). Confocal laser scanning electron microscopy showed metal accumulation in the tip cells in trichomes. The chemical forms of the exudated grains were identified as metal-substituted calcite (calcium carbonate) by using synchrotron-based X-ray microanalyses (Sarret et al. 2006, 2007). Observation by VP-SEM indicated that large crystals of 150 μm in size were formed on head cells of both short and long trichomes. An EDX analysis system fitted with VP-SEM revealed the crystals to contain amounts of Cd and Ca at much higher concentrations than in the head cells themselves.

TEM demonstrated crystal formation in amorphous osmiophilic deposits in vacuoles in tobacco (Choi et al. 2001). The majority of Ni is stored either in *Alyssum* leaf epidermal cell vacuoles or in the basal portions only of the numerous stellate trichomes. Broadhurst et al. (2004) reported simultaneous and region-specific localization of high levels of Ni, Mn, and Ca within *Alyssum* trichomes as determined by SEM/EDX. The metal concentration in the trichome basal compartment was about 15–20% dry weight, the highest ever reported for healthy vascular plant tissue (Broadhurst et al. 2004). In aerial parts, Zn was predominantly octahedral coordinated and complexed to malate.

In *A. halleri*, secondary organic species were identified in the bases of the trichomes, which contained elevated Zn concentrations, and in which Zn was tetrahedrally coordinated and complexed to carboxyl and/or hydroxyl functional groups (Sarret et al. 2002). In *A. halleri* leaves, the trichomes had by far the largest concentration of Zn and Cd. Inside the trichomes, there was a striking subcellular compartmentation, with almost all the Zn and Cd being accumulated in a narrow ring in the trichome base. Another phenomenon is that the epidermal cells other than trichomes were very small and contained lower concentrations of Zn and Cd than mesophyll cells. In particular, the concentrations of Cd and Zn in the mesophyll cells increased markedly in response to increasing Zn and Cd concentrations in the nutrient solution. This indicates that the mesophyll cells in the leaves of

A. halleri are the major storage site for Zn and Cd, and play an important role in their hyperaccumulation (Kupper et al. 2000). In contrast, Cd was detected in tracheids of *A. thaliana* but not in the mesophyll tissue (Van Belleghem et al. 2007). In *Potentilla griffithii* leaves, Zn and Cd shared a similar distribution pattern, and both were mostly accumulated in epidermis and bundle sheath. However, in leaves of 40 mg L^{-1} Cd treatment, which caused the phytotoxicity, Cd was also found in the mesophyll cells. The major storage site for Zn and Cd in leaves of *P. griffithii* was vacuoles, and to a lesser extent cell walls or cytosol. The present study demonstrates that the predominant sequestration of Zn and Cd in vacuoles of epidermis and bundle sheath of leaves may play a major role in strong tolerance and hyperaccumulation of Zn and Cd in *P. griffithii* (Hu et al. 2009).

4 Conclusion

It is obvious that plants utilize various types of biological ligands to conjugate, co-transport and partition heavy metal elements (Fig. 2). The biochemical and genetic bases of the Cd-tolerance phenotypes of plants may involve both the PC-dependent and -independent processes. The former involves several different processes: the activation of PC synthase, GSH biosynthesis, and accumulation of acid-labile sulfide, sulfur assimilation and transport of the Cd-PC complexes into vacuoles. All these would be required for the formation of the stable and nontoxic Cd–complexes in the vacuole or other sites in the cells of most plants, where the PC synthase is a key factor for the tolerance phenotypes to Cd and other heavy metals. The PC-independent mechanisms are apparently present in more differentiated higher plants that habituate on terrestrial system. Their hyperaccumulation phenotype of metal/metalloids from soil and water can be attributed to the highly developed apoplastic transport systems. The low constant pH condition and changeable solute components in the xylem sap and other apoplastic sites may allow more variable and more complicated interactions between the metal and biological ligands in plants. This might be a potential for the differentiation and specification of a unique hyperaccumulator to be evolved on ground. Readjustment of both the symplastic and apoplastic activities including the formations of PC-dependent and -independent metal-binding ligands and their transport systems can be beneficial for more effective and intentional approaches to conduct the remediation technique under contaminated soil and water environments.

References

Ayres RU (1992) Toxic heavy metals: materials cycle optimization. Proc Natl Acad Sci U S A 89:815–820

Bialonska D, Zobel AM, Kuras M, Tykarska T, Sawicka-Kapusta K (2007) Phenolic compounds and cell structure in bilberry leaves affected by emissions from a Zn-Pb smelter. Water Air Soil Poll 181:123–133

Bogs J, Bourbouloux A, Cagnac O, Wachter A, Rausch T, Delrot S (2003) Functional characterization and expression analysis of a glutathione transporter, BjGT1, from *Brassica juncea*: evidence for regulation by heavy metal exposure. Plant Cell Environ 26:1703–1711

Bovet L, Eggmann T, Meylan-Bettex M, Polier J, Kammer P, Marin E, Feller U, Martinoia E (2003) Transcript levels of AtMRPs after cadmium treatment: induction of AtMRP3. Plant Cell Environ 26:371–381

Bovet L, Rossi L, Lugon-Moulin N (2006) Cadmium partitioning and gene expression studies in *Nicotiana tabacum* and *Nicotiana rustica*. Physiol Plant 128:466–475

Broadhurst CL, Chaney RL, Angle JS, Maugel TK, Erbe EF, Murphy CA (2004) Simultaneous hyperaccumulation of nickel, manganese, and calcium in *Alyssum* leaf trichomes. Environ Sci Technol 38:5797–5802

Callahan DL, Baker AJM, Kolev SD, Wedd AG (2006) Metal ion ligands in hyperaccumulating plants. J Biol Inorg Chem 11:2–12

Callahan DL, Kolev SD, O'Hair RAJ, Salt DE, Baker AJM (2007) Relationships of nicotianamine and other amino acids with nickel, zinc and iron in *Thlaspi* hyperaccumulators. New Phytol 176:836–848

Chen J, Zhou J, Goldsbrough PB (1997) Characterization of phytochelatin synthase from tomato. Physiol Plant 101:165–172

Chiang PN, Wang MK, Chiu CY, Chou SY (2006) Effects of cadmium amendments on low-molecular-weight organic acid exudates in rhizosphere soils of tobacco and sunflower. Environ Toxicol 21:479–488

Choi YE, Harada E (2005) Roles of calcium and cadmium on Cd-containing intra-and extra cellular formation of Ca crystals in tobacco. J Plant Biol 48:113–119

Choi YE, Harada E, Wada M, Tsuboi H, Morita Y, Kusano T, Sano H (2001) Detoxification of cadmium in tobacco plants: formation and active excretion of crystals containing cadmium and calcium through trichomes. Planta 213:45–50

Clemens S (2006) Evolution and function of phytochelatin synthases. J Plant Physiol 163:319–332

Clemens S, Kim EJ, Neuman D, Schroeder JI (1999) Tolerance to toxic metals by a gene family of phytochelatin syntheses from plants and yeast. EMBO J 18:3325–3333

Cobbett CS (2000) Phytochelatins and their roles in heavy metal detoxification. Plant Physiol 123:825–832

Cobbett C, Goldsbrough P (2002) Phytochelatins and metallothioneins and Role in heavy metal detoxification and homeostasis. Annu Rev Plant Biol 53:159–182

Douchiche O, Rihouey C, Schaumann A, Driouich A, Morvan C (2007) Cadmium-induced alterations of the structural features of pectins in flax hypocotyl. Planta 225:1301–1312

Douchiche O, Soret-Morvan O, Chaibi W, Morvan C, Paynel F (2010) Characteristics of cadmium tolerance in 'Hermes' flax seedlings: contribution of cell walls. Chemosphere 81:1430–1436

Foy CD, Chaney RL, White MC (1978) The physiology of metal toxicity in plants. Annu Rev Plant Physiol 29:511–566

Gasic K, Korban SS (2007) Transgenic Indian mustard (*Brassica juncea*) plants expressing an *Arabidopsis* phytochelatin synthase (AtPCS1) exhibit enhanced As and Cd tolerance. Plant Mol Biol 64:361–369

Gong JM, Lee DA, Schroeder JI (2003) Long-distance root-to-shoot transport of phytochelatins and cadmium in *Arabidopsis*. Proc Natl Acad Sci U S A 100:10118–10123

Grill E, Winnacker EL, Zenk MH. (1987) Phytochelatins, a class of heavy-metal- inding peptides from plants, are functionally analogous to metallothioneins. Proc Natl Acad Sci U S A. 84:439–443

Grill E, Loffler S, Winnacker EL, Zink MH (1989) Phytochelatins, the heavy metal binding peptides of plants are synthesized from glutathione by a specific γ-glutamyl cysteine dipeptidyl transpeptidase (phytochelatin synthase). Proc Natl Acad Sci U S A 86:6838–6842

Guo J, Dai X, Xu W, Ma M (2008) Over-expression *GSH1* and *AsPCS1* simultaneously increases the tolerance and accumulation of cadmium and arsenic in *Arabidopsis thaliana*. Chemosphere 72:1020–1026

Gupta DK, Tohoyama H, Joho M, Inouhe M (2002) Possible roles of phytochelatins and glutathione metabolism in cadmium tolerance in chickpea roots. J Plant Res 115:429–437
Gupta DK, Tohoyama H, Joho M, Inouhe M (2004) Changes in the levels of phytochelatins and related metal binding peptides in chickpea seedlings exposed to arsenic and different heavy metal ions. J Plant Res 117:253–256
Gupta DK, Huang HG, Yang XE, Razafindrabe BHN, Inouhe M (2010) The detoxification of lead in *Sedum alfredii* H. is not related with phytochelatins but the glutathione. J Hazard Mater 177:437–444
Ha SB, Smith AP, Howden R, Dietrich WM, Bugg S, O'Connell MJ, Goldsbrough PB, Cobbett CS (1999) Phytochelatin synthase genes from *Arabidopsis* and the yeast *Schizosaccharomyces pombe*. Plant Cell 11:1153–1164
Hall JL (2002) Cellular mechanisms for heavy metal detoxification and tolerance. J Exp Bot 53:1–11
Han YL, Yuan HY, Huang SZ, Guo Z, Xia B, Gu JG (2007) Cadmium tolerance and accumulation by two species of Iris. Ecotoxicology 16:557–563
Harada E, Choi YE (2008) Investigation of metal exudates from tobacco glandular trichomes under heavy metal stresses using a variable pressure scanning electron microscopy system. Plant Biotechnol 25:407–411
Harada E, Choi YE, Tsuchisaka A, Obata H, Sano H (2001) Transgenic tobacco plants expressing a rice cysteine synthase gene are tolerant to toxic levels of cadmium. J Plant Physiol 158:655–661
Herbette S, Taconnat L, Hugouvieux V, Piette L, Magniette MLM, Cuine S, Auroy P, Richaud P, Forestier C, Bourguignon J, Renou JP, Vavasseur A, Leonhardt N (2006) Genome-wide transcriptome profiling of the early cadmium response of *Arabidopsis* roots and shoots. Biochimie 88:1751–1765
Haydon MJ, Cobbett CS (2006) A novel major facilitator superfamily protein at the tonoplast. Plant J 46: 861–879
Haydon MJ, Cobbett CS (2007) Transporters of ligands for essential metal ions in plants. New Phytol 174:499–506
Howden R, Cobbett CS (1992) Cadmium-sensitive mutants of *Arabidopsis thaliana*. Plant Physiol 99:100–107
Howden R, Goldsbrough PB, Andersen CR, Cobbett CS (1995) Cadmium-sensitive, *cad1* mutants of *Arabidopsis thaliana* are phytochelatin deficient. Plant Physiol 107:1059–1066
Hu PJ, Qiu RL, Senthilkumar P, Jiang D, Chen ZW, Tang YT, Liu FJ (2009) Tolerance, accumulation and distribution of zinc and cadmium in hyperaccumulator *Potentilla griffithii*. Environ Exp Bot 66:317–325
Inouhe M (2005) Phytochelatins. Braz J Plant Physiol 17:65–78
Inouhe M, Mitsumune M, Tohoyama H, Joho M, Murayama T (1991) Contributions of cell wall and metal-binding peptides in suspension-cultured cells of tomato. Bot Mag Tokyo 104:217–229
Inouhe M, Ninomiya S, Tohoyama H, Joho M, Murayama T (1994) Different characteristics of roots in the cadmium-tolerance and Cd-binding complex formation between mono- and dicotyledonous plants. J Plant Res 107:201–207
Inouhe M, Ito R, Ito S, Sasada N, Tohoyama H, Joho M (2000) Azuki bean cells are hypersensitive to cadmium and do not synthesize phytochelatins. Plant Physiol 123:1029–1036
Janas KM, Zielinska Tomaszewska J, Rybaczek D, Maszewski J, Posmyk MM, Amarowicz R, Kosinska A (2010) The impact of copper ions on growth, lipid peroxidation, and phenolic compound accumulation and localization in lentil (*Lens culinaris* Medic.) seedlings. J Plant Physiol 167:270–276
Jung C, Maeder V, Funk F, Frey B, Sticher H, Frossard E (2003) Release of phenols from *Lupinus albus* L. roots exposed to Cu and their possible role in Cu detoxification. Plant Soil 252:301–312

Kato M, Ishikawa S, Inagaki K, Chiba K, Hayashi H, Yanagisawa S, Yoneyama T (2010) Possible chemical forms of cadmium and varietal differences in cadmium concentrations in the phloem sap of rice plants. Soil Sci Plant Nutr 56:839–847

Khan DH, Duckett B, Frankland B, Kirkham JB (1984) An X-ray microanalytical study of the distribution of cadmium in roots of *Zea mays* L. J Plant Physiol 115:19–28

Klapheck S, Schlunz S, Bergmann L (1995) Synthesis of phytochelatins and homo-phytochelatins in *Pisum sativum* L. Plant Physiol 107:515–521

Kneer R, Zenk MH (1992) Phytochelatins protect plant enzymes from heavy metal poisoning. Phytochemistry 31:2663–2667

Kovacik J, Backor M, Kadukova J (2008) Physiological responses of *Matricaria chamomilla* to cadmium and copper excess. Environ Toxicol 23:123–130

Kramer U, Pickering IJ, Prince RC, Raskin I, Salt DE (2000) Subcellular localization and speciation of nickel in hyperaccumulator and non-accumulator *Thlaspi* species. Plant Physiol 122:1343–1353

Krzeslowska M, Lenartowska M, Mellerowicz EJ, Samardakiewicz S, Wozny A (2009) Pectinous cell wall thickenings formation-A response of moss protonemata cells to lead. Environ Exp Bot 65:119–131

Kupper H, Lombi E, Zhao FJ, McGrath SP (2000) Cellular compartmentation of cadmium and zinc in relation to other elements in the hyperaccumulator *Arabidopsis halleri*. Planta 212:75–84

Kupper H, Mijovilovich A, Meyer Klaucke W, Kroneck PMH (2004) Tissue and age dependent differences in the complexation of cadmium and zinc in the cadmium/zinc hyperaccumulator *Thlaspi caerulescens* (Ganges ecotype) revealed by X-ray absorption spectroscopy. Plant Physiol 134:748–757

Lafaye A, Junot C, Pereira Y, Lagniel G, Tabet JC, Ezan E, Labarre J (2005) Combined proteome and metabolite-profiling analyses reveal surprising insights into yeast sulfur metabolism. J Biol Chem 280:24723–24730

Leita L, DeNobili M, Cesco S, Mondini C (1996) Analysis of intercellular cadmium forms in roots and leaves of bush bean. J Plant Nutr 19:527–533

Leustek T (2002) Sulfate metabolism. In: Somerville CR, Meyerowitz EM (eds) The *Arabidopsis* book. American Chemical Society of Plant Biologists, Rockville

Li ZS, Lu YP, Thiele DJ, Rea PA (1997) A new pathway for vacuolar cadmium sequestration in *Saccharomyces cerevisiae*: YCF1-mediated transport of *bis* (glutathionato) cadmium. Proc Natl Acad Sci U S A 94:42–47

Li Y, Dhanker OP, Carreira L, Lee D, Chen A, Schroeder JI, Balish RS, Meagher RB (2004) Over expression of phytochelatin synthase in *Arabidopsis* leads to enhanced arsenic tolerance and cadmium hypersensitivity. Plant Cell Physiol 45:1787–1797

Loeffler S, Hochberger A, Grill E, Winnacker EL, Zenk MH (1989) Termination of phytochelatin synthase reaction through sequestration of heavy metals by the reaction product. FEBS Lett 258:42–46

Lugon Moulin N, Zhang M, Gadani F, Rossi L, Koller D, Krauss M, Wagner GJ (2004) Critical review of the science and options for reducing cadmium in tobacco (*Nicotiana tabacum* L.) and other plants. Adv Agron 83:111–180

Ma JF, Ueno D, Zhao FJ, McGrath SP (2005) Subcellular localisation of Cd and Zn in the leaves of a Cd-hyperaccumulating ecotype of *Thlaspi caerulescens*. Planta 220:731–736

Mari S, Gendre D, Pianelli K, Ouerdane L, Lobinski R, Briat JF, Lebrun M, Czernic P (2006) Root to shoot long distance circulation of nicotianamine and nicotianamine-nickel chelates in the metal hyperaccumulator *Thlaspi caerulescens*. J Exp Bot 57:4111–4122

Maruyama NA, Nakamura Y, Yamaya T, Takahashi H (2004) A novel regulatory pathway of sulfate uptake in *Arabidopsis* roots: implication of CRE1/WOL/AHK4-mediated cytokinin-dependent regulation. Plant J 38:779–789

Mehra RK, Winge DR (1991) Metal ion resistance in fungi: molecular mechanisms and their regulated expression. J Cell Biochem 45:1–11

Murasugi A, Wada C, Hayashi Y (1981) Cadmium-binding peptide induced in fission yeast, *Schizosaccharimyces pombe*. J Biochem 90:1561–1564

Nocito FF, Pirovano L, Cocucci M, Sacchi GA (2002) Cadmium-induced sulfate uptake in maize roots. Plant Physiol 129:1872–1879

Oshima T, Hayashi H, Chino M (1990) Collection and chemical composition of pure phloem sap from *Zea mays* L. Plant Cell Physiol 31:735–737

Ortiz DF, Ruscitli T, McCue KF, Ow DW (1995) Transport of metal binding peptides by HMT1: a fission yeast ABC-type vacuolar membrane protein. J Biol Chem 270:4721–4728

Pal R, Rai JPN (2010) Phytochelatins: peptides involved in heavy metal detoxification. Appl Biochem Biotechnol 160:945–963

Pianelli K, Mari S, Marques L, Lebrun M, Czernic P (2005) Nicotianamine over-accumulation confers resistance to nickel in *Arabidopsis thaliana*. Transgenic Res 14:739–748

Piechalak A, Tomaszewska B, Baralkiewicz D, Malecka A (2002) Accumulation and detoxification of lead ions in legumes. Phytochemistry 60:153–162

Pomponi M, Censi V, Di Girolamo, de Paolis A, di Sanit'a Toppi L, Aromolo R (2006) Overexpression of *Arabidopsis* phytochelatin synthase in tobacco plants enhances Cd^{2+} tolerance and accumulation but not translocation to the shoot. Planta 223:180–190

Rauser WE (1995) Phytochelatins and related peptides: structure, biosynthesis, and function. Plant Physiol 109:1141–1149

Rauser WE (1999) Structure and function of metal chelators produced by plants-The case for organic acids, amino acids, phytin, and metallothioneins. Cell Biochem Biophys 31:19–48

Romheld V, Awad F (2000) Significance of root exudates in acquisition of heavy metals from a contaminated calcareous soil by graminaceous species. J Plant Nutr 23:1857–1866

Sakihama Y, Cohen MF, Grace SC, Yamasaki H (2002) Plant phenolic antioxidant and prooxidant activities: phenolics-induced oxidative damage mediated by metals in plants. Toxicology 177:67–80

Salt DE, Prince RC, Baker AJM, Raskin I, Pickering IJ (1999) Zinc ligands in the metal hyperaccumulator *Thlaspi caerulescens* as determined using X-ray absorption spectroscopy. Environ Sci Technol 33:713–717

Santiago LJM, Louro RP, De Oliveira DE (2000) Compartmentation of phenolic compounds and phenylalanine ammonia-lyase in leaves of *Phyllanthus tenellus* Roxb. and their induction by copper sulphate. Ann Bot 86:1023–1032

Sarret G, Saumitou Laprade P, Bert V, Proux O, Hazemann JL, Traverse AS, Marcus MA, Manceau A (2002) Forms of zinc accumulated in the hyperaccumulator *Arabidopsis halleri*. Plant Physiol 130:1815–1826

Sarret G, Harada E, Choi YE, Isaure MP, Geoffroy N, Fakra S, Marcus MA, Birschwilks M, Clemens S, Manceau A (2006) Trichomes of tobacco excrete zinc as zinc-substituted calcium carbonate and other zinc-containing compounds. Plant Physiol 141:1021–1034

Sarret G, Isaure MP, Marcus MA, Harada E, Choi YE, Pairis S, Fakra S, Manceau A (2007) Chemical forms of calcium in Ca, Zn and Ca, Cd containing grains excreted by tobacco trichomes. Can J Chem 85:738–746

Sharma SS, Dietz KJ (2006) The significance of amino acids and amino acid-derived molecules in plant responses and adaptation to heavy metal stress. J Exp Bot 57:711–726

Suresh K, Subramanyam C (1998) Polyphenols are involved in copper binding to cell walls of *Neurospora crassa*. J Inorg Biochem 69:209–215

Takahasi M, Terada Y, Nakai I, Nakanishi H, Yoshimura E, Mori S, Nishizawa NK (2003) Role of nicotianamine in the intracellular delivery of metals and plant reproductive development. Plant Cell 14:1263–1280

Thumann J, Grill E, Winnacker EL, Zenk MH (1991) Reactivation of matal- requiring apoenzymes by phytochelatin metal complexes. FEBS Lett 284:66–69

Toppi LS, Gabbrielli R (1999) Response to cadmium in higher plants. Env Exp Bot 41:105–130

Tong YP, Kneer R, Zhu YG (2004) Vacuolar compartmentalization: a second-generation approach to engineering plants for phytoremediation. Trends Plant Sci 9:7–9

Tsuji N, Nishikori S, Iwabe O, Shiraki K, Miyasaka H, Takagi M, Hirata K, Miyamaoto K (2004) Characterization of phytochelatin synthase-like protein encoded by alr0975 from a prokaryote, *Nostoc* sp. PCC 7120. Biochem Biophys Res Commun 315:751–755

Ueno D, Iwashita T, Zhao FJ, Ma JF (2008) Characterization of Cd translocation and identification of the Cd form in xylem sap of the Cd-hyperaccumulator *Arabidopsis halleri*. Plant Cell Physiol 49:540–548

van Steveninck RFM, van Steveninck ME, Fernando DR (1992) Heavy-metal (Zn, Cd) tolerance in selected clones of duch weed (*Lemna minor*). Plant Soil 146:271–280

van Belleghem F, Cuypers A, Semane B, Smeets K, Vangronsveld J, d'Haen J, Valcke R (2007) Subcellular localization of cadmium in roots and leaves of *Arabidopsis thaliana*. New Phytol 173:495–508

Vatamaniuk OK, Mari S, Lu YP, Rea PA (1999) AtPCS1, a phytochelatin synthase from: isolation and *in vitro* reconstitution. Proc Natl Acad Sci U S A 96:7110–7115

Vatamaniuk OK, Bucher EA, Ward JT, Rea PA (2002) Worms take the 'phyto' out of 'phytochelatins'. Trends Biotechnol 20:61–64

Vazquez S, Goldsbrough P, Carpena RO (2006) Assessing the relative contributions of phytochelatins and the cell wall to cadmium resistance in white lupin. Physiol Plant 128:487–495

Verkleij JAC, Schat H (1990) Mechanisms of metal tolerance in higher plants. In: Shaw AJ (ed) Heavy metal tolerance in plants: evolutionary aspects. CRC, Boca Raton, pp 179–193

Veselov D, Kudoyarova G, Symonyan M, Veselov S (2003) Effect of cadmium onion uptake, transpiration and cytokinin content in wheat seedlings. Bulg J Plant Physiol special issue. In: Proceedings of the European workshop on environmental stress and sustainable agriculture, Bulgaria, 7–12 Sept 2002

Vollenweider P, Cosio C, Gunthardt Goerg MS, Keller C (2006) Localization and effects of cadmium in leaves of a cadmium-tolerant willow (*Salix viminalis* L.) part II microlocalization and cellular effects of cadmium. Environ Exper Bot 58:25–40

Wawrzyn'ski A, Kopera E, Wawrzyn'ska A, Kamin'ska J, Bal W, Sirko A (2006) Effects of simultaneous expression of heterologous genes involved in phytochelatin biosynthesis on thiol content and cadmium accumulation in tobacco plants. J Exp Bot 57:2173–2182

Wagner GJ (1993) Accumulation of cadmium in crop plants and its consequences to human health. Adv Agron Vol 51:173–212

Weber M, Harada E, Vess C, von Roepenack LE, Clemens S (2004) Comparative microarray analysis of *Arabidopsis thaliana* and *Arabidopsis halleri* roots identifies nicotianamine synthase, a ZIP transporter and other genes as potential metal hyperaccumulation factors. Plant J 37:269–281

Weber M, Trampczynska A, Clemens S (2006) Comparative transcriptome analysis of toxic metal responses in *Arabidopsis thaliana* and the Cd^{2+}-hypertolerant facultative metallophyte *Arabidopsis halleri*. Plant Cell Environ 29:950–963

Winter H, Lohaus G, Heldt HW (1992) Phloem transport of amino acids in relation to their cytosolic levels in barley leaves. Plant Physiol 99:996–1004

Wojcik M, Vangronsveld J, D'Haen J, Tukiendorf A (2005) Cadmium tolerance in *Thlaspi caerulescens*-II. Localization of cadmium in *Thlaspi caerulescens*. Environ Exp Bot 53:163–171

Woolhouse HW et al (1983) Toxicity and tolerance in the responses of plants to metals. In: Lange OL (ed) Encyclopedia of plant physiology, 12Cth edn. Springer, New York, pp 245–300

Wu FB, Dong J, Qian QQ, Zhang GP (2005) Subcellular distribution and chemical form of Cd and Cd-Zn interaction in different barley genotypes. Chemosphere 60:1437–1446

Xiang C, Oliver DJ (1998) Glutathione metabolic genes co-ordinately respond to heavy metals and jasmonic acid in *Arabidopsis*. Plant Cell 10:1539–1550

Yang XE, Feng Y, He ZL, Stoffella PJ (2005) Molecular mechanisms of heavy metal hyperaccumulation and phytoremediation. J Trace Elem Med Biol 18:339–353

Zenk MH (1996) Heavy metal detoxification in higher plants-a review. Gene 179:21–30

Zhao FJ, Lombi E, Breedon T, McGrath SP (2000) Zinc hyperaccumulation and cellular distribution in *Arabidopsis halleri*. Plant Cell Environ 23:507–514

Zhou YQ, Huang SZ, Yu SL, Gu JG, Zhao JZ, Han YL, Fu JJ (2010) The physiological response and sub-cellular localization of lead and cadmium in *Iris pseudacorus* L. Ecotoxicology 19:69–76

Zhu YL, Pilon Smits EAH, Tarun AS, Weber SU, Jouanin L, Terry N (1999a) Cadmium tolerance and accumulation in Indian mustard is enhanced by overexpressing – glutamylcysteine synthetase. Plant Physiol 121:1169–1177

Zhu YL, Pilon Smits EAH, Jouanin L, Terry N (1999b) Over expression of glutathione synthetase in Indian mustard enhances cadmium accumulation and tolerance. Plant Physiol 119:73–79

Heavy Metal Perception in a Microscale Environment: A Model System Using High Doses of Pollutants

Luis E. Hernández, Cristina Ortega-Villasante, M. Belén Montero-Palmero, Carolina Escobar, and Ramón O. Carpena

Abstract The characterization of the mechanisms of heavy metal detoxification has been undertaken through several experimental approaches, where high metal concentrations have been frequently used. A *microscale* hydroponic system was used to discriminate between the direct and indirect phytotoxic effects that may occur under heavy metal stress at short exposure times. Induction of oxidative stress and generation of stress signaling molecules are some of the physiological responses triggered soon after the exposure of plant cells to heavy metals, which might be part of stress perception mechanisms. The generation of reactive oxygen species, in particular H_2O_2, ethylene or jasmonate are envisaged as messengers in signaling pathways that may result ultimately in cell senescence and growth inhibition.

L.E. Hernández • C. Ortega-Villasante
Laboratory of Plant Physiology, Department of Biology, Universidad Autónoma de Madrid, 28049 Madrid, Spain
e-mail: luise.hernandez@uam.es; cristina.ortega@uam.es

M.B. Montero-Palmero
Laboratory of Plant Physiology, Department of Biology, Universidad Autónoma de Madrid, 28049 Madrid, Spain

Department of Environmental Sciences, Universidad de Castilla la Mancha, Campus Fábrica de Armas, 45070 Toledo, Spain
e-mail: mariabelen.montero@uam.es

C. Escobar
Department of Environmental Sciences, Universidad de Castilla la Mancha, Campus Fábrica de Armas, 45070 Toledo, Spain
e-mail: Carolina.Escobar@uclm.es

R.O. Carpena (✉)
Department of Agricultural Chemistry, Universidad Autónoma de Madrid, 28049 Madrid, Spain
e-mail: ramon.carpena@uam.es

D.K. Gupta and L.M. Sandalio (eds.), *Metal Toxicity in Plants: Perception, Signaling and Remediation*, DOI 10.1007/978-3-642-22081-4_2,

Abbreviations

ACC Amino-cyclopropane-1-carboxylic acid
APX Ascorbate peroxidase
ECS Gamma glutamylcysteine synthetase
GR Glutathione reductase
GS Glutathione synthase
JA Jasmonic acid
PCS Phytochelatine synthase
ROS Reactive oxygen species
SA Salicylic acid, SOD, superoxide dismutase

1 Introduction

The accumulation of heavy metals in some ecosystems as a consequence of several contaminating human activities, such as mining or melting, poses a relevant environmental risk (Patra et al. 2004). Heavy metals that accumulate in soils are taken up by plants, and through a biomagnification process in the trophic chain, they can constitute a serious health problem for animals and humans. Due to their sessile nature, terrestrial plants have restricted mechanisms for stress avoidance, but during the course of evolution some plant species have developed mechanisms to cope with environmental stresses (Pastori and Foyer 2002). These tolerance mechanisms rely on the activation of complex processes of perception, transduction and transmission of stress stimuli (Mittler et al. 2004). In recent years, the understanding of physiological responses to heavy metal stress has been the subject of many studies, with the aim of discovering mechanisms of tolerance. The contribution of signaling molecules like ethylene, jasmonate or reactive oxygen species (ROS) has recently been described (Maksymiec 2007). However, to identify the signaling components a plethora of experiments have been carried out, using different plant species, doses of metals and exposure times, as thoroughly reviewed by Schützendübel and Polle (2002). In most studies, treatments are long enough to provoke substantial metabolic changes, such as the onset of oxidative stress symptoms (Gratão et al. 2005). Upon a stress situation, an unbalance of the cellular redox status occurs due to the accumulation of ROS and alterations in the antioxidant defences. Superoxide ($O_2^{\bullet -}$) and hydrogen peroxide (H_2O_2) are scavenged by the action of superoxide dismutase (SOD), ascorbate peroxidase (APX) and catalase (CAT) enzymes, which use the soluble antioxidants ascorbate and glutathione (GSH; Noctor and Foyer 1998). The cellular thiol status apparently plays a central role in redox homeostasis and cell function, in which the concentration of GSH and the balance with its oxidized counterpart (GSSG) is kept at a constant level (Noctor 2006). It is expected that components of the antioxidant system may play a relevant role in heavy metal stress perception, as alterations in the redox cellular homeostasis is well documented (Hall 2002; Schützendübel and Polle 2002; Gratão et al. 2005; Sharma and Dietz 2009).

The responses might differ as a function of doses, plant species, growing conditions and phenology status (Sanitá di Toppi and Gabbrielli 1999). Therefore, in many experiments it is extremely difficult to distinguish between direct and indirect responses if metal concentrations or treatment interval are too high or excessively prolonged. In such experiments, the metabolic alterations observed might reflect general failure of plant metabolism; but little is known about the earlier stages. Therefore, the characterization of heavy metal stress perception mechanisms should be undertaken in adequate experimental conditions, where we could learn about the primary cellular components involved. Little is known about the early effects of metal treatments, mainly because of the difficulty of detecting ROS accumulation or oxidative damage at the cellular level. Such a task could be accomplished by using highly sensitive fluorescent probes that react with ROS (Olmos et al. 2003; Garnier et al. 2006), or which are capable of tracing cellular integrity (Ortega-Villasante et al. 2005). This kind of experiments is paving the way to understanding in detail the dynamic aspects of plant cellular responses to heavy metals.

2 Microscale Versus Macroscale Analysis: Time Resolved Responses

To identify the specific physiological responses to metal stress it is important to fix the time of exposure and dose, avoiding acute metabolic alterations. The dose–response relationship is a complex phenomenon in which toxic metals produce several events at molecular, physiological and morphological levels. Moreover, the dose that is considered toxic depends on the specific heavy metal and is characteristic for the different plant species (Schützendübel and Polle 2002). In fact, the dose–response curves of essential elements have three phases: deficiency, tolerance and toxicity, while non-essential elements do not present a deficiency phase (Hagemeyer 2004). Early and direct phytotoxic symptoms of heavy metals could be followed by analyzing several physiological parameters. Thus, the reduction in cell proliferation and inhibition of growth correlates well with metal intoxication, and is frequently used as a phytotoxic index (Schützendübel et al. 2001). Another parameter is the rapid alteration of nutrient uptake and accumulation, as heavy metals provoke damages in the plasma membrane of exposed cells (Hernández and Cooke 1997), affecting water uptake and solute permeability (Hernández et al. 1997). A number of different studies have also shown that such early alterations could be caused by the induction of oxidative stress, which leads to lipid peroxidation (Lozano-Rodríguez et al. 1997; Chaoui et al. 1997) and cell senescence (Ortega-Villasante et al. 2005, 2007).

An extensive array of experiments, described in the literature, aimed to identify the direct effects of heavy metals on plant cell metabolism. In the vast majority of such experiments, plants under study were treated in a pure hydroponic culture for a certain interval to several doses of heavy metals. As an example of the experiments carried out to analyze the short-term effects of heavy metals, we show in Fig. 1a the *macroscale* system used to test the influence of Cd on nitrate, potassium and manganese uptake in *Pisum sativum* plantlets during 24–96 h (Hernández et al. 1997, 1998). A similar approach can be followed by treating the plants in a semi-hydroponic culture system, where plants are grown on an almost inert substrate (perlite) moistened by the nutrient solution (Fig. 1b). In these conditions, the growth and development of roots resembles the pattern typical of plants grown in soil, where nutrient and metal availability is more restricted than in pure hydroponics (Vázquez and Carpena-Ruiz 2005). As a consequence, remarkable differences in the responses to Cd and Hg were found with plants treated in pure hydroponic systems, inferring that each kind of growing condition affects the perception of heavy metals (Sobrino-Plata et al. 2009).

In spite of the valuable information that is provided by both kinds of macroscale experimental settings described above, they are unsuitable when very short and direct effects are to be characterized, since the dynamic responses might be poor with particular toxic concentrations of heavy metals. For example, alfalfa plants treated with 30 μM Hg showed a saturated toxic response of growth inhibition and lipid peroxidation after just 24 h exposure, whereas both parameters increased linearly with time in plants treated with 30 μM Cd (Ortega-Villasante et al. 2005). To overcome the saturation phase under certain experimental conditions, and to monitor minute and rapid changes in plant metabolism, alternative plant materials and experimental conditions have been used. A viable option is the use of plant cell cultures or protoplasts isolated from leaves (see protoplasts from alfalfa in Fig. 1c), which can be kept in a suspension culture to which the metals can be added directly (Sobkowiak and Deckert 2003). DNA damage was observed in protoplasts exposed to heavy metals (Zhigang et al. 2009). Similar negative effects on DNA stability and cell apoptosis induction were found in tobacco BY2 cells treated with Cd, and, interestingly, a DNA repair mechanism was induced when Cd was removed from the culture medium (Fojtová et al. 2002). In this experimental design, tobacco BY2 cells suffered an abrupt accumulation of H_2O_2 when exposed to extreme high concentration of Cd (1–5 mm) after just 15 min (Olmos et al. 2003), or after 24 h (Garnier et al. 2006). Such treatments also caused a severe depletion of GSH in *Arabidopsis thaliana* cultured cells after 24 h under 50 and 200 μM Cd (Sarry et al. 2006). Therefore, it is possible to gain some information about the early responses of plant cells to heavy metals, albeit the behavior might be completely different from root cells, since this organ accumulates higher concentrations of pollutants in the excluder plants. This problem was solved by using the *microscale* hydroponic system proposed by Ortega-Villasante et al. (2005, 2007) shown in Fig. 1d. We have successfully combined the *microscale* system with fluorescent probes to detect in vivo parameters related with heavy metal stress in intact alfalfa seedlings with a moderate supply of Cd or Hg, in the range 3–30 μM:

Fig. 1 Different plant materials used to study the direct effects of heavy metals. (**a**) Pure hydroponic culture using a nutrient solution with reduced volume, used to study short-term responses of pea plants. (**b**) Semi-hydroponic system where alfalfa plants are grown in an inert substrate. (**c**) Protoplast isolated from alfalfa leaves after digestion with cell wall-degrading enzymes. (**d**) Microscale hydroponic system, which allows very precise control of exposure times and in vivo visualization of heavy metal stress in alfalfa seedlings

(a) 2′,7′-dichlorofluorescin diacetate (H_2DCFDA) to visualize oxidative stress; (b) monochlorobimane (MCB) to detect the cellular concentrations of GSH and homoglutathione (hGSH; homologous tripeptide to GSH that accumulates in alfalfa); and (c) propidium iodide (PI) to estimate the amount of cellular death. In addition, we used Amplex Red to monitor in situ the secretion of H_2O_2 from alfalfa root tips using a fluorescence titer-plate reader (Ortega-Villasante et al. 2007).

Epifluorescence or laser confocal fluorescence microscopy can be used to visualize the changes in the above mentioned parameters in *Medicago sativa* root epidermal cells (Fig. 2). Cadmium and Hg caused a depletion of the GSH/hGSH fluorescence signal in the epidermal cells of the alfalfa root (Fig. 2c, d). Buthionine sulfoximine (BSO) is a potent inhibitor of GSH/hGSH synthesis, and was used as a negative control to discriminate auto-fluorescence cell epidermis (Fig. 2b), which in turn permits to visualize the hazardous effect of GSH/hGSH depletion in cellular redox homeostasis. In parallel, oxidative stress was detected in BSO and Cd treated seedlings using H_2DCFDA (Fig. 2f, g). Interestingly, oxidative stress appeared scattered under Cd stress, implying that different epidermal cells accumulated the metal depending on their particular physiological status, as we could confirm in a kinetic experiment after very short exposure times (Ortega-Villasante et al. 2007). Remarkably, Hg caused an extremely high rate of cell death at the same dose of

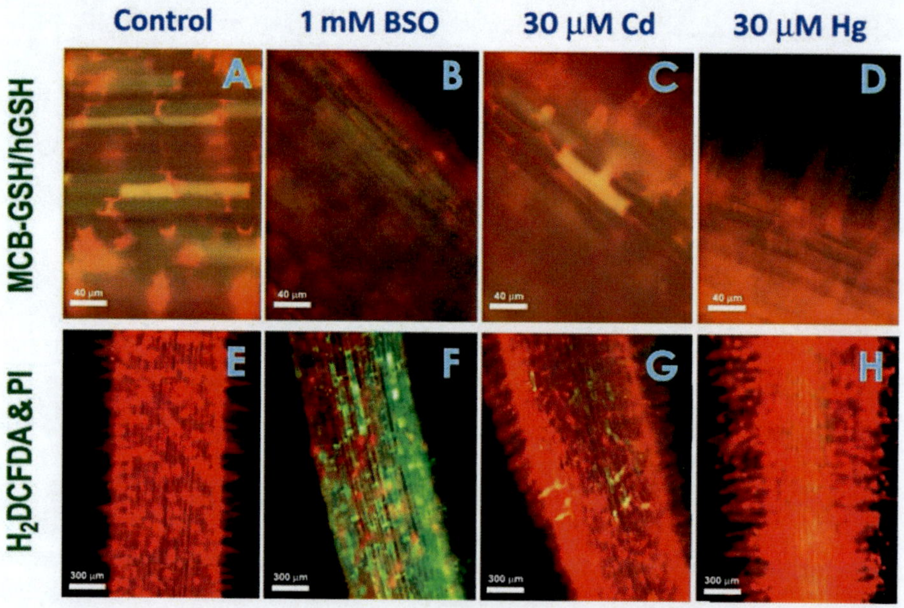

Fig. 2 Fluorescence probes that can be used to trace alteration in the pool of glutathione (monochlorobimane-MCB), oxidative stress (2′,7′-dichlorofluorescin diacetate-H_2DCFDA), and cell death (propidium iodide-PI). MCB (*green*) was visualized with epifluorescence microscopy (**a–d**), and H_2DCFDA (*green*) and IP (red) with confocal microscopy (**e–h**). Alfalfa seedlings were treated with Cd, Hg and buthionine sulfoximine (*BSO*) for 24 h prior to staining with the different probes

30 μM (observed as red fluorescence of PI in condensed nuclei; Fig. 2h), suggesting that oxidative stress only occurs in still viable cells with functional metabolism (Ortega-Villasante et al. 2005). A similar experiment with barley root tips showed that after 3–6 h exposure to 1 mM Cd or 0.5 mM Hg caused the overexpression of several genes encoding aquaporins and dehydrins, suggesting the onset of dehydration stress by heavy metal. These responses were accompanied by a significant inhibition of root growth and induction of oxidative stress (Tamás et al. 2010).

Another experimental alternative used frequently to study the direct effects of heavy metals on particular cellular components is the extraction of such materials from untreated plants, which are then tested with different treatments of metals in vitro. For example, it has been observed that direct exposure of tylakoid membranes caused the release of their components, especially proteins of the splitting water system and galactolipids probably connected with photosystem I (PSI) inhibition after heavy metals exposure (Skórzyńska and Baszyński 1993; Nouari et al. 2006). These experiments showed that heavy metals bind to membranes through oxygen atoms or aminoacids such as histidine, tryptophan or tyrosine in proteins, leading to electron flow disturbance in photosystem II after illumination (Maksymiec 1997). Other studies indicated that high heavy metal concentration lead to substitution of the Mg in the chlorophyll molecules (Kowalewska et al. 1987). Similarly, Hg substitutes Cu in plastocyanin molecule, blocking electron passage to PSI (Radmer and Kok 1974), and Cd, Hg, and Pb may also bind to LHCII producing conformational changes. Inhibition of enzymes involved in chlorophyll production also produces a decrease in its synthesis (Böddi et al. 1995). In spite of these findings, it is not clear that alterations in the photosynthetic electron transport and dark phase reactions observed in vivo are the direct effect of metals, as these effects appear only after several days of exposure (Wang et al. 2009).

3 ROS Signaling and Antioxidant Responses

An increase in H_2O_2 production has been reported in plant cells treated with several heavy metals, even those that have no direct or very little redox activity such as Cd or Hg. Cadmium is one of the heavy metals most widely studied, and relevant information has been provided by Olmos et al. (2003), Garnier et al. (2006), Cho and Seo (2005), and Romero-Puertas et al. (2002, 2004). Mercury is also a potent H_2O_2 inductor, as shown by Cho and Park (2000) and Ortega-Villasante et al. (2007). Similar responses were found for Cu (Xiang and Oliver 1998; Maksymiec and Krupa 2006) and Mn (Demirevska-Kepova et al. 2004), although at much higher doses due to their essential nature, which depends on the threshold of toxicity for each plant species. This production of ROS usually leads to damage in several cellular components, causing membrane lipid peroxidation (Lozano-Rodríguez et al. 1997), alteration of nucleic acids structure (Fojtová et al. 2002), or oxidation of proteins (Romero-Puertas et al. 2002; Rellán-Álvarez et al. 2006) and photosynthetic pigments (Somashekaraiah et al. 1992; Weckx and Clijsters

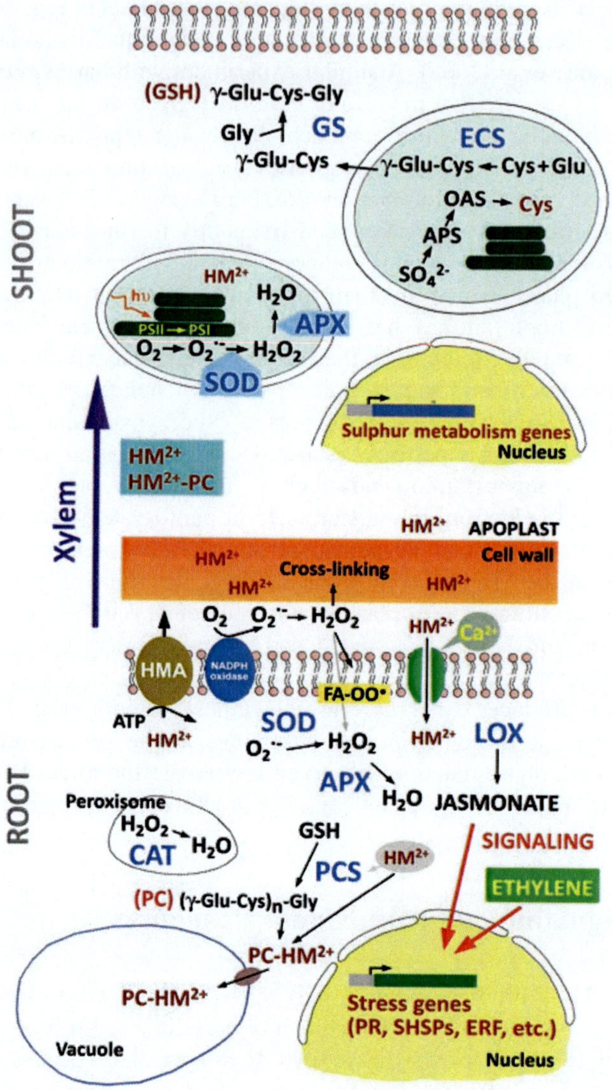

Fig. 3 Major physiological changes described in plants treated with heavy metals (*HM*$^{2+}$). In excluder plants metals accumulate in the root, where most of the alterations occur. The cell wall can act as a sink, but when this barrier is overridden, metals enter the protoplast probably through ion channels (for example Ca^{2+}-channel). Part of the uptaken metal can be expelled by transporters of the HMA, possibly loading them in the xylem. Plasma membrane NADPH-oxidases are activated to generate $O_2^{\bullet -}$, which is converted to H_2O_2 in the apoplast to serve as substrate of cell wall peroxidises to stiffen cell wall polymers. H_2O_2 can also oxidize membrane lipids, generating peroxide lipid radicals (FA-OO$^{\bullet}$), or permeate to the cytoplasm where antioxidant enzymes must control cellular levels of ROS (*SOD*, *CAT* and *APX*). On the other hand, metals would activate phytochelatin synthase (*PCS*) triggering the synthesis of phytochelatins (*PC*). These biothiols would form complexes that are thought to be transported to the vacuoles.

1997; Drążkiewicz et al. 2004). Such accumulation of H_2O_2 is usually connected with changes in the cellular redox status, which is considered as a component of stress signaling processes (Lamb and Dixon 1997; Pastori and Foyer 2002; De Gara et al. 2010), being part of a complex redox sensing network between several cellular comparments (Foyer and Noctor 2003). Recent evidence also indicates that H_2O_2 accumulation interacts with other relevant signaling molecules like Ca^{2+} (Rentel and Knight 2004). Indeed, Rivetta et al. (1997) described the involvement of Ca^{2+}-calmodulin in Cd toxicity during the early phases of radish seed germination. The relationship between heavy metals and Ca could also be explained by the possible permeation of Cd^{2+} through Ca^{2+} channels (Fig. 3), as was observed in protoplasts prepared from *Vicia faba* stomata cells (Perfus-Barbeoch et al. 2002).

NADPH oxidase is probably the major source of H_2O_2 under heavy metal stress, which accumulated principally in the apoplast after the generation of $O_2^{\bullet -}$ (Romero-Puertas et al. 2004; Rodriguez-Serrano et al. 2009; Fig. 3). Similarly, H_2O_2 was exudated in alfalfa seedling roots shortly after Cd and Hg exposure (90 min), which was partially inhibited by the addition of diphenyleneiodonium chloride (DPI), a NADPH-oxidase inhibitor (Ortega-Villasante et al. 2007). Using the microscale system, we have also observed that NADPH-oxidase activity augmented in alfalfa roots in particular under Cd stress (3 and 10 μM) after 6 h of treatment, confirming the involvement of this plasma membrane enzyme in ROS generation under heavy metal stress (Hernández et al. personal communication). The accumulation of H_2O_2 in the apoplast could also increase the activity of cell wall peroxidases, which would be followed by cell wall stiffening due to cross-linking reactions among cell wall polysaccharides and proteins, as described under Cu stress (Lin et al. 2005). Extensins (hydroxyproline-rich glycoprotein), a group of cell wall structural protein prone to suffer cross-linking reactions, are induced under heavy metal stress (Chai et al. 1998). It has been proposed that, under stress conditions, basic peroxidases could render the cell wall more rigid, which could be correlated with root growth inhibition under Cd stress (Tamás et al. 2007; Sobrino-Plata et al. 2009). In addition, H_2O_2 can be a diffusible signal to trigger at least partially some of the physiological responses to heavy metals (Miller et al. 2008). Tamás et al. (2010) showed in barley root tips that the supply of 10 mM H_2O_2 mimicked the transcriptional response of Cd and Hg, such as the overexpression of stress genes like a cell wall peroxidase.

To counteract the accumulation in ROS caused by heavy metal treatments the plant cells possess versatile antioxidant defence mechanisms in which GSH plays a central role (Schützendübel and Polle 2002; Sharma and Dietz 2009). In addition to

Fig. 3 (continued) Alternatively, some HM^{2+}–PC complexes could be transferred to the shoot via xylem long transport. Signaling molecules like ethylene or jasmonate accumulate, which would cause the overexpression of stress related genes (PRs, SHSPs, ERF, etc.). In shoots, the most significant changes would be the modification of important metabolic processes, such as photochemical reactions and electron transport in the tylakoid. In addition, the overexpression of sulfur metabolism genes and activation of GSH synthesis occur, which are key metabolites in heavy metal detoxification and tolerance, for example ECS this function, GSH is required to tolerate heavy metals via the synthesis of phytochelatins (PCs; Cobbett and Goldsbrough 2002), and it is also known to be conjugated with xenobiotics by the action of GSH S-transferases or to serve as the substrate of GSH peroxidases (Edwards et al. 2000). The redox couple formed by GSH and its oxidised form GSSG (GSH/GSSG ratio), along with ascorbate/dehydroascorbate (AA/DHA), is important to control the cellular redox homeostasis (Noctor 2006). Only under acute Hg-stress conditions characterised by high rates of cell death in alfalfa seedlings grown in the microscale hydroponic system, we could detect a displacement of the GSH/GSSG equilibrium towards the accumulation of GSSG (Ortega-Villsante et al. 2007). After just 24 h of treatment with 50 μM Cu there was an impairment of the GSH/GSSG and AA/DHA redox pairs in *Arabidopsis leaves* (Cuypers et al. 2000). Similarly, *Arabidopsis* treated with 100 μM Cd or Cu for 24 h suffered a remarkable accumulation of GSSG under extensive oxidative stress (Xiang and Oliver 1998). Interestingly, there was in parallel an overexpression of GR gene, possibly as a mechanism exerted in highly stressed alfalfa seedlings to restore original GSH cellular levels (Ortega-Villasante et al. 2007). These results imply that the cellular level of GSH is tightly controlled relative to GSSG, by means of higher glutathione reductase (GR) activity or by increased synthesis of GSH. Indeed, the maintenance of a cellular GSH threshold is fundamental to direct many of the processes of heavy metal detoxification and plant defence against stress (Maughan et al. 2010), and for normal development of plant cells (Pasternak et al. 2008). *Arabidopsis thaliana* mutants defective in GSH accumulation, such as *rml1* or *cad2-1*, or plants incubated with BSO suffered a diminution in cell division at the root meristem, which results in abnormal root growth after germination under a low level of cytosolic GSH (Vernoux et al. 2000).

Glutathione is synthesized in a two-step reaction: the first catalyzed by the rate limiting glutamate-cysteine ligase, or γ-glutamyl cysteine synthetase (ECS), and the second catalyzed by glutathione synthase (GS). In *Arabidopsis*, ECS is located exclusively in the chloroplast, whereas GS is found in the chloroplast and in the cytosol (Wachter et al. 2005). Plants treated with BSO, which precisely inhibits ECS and causes a similar phenotypic response to stress than as *Arabidopsis thaliana rml1* or *cad2-1* mutants, suffered from redox cellular imbalance (Fig. 2f; Ortega-Villasante et al. 2005). One common response of plants to heavy metals is the activation of the sulphur metabolism, increasing the assimilation of sulfate and the production of Cys and GSH (Ernst et al. 2008). Indeed, the early overexpression of ECS has been found in a number of experiments, with *Arabidopsis thaliana* cultured cells upon Cd stress (Sarry et al. 2006), in the roots of Cd-treated rice (Aina et al. 2007) or in alfalfa seedlings exposed to Cd and Hg for up to 24 h (Ortega-Villasante et al. 2007). It is feasible that the accumulation of GSH could be regulated by the redox cellular status, as it has been reported recently that the genes encoding the enzymes involved in GSH synthesis are up-regulated by H_2O_2, implying that the accumulation of ROS may trigger GSH accumulation (Queval et al. 2009). In addition, plant ECS responds to the redox cellular status, so under oxidizing conditions ECS forms a homodimer that enhances its activity three-fold, augmenting the level of GSH in the cells (Gromes et al. 2008). As commented, phytochelatins are a large family of biothiols synthesized from GSH, known to bind Cd, Hg and other toxic elements by means of sulfhydryl residues (Cobbett and Goldsbrough 2002). A transient depletion of GSH has been observed under heavy metal stress due to the synthesis of PC, which might result in poorer cellular redox homeostasis, and may probably alter other metabolic processes driven by GSH (Saito 2004). The phytochelatin synthase enzyme (PCS) catalyses the condensation of the γ-Glu-Cys moiety of GSH with the Glu residue of a second GSH, releasing Gly and increasing the length of the PC molecule (Vatamaniuk et al. 2004; Clemens 2006). Interestingly, the PCS defective *Arabidopsis cad1-3* mutant suffered a remarkable and rapid increase in GSH concentration after short-term Cd and Hg treatments, which was not used eventually to synthesize PCs. Therefore, this implies that the GSH synthesis pathway was induced under heavy metal stress, and might be the limiting step in metal tolerance (Carrasco-Gil et al. 2011). The metal–PC complex formed is mainly transported to the vacuole where is safely stored, although it cannot be discounted that a portion of the metal might be secreted to the xylem, permitting the translocation of the metal to the shoots (Saathoff et al. 2011). However, *Arabidopsis cad1-3* mutants were capable of translocating more Cd to the shoot, indicating that transporters of the HMA family would have more free cytosolic Cd as transportable solute (Wong and Cobbett 2009).

4 Phytohormone Signaling Pathways

The accumulation of H_2O_2 driven by plasma membrane NADPH-oxidases are thought to induce the formation of lipid peroxides (i.e., fatty acid radicals or FA-OO; Neil et al. 2002). This oxidative process, together with the activity of several classes of lipoxigenases, generate oxylipins, which are precursors of jasmonate (JA; Turner et al. 2002; Foreman et al. 2003). Several short-term experiments have shown that heavy metals can induce the overexpression of several genes encoding LOXs. Thus, the exposure to μM levels of Cd and Cu for 24 h caused the overexpression of LOX in *Arabidopsis* (Remans et al. 2010). Similar results were observed by Maksymiec et al. (2005) in *Arabidopsis* treated with excess of Cd or Cu, seedlings that accumulated JA after just 6–7 h of incubation. In addition, Rakwal et al. (1996) showed that Cu induced a rapid increase of JA content in *Oryza sativa* excised leaves. A biphasic JA accumulation was found in Cu or Cd stressed plants, e.g., a rapid one within a few hours, and a slow one after several days (Maksymiec et al. 2005) may be the reason of the usually observed rapid decrease in growth processes and senescence intensification of plants after prolonged exposure to the heavy metals.

Ethylene could also be part of the signaling process triggered by the exposure of plants to heavy metals (Sanitá di Toppi and Gabbrielli 1999). The first evidence came from the studies of Sandmann and Böger (1980), which showed that Cu induced the synthesis of ethylene, leading to cell senescence. It is feasible that the growth inhibition observed in dicotyledoneous plants treated with Cu may be mediated by ethylene (Maksymiec and Krupa 2006). Cu stimulated ethylene production possibly via an augmentation of ACC synthase enzymatic activity in *Arabidopsis plants* (Arteca and Arteca 2007). Similarly, pea plants exposed to 50 μM Cd for 14 days accumulated a remarkable amount of ethylene in leaves (Rodriguez-Serrano et al. 2009), which would be part of the signaling process involved in the defence responses to heavy metals (Guo and Ecker 2004).We have concluded a transcriptomic analysis of the early responses of alfalfa seedlings to 3 μM Hg using the microscale hydroponic system. Our results support the idea that ethylene might be part of the signalling molecules that are induced by heavy metals, and we found that transcription factors responding to ethylene (such as ERF1) are up-regulated after just 3 h of treatment (Montero-Palmero et al., personal communication). Ethylene acts by the activation of the mitogen-activated protein kinase (MAPK) cascade (Guo and Ecker 2004), signal transduction process that is also activated by Cd and Cu (Yeh et al. 2003; Jonak et al. 2004), suggesting the ethylene-mediated signaling processes might be induced in plant cells exposed to heavy metals. There might be interactions between different phytohormones, as has been found that there was a cross-talk between JA, ethylene and other phytohormones to mediate the expression of genes like cytochrome P450 under several biotic and abiotic stresses (Narusaka et al. 2004). On the other hand, Metwally et al. (2003) observed that Cd toxicity might be alleviated by salicylic acid, which is a known as a signaling factor that often blocks the JA pathway (Gupta et al. 2000). Some parallels have been found in the responses in plants exposed to heavy metals and to JA, for example, at the gene expression pattern. Xiang and Oliver (1998) found common responses between plants treated with Cd or Cu and JA in the transcription profile of genes involved in the GSH metabolism. A similar effect was observed in the induction of VSP2 (vegetative-storage protein) and MAPK transcription (Mira et al. 2002; Agrawal et al. 2003a, b; Kim et al. 2003). Secondary metabolites and pathogenesis-related proteins have been described in heavy metal-stressed plants (Cruz-Ortega and Ownby 1993; Rakwal et al. 1996; Pitta-Alvarez et al. 2000; Schützendübel and Polle 2002), in which transcription is mediated by JA and/or ethylene, implying that biotic and abiotc stresses may share some signaling processes (Mithöfer et al. 2004). Interestingly, Ghoshroy et al. (1998) and Mittra et al. (2004) showed that a pre-exposure to a mild dose of Cd increased plant resistance to both viral and fungus infection. Poschenrieder et al. (2006) suggested that some plants may increase tolerance to pathogens by taking advantage of the induction of defences like glucanases, chitinases or proteinases; which genes were overexpressed under heavy metal stress (Przymusiński and Gwóźdź 1999).

5 Conclusion

The information provided by short-term experiments using appropriate experimental settings indicates that several signalling processes are activated, in which H_2O_2, jasmonate and ethylene play a central role. The microscale hydroponic system might provide new data, as it is possible to study the changes that occur at the cellular level. Future work should be directed to describe in detail the components that permit the perception of the stress induced by heavy metals, and how different phytohormones and signaling components interact. Functional tools available (i.e. mutants) from the studies of biotic interactions could be used to the characterization of metal responsive components, due to the common responses found.

Acknowledgements This work was supported by Fundación Ramón Areces (http://www.fundacionareces.es), the Spanish Ministry of Science and Innovation (AGL2010-15151-PROBIOMET), Comunidad de Madrid (EIADES S2009/AMB-1478), and Junta de Comunidades de Castilla-La Mancha (FITOALMA2, POII10-0087–6458).

References

Agrawal GK, Jwa NS, Agrawal SK, Tamogami S, Iwahashi H, Rakwal R (2003a) Cloning of novel rice allene oxide cyclase (OsAOC): mRNA expression and comparative analysis with allene oxide synthase (OsAOS) gene provide insight into the transcriptional regulation of octadecanoid pathway biosynthetic genes in rice. Plant Sci 164:979–992

Agrawal GK, Tamogami S, Iwahashi H, Agrawal VP, Rakwal R (2003b) Transient regulation of jasmonic acid-inducible rice MAP kinase gene (OsBWMK1) by diverse biotic and abiotic stresses. Plant Physiol Biochem 41:355–361

Aina R, Labra M, Fumagalli P, Vannini MM, Cucchi U, Bracale M, Sgorbati S, Citterio S (2007) Thiol-peptide level and proteomic changes in response to cadmium toxicity in *Oryza sativa* L. Roots. Environ Exp Bot 59:381–392

Arteca RN, Arteca JM (2007) Heavy-metal-induced ethylene production in *Arabidopsis thaliana*. J Plant Physiol 164:1480–1488

Böddi B, Oravecz AR, Lehoczki E (1995) Effect of cadmium on organization and photoreduction of protochlorophyllide in dark-grow leaves and etioplast inner membrane preparations of wheat. Photosynthetica 31:411–420

Carrasco-Gil S, Álvarez-Fernández A, Sobrino-Plata J, Millán R, Carpena-Ruiz RO, Leduc DL, Andrews JC, Abadía J, Hernández LE (2011) Complexation of Hg with phytochelatins is important for plant Hg tolerance. Plant Cell Environ 34:778–791

Chai TY, Didierjean L, Burkard G, Genot G (1998) Expression of a green tissue-specific 11 kDa proline-rich protein gene in bean in response to heavy metals. Plant Sci 133:47–56

Chaoui A, Mazhoudi S, Ghorbal MN, Ferjani EE (1997) Cadmium and zinc induction of lipid peroxidation and effects on antioxidant enzymes activities in bean (*Phaseolus vulgaris* L.). Plant Sci 127:139–147

Cho UH, Park JO (2000) Mercury-induced oxidative stress in tomato seedlings. Plant Sci 156:1–9

Cho UH, Seo NH (2005) Oxidative stress in *Arabidopsis thaliana* exposed to cadmium is due to hydrogen peroxide accumulation. Plant Sci 168:113–120

Clemens S (2006) Evolution and function of phytochelatin synthases. J Plant Physiol 163:319–332

Cobbett CS, Goldsbrough P (2002) Phytochelatins and metallothioneins: roles in heavy metal detoxification and homeostasis. Annu Rev Plant Biol 53:159–182

Cruz-Ortega R, Ownby JD (1993) A protein similar to PR (pathogen-related) proteins is elicited by metal toxicity in wheat roots. Physiol Plant 89:211–219

Cuypers A, Vangronsveld J, Clijsters H (2000) Biphasic effect of copper on the ascorbate-glutathione pathway in primary leaves of *Phaseolus vulgaris* seedlings during the early stages of metal assimilation. Physiol Plant 110:512–517

De Gara L, Locato V, Dipierro S, De Pinto MC (2010) Redox homeostasis in plants. The challenge of living with endogenous oxygen production. Respir Physiol Neurobiol 173:S13–S19

Demirevska-Kepova K, Simova-Stoilova L, Stoyanova Z, Hölzer R, Feller U (2004) Biochemical changes in barley plants after excessive supply of copper and manganese. Environ Exp Bot 52:253–266

Drążkiewicz M, Skórzyńska-Polit E, Krupa Z (2004) Copper-induced oxidative stress and antioxidant defence in *Arabidopsis thaliana*. Biometals 17:379–387

Edwards R, Dixon RA, Walbot V (2000) Plant glutathione S-transferases: enzymes with multiple functions in sickness and in health. Trends Plant Sci 5:193–198

Ernst WHO, Krauss GJ, Verkleij JAC, Wesenberg D (2008) Interaction of heavy metals with the sulphur metabolism in angiosperms from an ecological point of view. Plant Cell Environ 31:123–143

Fojtová M, Fulnečková J, Fajkus J, Kovařík A (2002) Recovery of tobacco cells from cadmium stress is accompanied by DNA repair and increased telomerase activity. J Exp Bot 53:2151–2158

Foreman J, Demidchik V, Bothwell JHF, Mylona P, Miedema H, Torres MA, Linstead P, Costa S, Brownelee C, Jones JDG, Davies JM, Dolan L (2003) Reactive oxygen species produced by NADPH oxidase regulate plant cell growth. Nature 422:442–446

Foyer CH, Noctor G (2003) Redox sensing and signalling associated with reactive oxygen in chloroplasts, peroxisomes and mitochondria. Physiol Plant 119:355–364

Garnier L, Simon-Plas F, Thuleau P, Agnel JP, Blein JP, Ranjeva R, Montillet JL (2006) Cadmium affects tobacco cells by a series of three waves of reactive oxygen species that contribute to cytotoxicity. Plant Cell Environ 29:1956–1969

Ghoshroy S, Freedman K, Lartey R, Citovsky V (1998) Inhibition of plant viral systemic infection by non-toxic concentration of cadmium. Plant J 13:591–602

Gratão PL, Polle A, Lea PJ, Azevedo RA (2005) Making the life of heavy metal-stressed plants a little easier. Funct Plant Biol 32:481–494

Gromes R, Hothorn M, Lenherr ED, Rybin V, Scheffzek K, Rausch T (2008) The redox switch of γ-glutamylcysteine ligase via a reversible monomer-dimer transition is a mechanism unique to plants. Plant J 54:1063–1075

Guo H, Ecker JR (2004) The ethylene signaling pathway: new insights. Curr Opin Plant Biol 7:40–49

Gupta V, Willits MG, Galzebrook J (2000) *Arabidopsis thaliana* EDS4 contributes to salicylic acid (SA)-dependent expression of defense responses. Evidence for inhibition of jasmonic acid signaling by SA. Mol Plant Microbe Interact 13:503–511

Hall JL (2002) Cellular mechanisms for heavy metal detoxification and tolerance. J Exp Bot 53:1–11

Hagemeyer J (2004) Ecophysiology of plant growth under heavy metal stress. In: Prasad NMV (eds) Heavy metal stress in plants: from biomolecules to ecosystems. Springer, Berlin Heidelberg, pp 201–222

Hernández LE, Cooke DT (1997) Modification of the root plasma membrane lipid composition of cadmium-treated *Pisum sativum*. J Exp Bot 48:1375–1381

Hernández LE, Gárate A, Carpena-Ruiz RO (1997) Effects of cadmium on the uptake, distribution and assimilation of nitrate in *Pisum sativum*. Plant Soil 189:97–106

Hernández LE, Lozano-Rodríguez E, Gárate A, Carpena-Ruiz RO (1998) Influence of cadmium on the uptake, tissue accumulation and subcellular distribution of manganese in pea seedlings. Plant Sci 132:139–151

Jonak C, Nagkagami H, Hirt H (2004) Heavy metal stress. Activation of distinct mitogen-activated protein kinase pathways by copper and cadmium. Plant Physiol 136:3276–3283

Kim JA, Agrawal GK, Rakwal R, Han KS, Kim KN, Yun CH, Heu S, Park SY, Lee YH, Jwa NS (2003) Molecular cloning and mRNA expression analysis of a novel rice (*Oryza sativa* L.) MAPK kinase kinase, *OsEDR1*, an ortholog of *Arabidopsis AtEDR1*, reveal its role in defense/stress signalling pathways and development. Biochem Biophys Res Com 300:868–876

Kowalewska G, Falkowski L, Hoffmann SK, Szczepaniak L (1987) Replacement of magnesium by copper (II) in the chlorophyll porphyrin ring of planktonic algae. Acta Physiol Plant 9:43–52

Lamb C, Dixon R (1997) The oxidative burst in plant disease resistance. Annu Rev Plant Physiol Plant Mol Biol 48:251–275
Lin CH, Chen LM, Liu ZH (2005) Rapid effect of copper on lignin biosynthesis in soybean roots. Plant Sci 168:855–861
Lozano-Rodríguez E, Hernández LE, Bonay P, Carpena-Ruiz RO (1997) Distribution of cadmium in shoots and root tissues of maize and pea plants. Physiological disturbances. J Exp Bot 48:123–128
Maksymiec W (2007) Signaling responses in plants to heavy metal stress. Acta Physiol Plant 29:177–187
Maksymiec W, Krupa Z (2006) The effects of short-term exposition to Cd, excess Cu ions and jasmonate on oxidative stress appearing in *Arabidopsis thaliana*. Environ Exp Bot 57:187–194
Maksymiec W, Wianowska D, Dawidowicz AL, Radkiewicz S, Mardarowicz M, Krupa Z (2005) The level of jasmonic acid in *Arabidopsis thaliana* and *Phaseolus coccineus* plants under heavy metal stress. J Plant Physiol 162:1338–1346
Maksymiec W (1997) Effect of copper on cellular processes in higher plants. Photosynthetica 34: 321–342
Maughan SC, Pasternak M, Cairns N, Kiddle G, Brach T, Jarvis R, Hass F, Nieuwland J, Lim B, Müller C, Salcedo-Sora E, Kruse C, Orsel M, Hell R, Miller AJ, Bray P, Foyer CH, Murray JAH, Meyer AJ, Cobbett CS (2010) Plant homologs of the *plasmodium falciparum* chloroquine-resistance transporter, *Pf*CRT, are required for glutathione homeostasis and stress responses. Proc Natl Acad Sci U S A 107:2331–2336
Metwally A, Finkemeier I, Georgi M, Dietz KJ (2003) Salicylic acid alleviates the cadmium toxicity in barley seedlings. Plant Physiol 132:272–281
Miller G, Shulaev V, Mittler R (2008) Reactive oxygen signalling and abiotic stress. Physiol Plant 133:481–489
Mira H, Martinez N, Peñarrubia L (2002) Expression of vegetative-storage protein gene from Arabidopsis is regulated by copper, senescence and ozone. Planta 214:939–946
Mithöfer A, Schulze B, Boland W (2004) Biotic and heavy metal stress response in plants: evidence for common signals. FEBS Lett 566:1–5
Mittler R, Vanderauwera S, Gollery M, Van Breusegem F (2004) Reactive oxygen gene network of plants. Trends Plant Sci 9:490–498
Mittra B, Ghosh P, Henry SL, Mishra J, Das TK, Ghosh S, Babu CR, Mohanty P (2004) Novel mode of resistance to *Fusarium* infection by a mild dose pre-exposure of cadmium in wheat. Plant Physiol Biochem 42:781–787
Narusaka Y, Narusaka M, Seki M, Umezawa T, Ishida J, Nakajima M, Enju A, Shinozaki K (2004) Crosstalk in the responses to abiotic and biotic stresses in Arabidopsis: analysis of gene expression in cytochrome P450 gene superfamily by cDNA microarray. Plant Mol Biol 55:327–342
Neill SJ, Desikan R, Clarke A, Hurs RD, Hancock JT (2002) Hydrogen peroxide and nitric oxide as signaling molecules in plants. J Exp Bot 53:1237–1247
Noctor G (2006) Metabolic signalling in defence and stress: the central roles of soluble redox couples. Plant Cell Environ 29:409–425
Noctor G, Foyer CH (1998) Ascorbate and glutathione: keeping active oxygen under control. Annu Rev Plant Physiol Plant Mol Biol 49:249–279
Nouari I, Ammar WB, Yousef NB, Daoud DBM, Ghorbal NH, Zarrouk M (2006) Comparative study of cadmium effects on membrane lipid composition of *Brassica juncea* and *Brassica napus* leaves. Plant Sci 170:511–519
Olmos E, Martínez-Solano JR, Piqueras A, Hellín E (2003) Early steps in the oxidative burst induced by cadmium in cultured tobacco cells (BY-2 line). J Exp Bot 54:291–301
Ortega-Villasante C, Hernández LE, Rellán-Álvarez R, Del Campo FF, Carpena-Ruiz RO (2007) Rapid alteration of cellular redox homeostasis upon exposure to cadmium and mercury in alfalfa seedlings. New Phytol 176:96–107

Ortega-Villasante C, Rellán-Álvarez R, Del Campo FF, Carpena-Ruiz RO, Hernández LE (2005) Cellular damage induced by cadmium and mercury in *Medicago sativa*. J Exp Bot 56:2239–2251

Pasternak M, Lim B, Wirtz M, Hell R, Cobbett CS, Meyer AJ (2008) Restricting glutathione biosynthesis to the cytosol is sufficient for normal plant development. Plant J 53:999–1012

Pastori GM, Foyer CH (2002) Common components, networks, and pathways of cross-tolerance to stress. The central role of "redox" and abscisic acid-mediated controls. Plant Physiol 129:460–468

Patra M, Bhowmik N, Bandopadhyay B, Sharma A (2004) Comparison of mercury, lead and arsenic with respect to genotoxic effects on plant systems and the development of genetic tolerance. Environ Exp Bot 52:199–223

Perfus-Barbeoch L, Leonhardt N, Vavasseur A, Forestier C (2002) Heavy metal toxicity: cadmium permeates through calcium channels and disturbs the plant water status. Plant J 32:539–548

Pitta–Álvarez SI, Spollansky TC, Giulietti AM (2000) The influence of different biotic and abiotic elicitors on the production and profile of tropane alkaloids in hairy root cultures of *Brugmansia candida*. Enzyme Microbial Tech 26, 252–258

Poschenrieder C, Tolrà R, Barceló J (2006) Can metals defend plants against biotic stress? Trends Plant Sci 11:288–295

Przymusiński R, Gwóźdź EA (1999) Heavy metal-induced polypeptides in lupin roots are similar to pathogenesis-related proteins. J Plant Physiol 154:703–708

Queval G, Thominet D, Vanacker H, Miginiac-Maslow M, Gakičre B, Noctor G (2009) H_2O_2-Activated up-regulation of glutathione in *Arabidopsis* involves induction of genes encoding enzymes involved in cysteine synthesis in the chloroplast. Mol Plant 2:344–356

Rakwal R, Tamogami S, Kodama O (1996) Role of jasmonic acid as a signaling molecule in copper chloride-elicited rice phytoalexin production. Biosci Biotech Biochem 60:1046–1048

Radmer R, Kok B (1974) Kinetic observation of the System II electron acceptor pool isolated by mercuric ion. Biochim. Biophys. Acta 357:177–180

Rellán-Álvarez R, Ortega-Villasante C, Álvarez-Fernández A, del Campo FF, Hernández LE (2006) Stress responses of *Zea mays* to cadmium and mercury. Plant Soil 279:41–50

Remans T, Opdenakker K, Smeets K, Mathijsen D, Vangronsveld J, Cuypers A (2010) Metal-specific and NADPH oxidase dependent changes in lipoxygenase and NADPH oxidase gene expression in *Arabidopsis thaliana* exposed to cadmium or excess copper. Funct Plant Biol 37:532–544

Rentel M, Knight MR (2004) Oxidative stress-induced calcium signaling in *Arabidopsis*. Plant Physiol 135:1471–1479

Rivetta A, Negrini N, Cocucci M (1997) Involvement of Ca^{2+}-calmodulin in Cd^{2+} toxicity during the early phases of radish (*Raphanus sativus* L.) seed germination. Plant Cell Environ 20:600–608

Rodriguez-Serrano M, Romero-Puertas MC, Pazmiño DM, Testillano PS, Risueño MC, del Río LA, Sandalio LM (2009) Cellular response of pea plants to cadmium toxicity: cross talk between reactive oxygen species, nitric oxide, and calcium. Plant Physiol 150:229–243

Romero-Puertas MC, Palma JM, Gómez M, del Rio LA, Sandalio LM (2002) Cadmium causes the oxidative modification of proteins in pea plants. Plant Cell Environ 25:677–686

Romero-Puertas MC, Rodriquez-Serrano M, Corpas FJ, Gómez M, Del Rio LA, Sandalio LM (2004) Cadmium-induced subcellular accumulation of O_2^{2-} and H_2O_2 in pea leaves. Plant Cell Environ 27:1122–1134

Saathoff AJ, Ahner B, Spanswick RM, Walker LP (2011) Detection of phytochelatin in the xylem sap of *Brassica napus*. Environ Eng Sci 28:103–111

Saito K (2004) Sulfur assimilatory metabolism: the long and smelling road. Plant Physiol 136:2443–2450

Sandmann G, Böger P (1980) Copper-mediated lipid peroxidation processes in photosynthetic membranes. Plant Physiol 66:797–800

Sanitá di Toppi L, Gabbrielli R (1999) Response to cadmium in higher plants. Environ Exp Bot 41:105–130

Sarry JE, Kuhn L, Ducruix C, Lafaye A, Junot C, Hugouvieux V, Jourdain A, Bastien O, Fievet JB, Vailhen D, Amekraz B, Moulin C, Ezan E, Garin J, Bourguignon J (2006) The early responses of *Arabidopsis thaliana* cells to cadmium exposure explored by protein and metabolite profiling analyses. Proteomics 6:2180–2198

Schützendübel A, Polle A (2002) Plant responses to abiotic stresses: heavy metal-induced oxidative stress and protection by mycorrhization. J Exp Bot 53:1351–1365

Schützendübel A, Schwanz P, Teichmann T, Gross K, Langenfeld-Heyser R, Goldbold DL, Polle A (2001) Cadmium-induced changes in antioxidant systems, hydrogen peroxide content and differentiation in Scot pine roots. Plant Physiol 127:887–898

Sharma SS, Dietz KJ (2009) The relationship between metal toxicity and cellular redox imbalance. Trends Plant Sci 14:43–50

Skórzyńska E, Baszyński T (1993) The changes in PSII complex polypeptides under cadmium treatment are they of direct or indirect nature. Acta Physiol Plant 15:263–269

Sobkowiak R, Deckert J (2003) Cadmium-induced changes in growth and cell cycle gene expression in suspension-culture cells of soybean. Plant Physiol Biochem 41:767–772

Sobrino-Plata J, Ortega-Villasante C, Flores-Cáceres ML, Escobar C, Del Campo FF, Hernández LE (2009) Differential alterations of antioxidant defenses as bioindicators of mercury and cadmium toxicity in alfalfa. Chemosphere 77:946–954

Somashekaraiah BV, Padmaja K, Prasad ARK (1992) Phytotoxicity of cadmium ions on germinating seedlings of mung bean (*Phaseolus vulgaris*): involvement of lipid peroxides in chlorophyll degradation. Physiol Plant 85:85–89

Tamás L, Mistrík I, Huttová J, Halušková L, Valentovičová K, Zelinová V (2010) Role of reactive oxygen species-generating enzymes and hydrogen peroxide during cadmium, mercury and osmotic stresses in barley root tip. Planta 231:221–231

Tamás L, Ďurčeková K, Halušková L, Huttová J, Mistrík I, Ollé M (2007) Rhizosphere localized cationic peroxidase from barley roots is strongly activated by cadmium and correlated with root growth inhibition. Chemosphere 66:1292–1300

Turner JG, Ellis CH, Devoto A (2002) The jasmonate signal pathway. Plant Cell 14S:153–164

Vatamaniuk OK, Mari S, Lang A, Chalasani S, Demkiv LO, Rea P (2004) Phytochelatin synthase, a dipeptidyl transferase that undergoes multisite acylation with gammaglutamylcysteine during catalysis. Stoichiometric and site-directed mutagenic analysis of AtPCS1-catalyzed phytochelatin synthesis. J Biol Chem 279:22449–22460

Vázquez S, Carpena-Ruiz RO (2005) Use of perlite in cadmium plant studies: an approach to polluted soil conditions. J Environ Monit 7:1355–1358

Vernoux T, Wilson RC, Seeley KA, Reichheld JP, Muroy S, Brown S, Maughan SC, Cobbett CS, Van Montagu M, Inzé D, May MJ, Sung ZR (2000) The root meristemless1/cadmium sensitive2 gene defines a glutathione-dependent pathway involved in initiation and maintenance of cell division during postembryonic root development. Plant Cell 12:97–109

Wang H, Zhao SC, Liu RL, Zhou W, Jin JY (2009) Changes of photosynthetic activities of maize (*Zea mays* L.) seedlings in response to cadmium stress. Photosythetica 47:277–283

Wachter A, Wolf S, Steininger H, Bogs J, Rausch T (2005) Differential targeting of GSH1 and GSH2 is achieved by multiple transcription initiation: implications for the compartmentation of glutathione biosynthesis in the Brassicaceae. Plant J 41:15–30

Weckx JEJ, Clijsters HMM (1997) Zn phytotoxicity induces oxidative stress in primary leaves of *Phaseolus vulgaris*. Plant Physiol Biochem 35:405–410

Wong CKE, Cobbett CS (2009) HMA type P-type ATPase transporters are the major for root to shoot Cd translocation in *Arabidopsis thaliana*. New Phytol 181:71–81

Xiang C, Oliver DJ (1998) Glutathione metabolic genes coordinately respond to heavy metals and jasmonic acid in *Arabidopsis*. Plant Cell 10:1539–1550

Yeh CHM, Hung WCH, Huang HJ (2003) Copper treatment activates mitogen-activated protein kinase signalling in rice. Physiol Plant 119:392–399

Zhigang N, Wang Y, Shaoshan L (2009) Heavy metal-induced DNA damage in *Arabidopsis thaliana* protoplasts measured by single-cell gel electrophoresis. Chin Bull Bot 44:117–123

Genetic and Molecular Aspects of Metal Tolerance and Hyperaccumulation

Elena Maestri and Marta Marmiroli

Abstract Metals in the environment constitute a stress and a selective factor for plants, since they are genotoxic and toxic at high concentration. Plant responses to metals demonstrate the existence of different mechanisms for resistance, tolerance, accumulation and hyperaccumulation. This chapter analyses the most recent literature on the subject to highlight the different approaches which have been applied to elucidate the genetic and molecular bases of tolerance and accumulation. The data show that our knowledge of these events is not yet complete and that new research areas are needed.

1 Introduction

Metals and metalloids are common features in our environment. Some elements are essential nutrients for animals and for plants, whereas others have no known biological function. The capacity for accumulating metals in some plant species is an interesting evolutionary problem, also relevant for its practical application in phytoremediation of metal-contaminated soils and for the effects on mineral nutrition of humans. The study of the genetic and molecular bases of the plant response to metals in the environment is an active scientific field, in which observations from wild plants are linked to observations in crop plants, and in which information from "omics" technologies are used to improve the knowledge gathered through classical genetics. This chapter will summarise the most important results and highlight the diversity of experimental approaches.

E. Maestri (✉) • M. Marmiroli
Department of Environmental Sciences, University of Parma, Parma, Italy
e-mail: elena.maestri@unipr.it; marta.marmiroli@unipr.it

D.K. Gupta and L.M. Sandalio (eds.), *Metal Toxicity in Plants: Perception, Signaling and Remediation*, DOI 10.1007/978-3-642-22081-4_3,

1.1 Metals as Toxicants

At high concentrations many metals are toxic, as also are those metals which are essential for mineral nutrition such as Cu, Zn and Fe (Marmiroli and Maestri 2008). The most interesting feature of metal ions is the high affinity for S, N and O atoms in organic molecules: binding to proteins and nucleic acids can determine toxic effects in animal and plant cells. Some metal ions are active in redox reactions, and they can lead to the formation of radicals. Additionally, metal ions deplete the cellular stores of antioxidant molecules, especially glutathione (GSH), facilitating oxidative stress (Schuetzenduebel and Polle 2002). Sensitive sites in plant cells are the photosynthetic apparatus and the mitochondrial respiration (Clemens 2006; Maksymiec 2007). Their toxicity mechanisms partly explain their mutagenic effect, which is exerted indirectly through production of radicals (Clemens 2006). Metals also interfere with DNA repair mechanisms, essentially by competing with metals such as Mn or Zn essential for the functioning of enzymes (Hartwig 1998).

1.2 Metals as Stressors

It is recognised that high environmental concentration of metals are a factor of stress for plants. Metal stress occurs with other abiotic stresses, usually as secondary or tertiary stress. Metal stress is associated with oxidative stress (Clemens 2006) and water stress (Fusco et al. 2005). Protein denaturation is also a common feature for several abiotic stresses including metal excess (Suzuki et al. 2001). A recent review by Maksymiec (2007) lists a series of proteins which are induced by metals and by other environmental stresses: the list includes heat shock proteins, pathogen responsive proteins, and drought responsive proteins. Interaction at cellular level between different stress factors is mediated by signalling pathways, involving common molecules such as ethylene, jasmonic acid and abscisic acid. Nitric oxide is one of these mediators, even though the results relating to its role are still preliminary (Xiong et al. 2010). The possibility of interaction among metal-induced stress and other abiotic and biotic stresses is also becoming relevant also due the global climate change, which will lead to an increase of many stresses: temperature increase, floods and air pollution will exacerbate their effects when there is metal excess in the environment. Crop plants seem to be particularly exposed (Hashiguchi et al. 2010).

1.3 Defining Metal Tolerance

Tolerance to metals is a term used to indicate that some plants can survive in the presence of metal concentrations which are lethal or in some way noxious to other plants of the same or of different species; a comprehensive definition of hypotolerant, tolerant and hypertolerant plants has been provided by Ernst et al. (2008). As already recognised by Baker (1987), survival and reproduction of a plant in the presence of metals is defined as resistance. Resistance is achieved by two basic mechanisms, avoidance and tolerance. Avoidance is a feature by which plants do not really become exposed in their tissues to the high concentrations of metals in the environment, whereas tolerance is manifested by those plants which can survive even though metal concentration in tissues and organs becomes exceedingly high. In all cases, resistance mechanisms show heritability and variability, and their performance is attributed to genes and to gene products. According to Baker (1987), plants endowed with real tolerance will behave as metallophytes and pseudometallophytes; the former growing exclusively in metal contaminated areas, the latter also growing in non-contaminated areas. On the other hand, plants endowed with avoidance mechanisms will behave as accidental metallophytes, appearing only sporadically in contaminated areas. It can be hypothesised that a high concentration of toxic metals can act on a natural population of plants by selecting for survival and reproduction only those genotypes and variants which are tolerant to the metals. This leads to a bottleneck in which only a few individuals survive to produce the next generations, and these will be mostly tolerant to the metals. The testable prediction is that metal tolerance in populations can evolve rapidly after an event of contamination in the natural environment. Indeed, this has been shown to be the case, occurring in plant populations for which variability in the degree of tolerance was already existing prior to contamination (reviewed in Baker 1987). A recent report suggests that environmental factors related to metal content of soils have played a role in domestication of *Zea mays* L., the maize plant. Evidence of selection on genes encoding for metal transporters have been found, and the same genes map into genomic regions associated with domestication (Vielle-Calzada et al. 2009).

An inducible tolerance has been demonstrated in several cases of abiotic stresses. For instance, thermotolerance to lethal extreme temperatures can be induced in plants by previous exposure to moderately high temperatures (Vierling 1991; Marmiroli et al. 1997). In the case of metals, induction of tolerance was reported in the 1980s but it is not cited in recent literature. An interesting phenomenon described in yeast is "zinc shock" occurring when Zn-deficient cells are supplied with small amounts of Zn. Zinc is rapidly accumulated in the cells due to high activity of plasma membrane transporters, and resistance to the "shock" is obtained through vacuolar sequestration. Genes responsible for vacuolar sequestration, such as ZRC1, can be transcribed during Zn deficiency by a transcription factor which contains Zn atoms (McDiarmid et al. 2003). The induction of ZRC1 transcription at low Zn levels is necessary to confer tolerance to Zn shock. A similar mechanism may also exist in plants. On the other hand, constitutive tolerance refers to cases where an entire species or genus is uniformly tolerant to metals. For instance, *Thlaspi caerulescens* (now reclassified as *Noccaea caerulescens,* J. Presl & C. Presl; F.K. Mey) is endowed with constitutional tolerance to Zn, Cd, Ni, and this is also evident in ecotypes sampled from non-metalliferous soils. However, ecotypes from metalliferous soils show a greater degree of tolerance (Meerts and Van Isacker 1997).

Another issue concerning metal resistance (both avoidance and tolerance) is the case of multiple resistances, when a species or population is resistant to more than one metal at the same time. This has also been described as co-tolerance. A clear example is provided by accessions of *T. caerulescens*: individuals from sites La Calamine and Monte Prinzera exhibit tolerance to both Zn and Ni, which are both present in their environment (Assunçao et al. 2001).

Interaction of resistance to metals and resistance to other stresses has not been thoroughly explored; however, in the past, many studies on heat shock and drought stress did consider possible interactions with the response to heavy metals (see, for example, Gullì et al. 2005). Zhang et al. (2008) identified a correlation between resistance to water deficit, oxidative stress and metals by using transgenic plants overexpressing aquaporin PIP1. The authors postulated that the improved water status of the plant can explain the pleiotropic effect. The possible interactions between metals and organic pollutants in plants has been highlighted in a review (Verkleij et al. 2009), and metal-binding molecules such as glutathione and other antioxidative defences can act simultaneously on both types of pollutants (Rausch et al. 2007).

1.4 Defining Metal Accumulation

Metals move from soil and water into plants via their roots. Many excellent reviews have recently been written to explain our current knowledge on the uptake of metals through the root tissues and transport to the aerial parts (see, for example, other chapters in this book, and also Milner and Kochian 2008; Verbruggen et al. 2009; Verkleij et al. 2009). Some of the issues are discussed later in this chapter. Essentially, a combination of metal transporters, chelator molecules and other physiological features of the plants can lead to a phenomenon termed accumulation in which the concentration of a specific metal inside the plant, on a dry weight basis, exceeds the concentration of that metal in the substrate in which the plant is growing. Additionally, some plant species manifest a feature called hyperaccumulation. This is not easy to define, even though the name conveys the idea that extremely high metal concentrations are reached in the tissues. Internal metal thresholds for defining hyperaccumulators can be found in many papers (e.g., Verbruggen et al. 2009; Kraemer 2010; Maestri et al. 2010) and in chapters in this book. For the definition of hyperaccumulation it is essential that the shoot to root ratio of metal concentration was at least greater than one (Kraemer 2010). The evolution of the hyperaccumulation phenomenon has not yet been clarified; it is speculated that hyperaccumulating plants can benefit from the high metal concentration by gaining protection against predators, herbivores and pathogens (see, for example, Fones et al. 2010). Similar to tolerance, accumulation can exhibit variability among individuals in a population or species. *T. caerulescens* ecotypes show accumulation of Zn and Pb, but the level of accumulation was shown to be higher in ecotypes from non-contaminated soils (Meerts and Van Isacker 1997). Accessions of *T. caerulescens* from different metalliferous sites showed variation in Zn and Ni tolerance, accumulation and root to shoot ratio (Assunçao et al. 2001). On the other hand, analysis of metallicolous and nonmetallicolous populations of *Arabidopsis halleri* (L.) O'Kane & Al-Shehbaz from different sites showed that Zn and Cd accumulation is a constitutive property of the whole species (Bert et al. 2002).

In several reviews, it has been pointed out that hyperaccumulator species occur within many different taxa, suggesting independent evolution in the different species (e.g., Broadley et al. 2001). Hyperaccumulators are, however, more frequent in some plant families, notably the Brassicaceae (Kraemer 2010). Considering the numerosity of the class, the scarcity of hyperaccumulators in monocots seems quite surprising and should be investigated further. Some plant species can accumulate or hyperaccumulate just one metal, while others accumulate more than one. Broadley et al. (2001) analysing data from the literature concerning hundreds of species showed a positive correlation for accumulation between Cu and Cd, Zn and Cd, Zn and Cr, Ni and Cr, whereas Pb accumulation was not associated. This can be taken as a good indication of a common mechanism of uptake and transportation of metal ions. Baker and Brooks (1989) classified some plants in a third category of indicator plants, when they accumulate metals in proportion to the metal content in soils, without showing tolerance to toxic levels. One such plant is *Arabis alpina* L., which accumulates Cd in roots and does not tolerate Cd above 10 μM (Bovet et al. 2006).

2 Genetic Aspects of Tolerance and Accumulation

Metal tolerance and accumulation are complex genetic systems. Data from many species show that different genes organised in complex regulatory networks are responsible for these traits. Nonetheless, experiments with mutants, crosses, and reverse genetics approaches also demonstrate that specific genes can play major roles in determining the traits. These experiments have elucidated the specific roles of gene functions and of gene products, thus contributing to our understanding of the complex phenomenon.

2.1 *Evidence from Classical Mendelian Genetics and Mutants*

The genetic basis of tolerance and accumulation has been studied first through genetic approaches, analysing the segregating progenies of crosses between individuals with contrasting phenotypes. Interspecific crosses are sometimes required to study the hyperaccumulation phenomenon, which is found in different species of the same genus. One such example is the cross between *A. halleri* and *A. lyrata* (ssp. *petraea*) (L.) O'Kane & Al-Shehbaz, divergent for Zn and Cd hyperaccumulation. In the progeny, the traits of hyperaccumulation and tolerance segregate independently for both Zn (Macnair et al. 1999) and Cd (Bert et al. 2003).

Similar conclusions were reached by crossing *T. caerulescens* accessions differing for accumulation and tolerance to Zn (Assunção et al. 2003) and Cd (Zha et al. 2004). In this specific case, non-tolerant plants accumulated more Zn than tolerant plants: this negative correlation can be explained by a pleiotropism of genetic determinants. In these experiments the conclusion is usually that tolerance and accumulation are regulated by one or few major genes, although they might interact in a pleiotropic fashion (Bert et al. 2003; Frérot et al. 2003). Segregating populations for Ni and Zn response demonstrated that for Ni, tolerance and accumulation are also independently inherited gene traits, whereas Ni and Zn accumulation are genetically correlated (Richau and Schat 2009). All authors have therefore concluded that metal hyperaccumulation cannot be considered simply a tolerance strategy. Similar approaches have been applied to search for co-segregation with specific genes. Hassinen et al. (2009) analysed segregation of allelic variants of genes for metallothioneins MT2a and MT3 in *T. caerulescens*. These alleles derived from the metallicolous parental accession had a higher level of expression, but this did not co-segregate with Zn accumulation in the progeny. Mutant analysis is a powerful instrument for genetics. The literature on metal tolerance and accumulation provides examples mainly for the model plant *Arabidopsis thaliana* (L.) Heynh. Mutants for metal tolerance can be selected by screening for increased or decreased survival at high metal concentrations. The selection of Cd sensitive mutants in *A. thaliana* have led to the discovery that one gene, phytochelatin synthase, was responsible for the synthesis of metal ligands (Ha et al. 1999). An *A. thaliana* T-DNA insertional mutant in the gene AtHMA4 shows a lower accumulation of Zn and Cd in leaves as compared to the wild-type plants; at the same time accumulation in roots is higher (Verret et al. 2004). In the same plants, mutants defective for HMA2 and HMA4 show increased sensitivity to Cd and a decrease in Cd shoot translocation; in the double mutant *hma2, hma4* Cd translocation from roots to shoots was completely lost. Therefore, both HMA2 and HMA4 have a role in Cd translocation in *A. thaliana* (Wong and Cobbett 2009). Zinc-sensitive T-DNA insertional mutants of *A. thaliana* led to the identification of gene ZIF1, encoding for a tonoplast protein involved in vacuolar sequestration of Zn (Haydon and Cobbett 2007a). These mutants impaired in ZIF1 expression also show increased content of Zn in shoots, which is consistent with a reduced sequestration of Zn in the vacuoles. Caesium resistant mutants of *A. thaliana* have been isolated taking advantage of the collections of T-DNA insertional mutants (Marmiroli et al. 2009). Insertional mutants offer the possibility of cloning and identifying the mutated gene, but especially of identifying the putative functions involved in the phenotype. In this case, it was demonstrated with molecular and physical techniques that a single gene mutation impairs Cs uptake and translocation, as well as K and Ca homeostasis. Occurrence of mutants is often complicated by the redundancy of gene functions in plants. Guo et al. (2008b) studied *A. thaliana* mutants defective in metallothionein genes MT1a and MT2b and found no decrease in metal tolerance. On the other hand, a triple mutant also defective for phytochelatin synthase is more sensitive to Cu and Cd as compared to the single mutants. Such experiments showed a role of metallothioneins in metal homeostasis to be possibly redundant with the role of phytochelatins.

2.2 Evidence from Quantitative Genetics and Mapping

Genetic crosses have highlighted how metal tolerance was controlled by few genes. Following these first indications, some groups attempted the identification of Quantitative Trait Loci (QTLs) in segregating progenies of crosses between tolerant and non-tolerant accessions. As already mentioned, in the case of *A. halleri*, the absence of contrasting phenotypes within the species, prompted interspecific crosses to obtain segregating progeny; *A. lyrata* was the partner chosen for these studies. The comparative mapping strategy consisted in mapping QTLs in Zn tolerant *A. halleri* (Willems et al. 2007) and then with syntenic information from *A. thaliana* dissecting the QTLs (Roosens et al. 2008b), taking advantage of the genomic resources available for this model species (www.arabidopsis.org) and using common genetic markers. The chromosome location of the QTLs was compared with the position of known gene sequences which had been identified as involved in metal homeostasis (candidate genes). The results showed that two loci for MTP1 (see also 3.2.1) map within QTLs on different chromosomes. Other candidate genes associated to QTLs were HMA4 (see also 3.2.1) and CCH for a copper chaperone protein. At the same time, 20 additional candidate genes showed no association with QTLs in this particular segregating progeny. Courbot et al. (2007) also demonstrated an association of a QTL for Cd tolerance and HMA4. In *T. caerulescens* the segregating progenies used for QTL analysis derived from parentals of different ecotypes; in this case the trait under analysis was hyperaccumulation (Assunçao et al. 2006; Deniau et al. 2006). Different QTLs have been identified for Zn and Cd accumulation in root and shoot using different crosses. Accumulation of Zn and Cd are correlated in the segregating progenies utilised by Deniau et al. (2006), in fact one QTL position accounted for both Zn and Cd concentration in roots. In QTL analysis, association between the segregating genes may vary in different progenies. For instance QTL mapping for Zn hyperaccumulation in *A. halleri* by Filatov et al. (2007) failed to evidence any association with candidate genes identified in other studies. However, it has been shown that one of the QTLs overlaps a QTL for Zn tolerance which contains gene MTP1a (Roosens et al. 2008a). The QTL approach has been applied also to rice, mapping a region of chromosome 7 as a QTL conferring increased accumulation of Cd to cultivar Cho-Ko-Koku (Miyadate et al. 2011). By fine QTL mapping the region was restricted to four coding sequences, including those for metal transporters ZIP8 and HMA3. The gene HMA3 was then shown to be responsible for the phenotype, due to a truncated recessive allele (see also 3.2.3). The utilisation of rice, a model plant with a wide range of research tools available is instrumental to the success of an approach based on exploitation of genetic variation, fine mapping and genetic engineering.

2.3 Evidence from Reverse Genetics and Genetic Engineering

Transgenic plants have been used to study the effect of single genes on metal tolerance and accumulation. Some of the experiments have indeed been performed with the aim of producing plants better suited for phytoremediation, but some interesting genetic knowledge was also gathered. Reviews have elucidated the recent state-of-the-art (for example Kotrba et al. 2009; Maestri and Marmiroli 2011, and references therein). Table 1 summarises some information about plant genes which have been shown to be relevant for tolerance or for accumulation. These include metal transporters, proteins involved in the synthesis of low molecular weight ligands, enzymes of sulphur metabolism.

2.4 Evidence from Natural Populations Variability

Genetic variation in natural populations has also been examined to understand the genetic bases of metal tolerance and accumulation. According to the hypothesis of metals acting as selective factors, it should be expected that populations from metalliferous sites show reduced genetic variability as compared with populations of the same species from non-metalliferous sites. But in *Silene paradoxa* L. this hypothesis was not confirmed: populations growing in copper mine deposits and serpentine outcrops have the same degree of genetic diversity as those from the neighbouring non-contaminated soils (Mengoni et al. 2000). The same was observed in populations of *T. caerulescens* from metal polluted and non-polluted sites analysed with isoenzyme genetic markers (Koch et al. 1998). Studying *A. halleri*, Macnair (2002) confirmed the existence of genetic variability in Zn accumulation both within and between populations from different sites; the variability was heritable, and it was not correlated with the level of Zn contamination in the soil of origin. In this type of study, it is highly important to take into account the reproductive biology of the species. Outcrossing species such as *A. halleri* are expected to have higher genetic diversity as compared to self-fertilising species. In fact, populations of *Thlaspi*/*Noccea* species endemic to serpentine soils showed no genetic variability for Zn and Ni hyperaccumulation (Taylor and Macnair 2006). In *A. halleri* natural populations, Zn tolerance has evolved separately in different areas, and this suggests that mechanisms can be different in different populations, even at the molecular level (Pauwels et al. 2008). The study of genetic variation in genes involved in adaptation and tolerance could be advantageous in this type of study. Swiss populations of *T. caerulescens*,

Table 1 Summary of some genes involved in metal tolerance and accumulation: data are based on experiments with transgenic plants

Gene	Function	Effect on tolerance and accumulation	Reference
MT	Metallothionein	Confers metal tolerance, in some cases increases metal content	Zhigang et al. (2006); Hassinen et al. (2009)
PCS1	Phytochelatin synthase	Pb tolerance–As tolerance and accumulation–Cd/As tolerance and accumulation	Couselo et al. (2010); Wojas et al. (2010); Guo et al. (2008a)
ZAT	Zn transporter	Confers Zn tolerance and accumulation in roots	van der Zaal et al. (1999)
NtCBP4	Ion transporter, Calmodulin binding protein	Confers Ni tolerance and decreased accumulation, Pb sensitivity and accumulation	Arizi et al. (1999)
NRAMP3	Plasma membrane metal transporters	Increases Cd sensitivity, no effect on Ni resistance	Wei et al. (2009)
APS	ATP sulfurylase	Confers Se tolerance and increased accumulation	Pilon-Smits et al. (1999)
CGS	Cystathionine gamma synthase	Increases Se tolerance and volatilisation	Van Huysen et al. (2003)
SMT	Selenocysteine methyltransferase	Confers tolerance to Se compounds, increased Se accumulation and volatilisation	LeDuc et al. (2004)
CpNifS	Selenocysteine lyase	Confers Se tolerance and accumulation	Van Hoewyk et al. (2005)
NAS	Nicotianamine synthase	Ni tolerance/Ni accumulation	Kim et al. (2005); Pianelli et al. (2005)
SAT	Serine acetyltransferase	Ni resistance	Freeman et al. (2004)
OAS-TL	Cysteine synthase	Cd/Ni/Se tolerance	Kawashima et al. (2004)
ATP-PRT	Histidine synthesis	Confers Ni tolerance but not accumulation	Ingle et al. (2005)
PDF	Defensin	Confers Zn tolerance	Mirouze et al. (2006)
MTP3	Metal tolerance protein	Increases Zn accumulation and tolerance	Arrivault et al. (2006)
PDR8	ABC transporter	Confers Pb/Cd tolerance, Cd accumulation	Kim et al. (2007)
ZIF1	Zinc induced facilitator	Confers Zn tolerance	Haydon and Cobbett (2007a)
ZIP4	Zn transporter	Increases Zn accumulation in roots, decreases in shoots	Ishimaru et al. (2007)
CAX2, CAX4	Divalent action/ proton antiporters	Confers Cd/Zn/Mn tolerance	Korenkov et al. (2007)
TSB1	Tryptophan synthase	Confers Cd tolerance without accumulation	Sanjaya et al. (2008)
PIP1	Aquaporin	Increases Cd resistance	Zhang et al. (2008)
MRP7	ABC transporter	Confers Cd tolerance, increases concentration in leaf vacuoles	Wojas et al. (2009)
ECS	Gamma-glutamylcysteine synthetase	Confers Cd tolerance and increased accumulation	Zhao et al. (2010)

(continued)

Table 1 (continued)

Gene	Function	Effect on tolerance and accumulation	Reference
AtPcrs1	Metal binding, unknown function	Confers Cd resistance and decrease in Cd content	Song et al. (2004)
HMA4	P-type ATPase, transporter	Confers specific Zn tissue distribution and tolerance	Hanikenne et al. (2008)
bHLH100	Transcription factor	Confers metal tolerance	van de Mortel et al. (2008)
MTP1	Tonoplast metal transporter	Expression in roots decreases Zn accumulation in shoots. Increases Zn tolerance	Gustin et al. (2009)
GI, GII	Glyoxalase I and II	Confers Zn tolerance, increased Zn content in roots	Singla-Pareek et al. (2006)

compared with genetic markers to evaluate genetic diversity in correlation to metal content in soils (Besnard et al. 2009), showed high genetic differentiation between sites. Higher genetic diversity, such as allele richness, was observed in locations with high Cd concentrations. These studies showed a selective pressure acting on loci potentially involved in adaptation to metal content, such as metal transporters IRT and ZNT.

3 Molecular Aspects of Tolerance and Accumulation

Information about the macromolecules involved in metal tolerance and accumulation were obtained in the course of genetic studies. Specific knowledge emerged by applying different approaches: from isolation of proteins and clones of genes, to the high throughput techniques based on transcriptomics and proteomics (see Chap. 8, this book). Molecular differences between tolerant and non tolerant plants, or between accumulators and non-accumulators, can thus be hypothesized at all steps of the cellular processes from gene to protein and from protein to metabolite. Some of these observations are summarised in Table 2.

3.1 Evidence from Physiology and Biochemistry

To explain the differences between contrasting phenotypes for tolerance and accumulation, physiological and biochemical features have been analysed. We report here a few relevant observations. Milner and Kochian (2008) report the results of a whole series of observations performed in *T. caerulescens*, a hyperaccumulator, in comparison to *Thlaspi arvense* L., a non-accumulator. Starting from the level of root Zn uptake, it was shown that Zn transporters do

Table 2 Processes and features which differentiate hyperaccumulator/tolerant plants from non accumulator/non tolerant plants

Process involved	Altered feature	Examples of specific genes
1.1. Gene structure	1.1.1. Number of loci	HMA4, MTP1, IRT1, ZIP3, ZIP6, ZIP9
	1.1.2. Allelic variants	MT3, MT2a
	1.1.3. Length variants and structural polymorphisms	IRT1, MTP1, HMA4, HMA3
1.2. Transcription	1.2.1. Promoter efficiency	ZNT1/ZIP4, HMA4, ATP-PRT, IRT1, NRAMP3
	1.2.2. Constitutive vs inducible	HMA4, MT1, MT2
	1.2.3. Response to metal excess/deficiency	ZNT1/ZIP4, MT2, IRT1, NRAMP3
	1.2.4. Tissue/cell specificity	ZNT1/ZIP4
	1.2.5. siRNA inhibition	Not reported
1.3. Post-transcriptional modification	1.3.1. Pre-mRNA splicing	MTP1
	1.3.2. Export to cytoplasm	Not reported
2.1. Translation	2.1.1. mRNA stability	Not reported
	2.1.2. Codon usage bias	Not reported
2.2. Post-translational modification and protein activity	2.2.1. Subcellular localisation	MTP1(?), MT3
	2.2.2. Protein stability	Not reported
	2.2.3. Protein functionality	HMA2, PCS, MT3
	2.2.4. Protein degradation	Not reported
	2.2.5. Direct interaction with metals	AtPcrs1

not differ in affinity for Zn but rather the hyperaccumulator has a higher density of transporters in the plasma membrane of root epidermis. Additionally, the affinity for Zn does not decrease at high Zn concentration in soils, as happens in the non-accumulator. These biochemical differences in the transporters' physiology should be explained at the molecular level by cloning and identifying the genes involved. By moving inside the root, the metals are sequestered into vacuoles with higher efficiency in the non-accumulator, thus limiting the transfer to shoots. Xylem loading on the contrary is much higher in the hyperaccumulator, and has been attributed to the P-type ATPase encoded by gene HMA4. But in the aerial parts, Zn accumulates in leaf epidermis and mesophyll cells. To explain the binding capacity for metals, organic acids, metal carriers and ligands have been studied. In the case of *T. caerulescens*, phytochelatins levels do not differ significantly from the non accumulator. A comprehensive review of the role of sulphur-containing compounds in metal tolerance has been given by Ernst et al. (2008), whereas Haydon and Cobbett (2007b) have reviewed the whole series of metal ligands of relevance in plants. Interesting aspects were the interactions of metals with organic acids, histidine, nicotianamine, phytochelatins, glucosinolates, metallothioneins and the regulation of enzymes involved in their synthesis. The synthesis of these compounds has in some cases been associated to an increased tolerance and accumulation, e.g. elevated levels of histidine in Ni hyperaccumulation by *Alyssum lesbiacum* Candargy (Kraemer et al. 1996). Differences in histidine levels were also described by comparing *T. caerulescens* and *T. arvense* for Ni hyperaccumulation (Richau et al. 2009). Cell ligands for metals may contain sulphur, and this has been suggested as a possible mechanism of the interactions between metal response and response to biotic stresses: in many species defence against pathogens and herbivores also relies on S-based compounds (Tolrà et al. 2006). GSH has a central role in the plant response against metals as also against pathogens and parasites (Rausch et al. 2007).

3.2 *Evidence from Gene Cloning*

Earlier studies on genes involved in plant metal tolerance and accumulation took advantage of the complementation of yeast mutants deficient in tolerance or in metal uptake. Zinc transporters (Grotz et al. 1998; Pence et al. 2000; Draeger et al. 2004), metallothioneins and metal-transporting P-type ATPase (Papoyan and Kochian 2004) were studied in this way. The availability of cloned genes allowed a series of interesting studies, as exemplified in the following sections.

3.2.1 Gene Copy Number

Southern blot analysis with cloned genes can provide indications on the copy number. In *A. halleri* two genes are present in multiple loci as compared to other species. The metal transporter gene AhMTP1, homologue of *A. thaliana* ZAT, is triplicated in the *A. halleri* genome but is in single copy in the genomes of *A. thaliana* and *A. lyrata* (Draeger et al. 2004). The loci in *A. halleri* are not linked; they co-segregate with Zn tolerance and justify the high level of expression. Recent evidence showed that the *A. halleri* genome contains five copies of the MTP1 gene (Shahzad et al. 2010). The gene encoding for the P-type ATPase HMA4 is also triplicated in *A. halleri* with 99% sequence identity (Hanikenne et al. 2008) and quadruplicated in *T. caerulescens* (O Lochlainn et al. 2011). At variance with MTP1, the multiple copies of HMA4 are arrayed in tandem. In *T. caerulescens*, the gene IRT1 is duplicated: a full-length sequence and a truncated variant (Plaza et al. 2007) are expressed, respectively, in the Ganges ecotype (Cd accumulating) and in the Prayon ecotype (non-accumulating). The predicted structure of the gene products demonstrates that the shorter protein lacks a metal-binding site and five transmembrane helices, lacking the ability to transport Cd. The suppression of active Cd transport could be seen as an adaptive response to high metal concentrations.

3.2.2 Gene Expression

Cloned gene sequences can be used to compare their expression patterns through northern blots, in species and individuals with contrasting phenotypes. Pence et al.

(2000) compared expression of the Zn transporter ZNT 1 in *T. caerulescens* and *T. arvense*, demonstrating that ZNT1 is highly expressed in roots and shoots in the hyperaccumulator, independently of the Zn supply. But the expression in the non-accumulator is up-regulated at low Zn concentrations. Metallothionein genes have been cloned and compared in ecotypes of *T. caerulescens* with different degree of Zn tolerance, showing higher expression of MT2 in shoots and roots of the Zn-tolerant accession (Hassinen et al. 2007). Tissue-specific expression can also affect the level of tolerance or accumulation. The ZNT1 gene of *T. caerulescens* has higher levels of expression in leaf cells, when compared to the *A. thaliana* homologue ZIP4, enhancing the difference in Zn accumulation in the aerial parts (Milner and Kochian 2008). Promoter efficiency can affect the level of tolerance and accumulation. This was demonstrated for HMA4 in *A. halleri* in transgenic plants: the promoter sequence of HMA4 from *A. thaliana* confers a lower level of transcription whereas the promoters from *A. halleri* are as effective as the cauliflower mosaic virus 35 S promoter (Hanikenne et al. 2008). Quite surprisingly, no attempt has been made to analyse the sequence of promoters in genes identified as relevant for metal tolerance and accumulation, dissecting their sequences and searching for consensus elements. Zientara et al. (2009) analysed the promoter of gene MRP3 from *A. thaliana*, describing its inducibility by Cd, Ni, As, Co, and Pb. However, this promoter does not come from a gene isolated from a tolerant or hyperaccumulator plant.

3.2.3 Sequence Variants

Sequences of cloned genes are available for bioinformatic analysis and alignment to construct phylogenetic trees, to search for variants in coding sequences, to search differences for those promoter sequences which may be related to differential gene expression. The alleles for ZNT1 cloned from the two accessions of *T. caerulescens* Prayon and La Calamine differ by 30 amino acids at the N-terminus and 5 amino acids inside the coding sequence (Assunçao et al. 2001). In a phylogenetic tree, ZIP transporters from angiosperms were compared, including ZNT and IRT genes (Plaza et al. 2007). Allelic variants among ecotypes of *T. caerulescens* are evidenced, pointing for a possible role in metal tolerance. To assess the importance of single amino acid residues, Song et al. (2004) compared the gene AtPcrs1, involved in tolerance to Cd, to other similar sequences present in *A. thaliana* and *Oryza sativa* L.. Modified gene sequences were checked for their capability in conferring Cd tolerance in mutant yeast. In this way, it was demonstrated that a complete Cys-rich region is important for the protein function in Cd tolerance. In *T. caerulescens*, the sequence of the MT3 gene for metallothionein shows differences in accessions with different tolerance and accumulation levels: the allele from a Cd tolerant accession has a modification in the Cys domain which can result in a larger cavity for chelation of metals (Hassinen et al. 2007). An interesting feature has been described for MTP1 in *Thlaspi goesingense* [now reclassified as *Noccaea goesingensis* (Halàcsy) F.K. Mey.], a Ni hyperaccumulator:

two mRNAs have been found as a result of differential splicing of a genomic sequence (Persans et al. 2001). The two resulting proteins are supposed to have a role in sequestration of different metals inside the vacuoles. Alternative splicing has been described as a mechanism for regulating gene expression during heat shock (Sinibaldi and Mettler 1992), and it could be a mechanism for differential accumulation of metals. In the case of metallothionein genes cloned from different accessions of *T. caerulescens*, the alleles differ in the coding sequences, in the length of the 3' untranslated region and in the length of introns (Hassinen et al. 2009). In one cultivar of *O. sativa* identified as Cd accumulator, the phenotype has been attributed to a specific allele of gene HMA3, encoding for a vacuolar metal transporter (Miyadate et al. 2011). The allele carries a deletion in one exon due to variation in the number of tandem repeats, and the accumulator cultivar translocates higher amounts of Cd from roots to shoots as compared to other cultivars.

3.2.4 Structural Information

In silico analysis of coding sequences and of predicted amino acid sequences can provide indications about the putative functions and cellular localisation of the gene products. The coding sequence of the gene AhMTP1 in *A. halleri* contains a microsatellite region, a series of tandem repeats of a 6-bp sequence (CATGAT coding for HisGlu) whose length differs in the alleles from individual accessions (Draeger et al. 2004). The impact of this polymorphism on Zn tolerance has not yet been elucidated. The protein encoded by gene TcHMA4 has the presence of a C-terminal stretch rich in His and Cys residues which confers tolerance to Cd, presumably by providing binding sites for metal ions: tolerance is correlated to the length of this stretch (Papoyan and Kochian 2004). Concerning the subcellular localization of the proteins, the MTP1 gene in *T. caerulescens* is expressed in the tonoplast, whereas the homologue in *T. goesingense*, Ni accumulator, is expressed in the plasma membrane (Milner and Kochian 2008). However, a later paper also showed localisation of MTP1 to the tonoplast in *T. goesingense* (Gustin et al. 2009).

3.3 *Evidence from Transcriptomic Analysis*

In the first decade after 2000, the availability of high throughput analytical techniques has provided exceptional opportunities for RNA profiling and for determining genetic differences between plants with contrasting phenotypes: tolerant versus sensitive, accumulators versus non-accumulators. The study of two species has particularly benefited from this approach: *T. caerulescens* and *A. halleri*, both Brassicaceae. Their high level of relatedness to *A. thaliana* has provided the opportunity to study the transcriptome with microarrays specific for this model plant. The main results from these experiments are summarised in Table 3. In 2004 Becher et al. analysed gene expression in *A. halleri* plants in

Table 3 Microarray transcriptomic analysis in different plants: tolerant, hyperaccumulator, non tolerant, non accumulator

Plants	Comparison	Main observation	Reference
A. halleri and *A. thaliana*	Zn-tolerant hyperaccumulator vs non-tolerant non-accumulator; different Zn concentrations	High constitutive expression of metal homeostasis genes and stress-protection genes in shoots; different response to Zn concentration decrease. Candidate genes confirmed by RealTime PCR analysis	Becher et al. (2004)
A. halleri and *A. thaliana* roots	Zn/Cd-tolerant hyperaccumulator vs non-tolerant non-accumulator at low Zn concentrations	High constitutive expression of metal homeostasis genes: metal transport, nicotianamine and cysteine synthesis	Weber et al. (2004)
A. halleri and *A. thaliana* roots	Cd-tolerant hyperaccumulator vs non-tolerant non-accumulator under Cu and Cd treatment	High constitutive expression of metal homeostasis genes	Weber et al. (2006)
A. halleri and *A. thaliana*	Zn/Cd-tolerant hyperaccumulator vs non-tolerant non-accumulator	High constitutive expression of metal homeostasis genes: metal transporters, metallothioneins, peroxidases	Chiang et al. (2006)
A. halleri and *A. thaliana*	Zn/Cd-tolerant hyperaccumulator vs non-tolerant non-accumulator at different Zn concentrations	High constitutive expression of metal homeostasis genes: metal transporters, metal chelation	Talke et al. (2006)
A. halleri and *A. petraea* and segregating progeny from their cross, roots and leaves	Zn-tolerant hyperaccumulator vs non-tolerant non-accumulator at high or low Zn concentrations	High constitutive expression in accumulator at high and low Zn concentrations of metal homeostasis genes: transporters, signal transduction, carbonic anhydrase, aconitase	Filatov et al. (2006)
T. caerulescens and *T. arvense* shoots	Zn hyperaccumulator vs non-accumulator	High constitutive expression in accumulator: defensin genes, Zn transporters, metallothioneins, GSH synthesis	Hammond et al. (2006)

(continued)

Table 3 (continued)

Plants	Comparison	Main observation	Reference
T. caerulescens roots from accessions Lellingen and La Calamine	Zn-tolerant ccession from metalliferous soil vs non-tolerant accession from non-metalliferous soil	Higher expression in Zn tolerant: Glycoprotein EP1, sulfur metabolism,	Plessl et al. (2005)
T. caerulescens and *A. thaliana* roots	Zn/Cd hyperaccumulator vs non-accumulator at different Zn concentrations	High constitutive expression in accumulator: metal homeostasis, abiotic stress response, lignin biosynthesis	Van de Mortel et al. (2006)
T. caerulescens and *A. thaliana* roots	Zn/Cd hyperaccumulator vs non-accumulator under Zn or Cd treatment	High constitutive expression in accumulator: metal transporters, lignin biosynthesis, GSH synthesis, sulphate metabolism	Van de Mortel et al. (2008)

response to low and high Zn concentrations. Further experiments by Filatov et al. (2006) examined the differential gene expression between *A. halleri* and *A. petraea*, extending the observations to F3 families of the segregating progenies. Genes with higher expression in the accumulator plant were also highly expressed in the accumulator progenies, identifying the heritability of this trait. Other transcriptomic studies which identified several interesting candidate genes differentially expressed in the hyperaccumulator *A. halleri* were performed by Chiang et al. (2006), Talke et al. (2006); Weber et al. (2004, 2006). Hammond et al. (2006) showed the transcriptomes of *T. caerulescens* and *T. arvense* could be effectively analysed with *A. thaliana* gene chips. They showed differential expression for thousands of genes, including many involved in metal homeostasis. In 2006 and 2008, Van de Mortel et al. analysed gene expression in *T. caerulescens* in comparison to *A. thaliana*, identifying high levels of expression for several genes for metal transport, as ten genes of the ZIP family. Other genes with high levels of expression included transcription factors, enzymes of lignin biosynthesis, glutathione biosynthesis and sulphur metabolism. In particular, they demonstrate several differences in the transcription response to metal excess or deficiency.

4 Conclusion

In a recent review, the state of the art knowledge of regulatory networks in abiotic stress response in plants was highlighted (Urano et al. 2010). It has been shown how transcriptomics, metabolomics and proteomics have all provided a large number of data useful to explain how plants adapt to stress through cellular, physiological and molecular mechanisms. The literature on metal tolerance and accumulation in plants, which has been reviewed here, shows how our knowledge has progressed taking advantage of many approaches. However, some aspects have not yet been fully addressed in relation to metal stress and response of the plants. One example is the lack of detailed description of the structure and of the regulation of promoter sequences from the genes identified as differentially induced. Other features which might become relevant are the synthesis of small RNAs (miRNAs) with impacts on transcriptional regulation, and the existence of post-translational processing of proteins. A feature that will deserve more attention is certainly the role of duplication of gene sequences in tolerant and accumulator plants, an interesting phenomenon which could help to explain the evolution of these phenotypes.

Acknowledgments We thank the authors of papers and reviews cited in the chapter, and apologise to the authors who have not been cited due to space constraints. We acknowledge financial support by PRIN contracts (Italian Ministry of University). We thank Prof. Nelson Marmiroli for critical reading of the manuscript and helpful suggestions.

References

Arizi T, Sunkar R, Kaplan B, Fromm H (1999) A tobacco plasma membrane calmodulin-binding transporter confers Ni^{2+} tolerance and Pb^{2+} hypersensitivity in transgenic plants. Plant J 20:171–182

Arrivault S, Senger T, Kraemer U (2006) The *Arabidopsis* metal tolerance protein AtMTP3 maintains metal homeostasis by mediating Zn exclusion from the shoot under Fe deficiency and Zn oversupply. Plant J 46:861–879

Assunçao AGL, Da Costa MP, de Folter S, Vooijs R, Schat H, Aarts MGM (2001) Elevated expression of metal transporter genes in three accesssons of the metal hyperaccumulator *Thlaspi caerulescens*. Plant Cell Environ 24:217–226

Assunçao AGL, Pieper B, Vromans J, Lindhout P, Aarts MG, Schat H (2006) Construction of a genetic linkage map of *Thlaspi caerulescens* and quantitative trait loci analysis of zinc accumulation. New Phytol 170:21–32

Assunção AGL, Ten Bookum WM, Nelissen HJM, Vooijs R, Schat H, Ernst WHO (2003) A cosegregation analysis of zinc (Zn) accumulation and Zn tolerance in the Zn hyperaccumulator *Thlaspi caerulescens*. New Phytol 159:383–390

Baker AJM (1987) Metal tolerance. New Phytol 106(Suppl s1):93–111

Baker AJM, Brooks RR (1989) Terrestrial higher plants which hyperaccumulate metallic elements. A review of their distribution, ecology and phytochemistry. Biorecovery 1:81–126

Becher M, Talke IN, Krall L, Kraemer U (2004) Cross-species microarray transcript profiling reveals high constitutive expression of metal homeostasis genes in shoots of the zinc hyperaccumulator *Arabidopsis halleri*. Plant J 37:251–268

Bert V, Bonnin I, Saumitou-Laprade P, de Laguérie P, Petit D (2002) Do *Arabidopsis halleri* from nonmetallicolous populations accumulate zinc and cadmium more effectively than those from metallicolous populations? New Phytol 155:47–57

Bert V, Meerts P, Saumitou-Laprade P, Salis P, Gruber W, Verbruggen N (2003) Genetic basis of Cd tolerance and hyperaccumulation in *Arabidopsis halleri*. Plant Soil 249:9–18

Besnard G, Basic N, Christin P-A, Savova-Bianchi D, Galland N (2009) *Thlaspi caerulescens* (Brassicaceae) population genetics in western Switzerland: is the genetic structure affected by natural variation of soil heavy metal concentration? New Phytol 181:974–984

Bovet L, Kammer PM, Meylan-Bettex M, Guadagnuolo R, Matera V (2006) Cadmium accumulation capacities of *Arabis alpina* under environmental conditions. Environ Exp Bot 57:80–88

Broadley MR, Willey NJ, Wilkins JC, Baker AJM, Mead A, White PJ (2001) Phylogenetic variation in heavy metal accumulation in angiosperms. New Phytol 152:9–27

Chiang HC, Lo JC, Yeh KC (2006) Genes associated with heavy metal tolerance and accumulation in Zn/Cd hyperaccumulator *Arabidopsis halleri*: a genomic survey with cDNA microarray. Environ Sci Technol 40:6792–6798

Clemens S (2006) Toxic metal accumulation, responses to exposure and mechanisms of tolerance in plants. Biochimie 88:1707–1719

Courbot M, Willems G, Motte P, Arvidsson S, Roosens N, Saumitou-Laprade P, Verbruggen N (2007) A major quantitative trait locus for cadmium tolerance in *Arabidopsis halleri* colocalizes with *HMA4*, a gene encoding a heavy metal ATPase. Plant Physiol 144:1052–1065

Couselo JL, Navarro-Avino J, Ballester A (2010) Expression of the phytochelatin synthase TaPCS1 in transgenic aspen, insight into the problems and qualities in phytoremediation of Pb. Int J Phytoremediation 12:358–370

Deniau AX, Pieper B, Ten Bookum WM, Lindhout P, Aarts MG, Schat H (2006) QTL analysis of cadmium and zinc accumulation in the heavy metal hyperaccumulator *Thlaspi caerulescens*. Theor Appl Genet 113:907–920

Draeger DB, Desbrosses-Fonrouge A-G, Krach C, Chardonnens AN, Meyer RC, Saumitou-Laprade P, Kraemer U (2004) Two genes encoding *Arabidopsis halleri* MTP1 metal transport proteins co-segregate with zinc tolerance and account for high *MTP1* transcript levels. Plant J 39:425–439

Ernst WHO, Krauss GJ, Verkleij JAC, Wesenberg D (2008) Interaction of heavy metals with the sulphur metabolism in angiosperms from an ecological point of view. Plant Cell Environ 31:123–143

Filatov V, Dowdle J, Smirnoff N, Ford-Lloyd B, Newbury HJ, Macnair M (2006) Comparison of gene expression in segregating families identifies genes and genomic regions involved in a novel adaptation, zinc hyperaccumulation. Mol Ecol 15:3045–3059

Filatov V, Dowdle J, Smirnoff N, Ford-Lloyd B, Newbury HJ, Macnair M (2007) A quantitative trait loci analysis of zinc hyperaccumulation in *Arabidopsis halleri*. New Phytol 174:580–590

Fones H, Davis CAR, Rico A, Fang F, Smith JAC, Preston GM (2010) Metal hyperaccumulation armors plants against disease. PLoS Pathog 6(9):e1001093

Freeman JL, Persans MW, Nieman K, Albrecht C, Peer W, Pickering IJ, Salt DE (2004) Increased glutathione biosynthesis plays a role in nickel tolerance in *Thlaspi* nickel hyperaccumulators. Plant Cell 16:2176–2191

Frérot H, Petit C, Lefèbvre C, Gruber W, Collin C, Escarré J (2003) Zinc and cadmium accumulation in controlled crosses between metallicolous and nonmetallicolous populations of *Thlaspi caerulescens* (Brassicaceae). New Phytol 157:643–648

Fusco N, Micheletto L, Dal Corso G, Borgato L, Furini A (2005) Identification of cadmium-regulated genes by cDNA-AFLP in the heavy metal accumulator *Brassica juncea* L. J Exp Bot 56:3017–3027

Grotz N, Fox T, Connolly E, Park W, Guerinot ML, Eide D (1998) Identification of a family of zinc transporter genes from *Arabidopsis* that respond to zinc deficiency. Proc Natl Acad Sci USA 95:7220–7224

Gullì M, Rampino P, Lupotto E, Marmiroli N, Perrotta C (2005) The effect of heat stress and cadmium ions on the expression of a small hsp gene in barley and maize. J Cereal Sci 42:25–31

Guo J, Dai X, Xu W, Ma M (2008a) Overexpressing GSH1 and AsPCS1 simultaneously increases the tolerance and accumulation of cadmium and arsenic in *Arabidopsis thaliana*. Chemosphere 72:1020–1026

Guo W-J, Meetam M, Goldsbrough PB (2008b) Examining the specific contribution of individual Arabidopsis metallothioneins to copper distribution and metal tolerance. Plant Physiol 146:1697–1706

Gustin JL, Loureiro ME, Kim D, Na G, Tikhonova M, Salt DE (2009) MTP1-dependent Zn sequestration into shoot vacuoles suggests dual roles in Zn tolerance and accumulation in Zn-hyperaccumulating plants. Plant J 57:1116–1127

Ha S-B, Smith AP, Howden R, Dietrich WM, Bugg S, O'Connell MJ, Goldsbrough PB, Cobbett CS (1999) Phytochelatin synthase genes from *Arabidopsis* and the yeast *Schizosaccharomyces pombe*. Plant Cell 11:1153–1164

Hammond JP, Bowen HC, White PJ, Mills V, Pyke KA, Baker AJM, Whiting SN, May ST, Broadley MR (2006) A comparison of the *Thlaspi caerulescens* and *Thlaspi arvense* shoot transcriptomes. New Phytol 170:239–260

Hanikenne M, Talke IN, Haydon MJ, Lanz C, Nolte A, Motte P, Kroymann J, Weigel D, Kraemer U (2008) Evolution of metal hyperaccumulation required *cis*-regulatory changes and triplication of *HMA4*. Nature 453:391–396

Hartwig A (1998) Carcinogenicity of metal compounds: possible role of DNA repair inhibition. Toxicol Lett 102–103:235–239

Hashiguchi A, Ahsan N, Komatsu S (2010) Proteomics application of crops in the context of climatic changes. Food Res Int 43:1803–1813

Hassinen VH, Tervahauta AI, Halimaa P, Plessl M, Peraniemi S, Schat H, Aarts MGM, Servomaa K, Karenlampi SO (2007) Isolation of Zn-responsive genes from two accessions of the hyperaccumulator plant *Thlaspi caerulescens*. Planta 225:977–989

Hassinen VH, Tuomainen M, Peraniemi S, Schat H, Karenlampi SO, Tervahauta AI (2009) Metallothioneins 2 and 3 contribute to the metal-adapted phenotype but are not directly linked to Zn accumulation in the metal hyperaccumulator, *Thlaspi caerulescens*. J Exp Bot 6:187–196

Haydon MJ, Cobbett CS (2007a) A novel major facilitator superfamily protein at the tonoplast influences zinc tolerance and accumulation in Arabidopsis. Plant Physiol 143:1705–1719

Haydon MJ, Cobbett CS (2007b) Transporters of ligands for essential metal ions in plants. New Phytol 174:499–506

Ingle RA, Mugford ST, Rees JD, Campbell MM, Smith JAC (2005) Constitutively high expression of the histidine biosynthetic pathway contributes to nickel tolerance in hyperaccumulator plants. Plant Cell 17:2089–2106

Ishimaru Y, Masuda H, Suzuki M, Bashir K, Takahashi M, Nakanishi H, Mori S, Nishizawa NK (2007) Overexpression of the OsZIP4 zinc transporter confer disarrangement of zinc distribution in rice plants. J Exp Bot 58:2909–2915

Kawashima CG, Noji M, Nakamura M, Ogra Y, Suzuki KT, Saito K (2004) Heavy metal tolerance of transgenic tobacco plants over-expressing cysteine synthase. Biotechnol Lett 26:153–157

Kim S, Takahashi M, Higuchi K, Tsunoda K, Nakanishi H, Yoshimura E, Mori S, Nishizawa NK (2005) Increased nicotianamine biosynthesis confers enhanced tolerance of high levels of metals, in particular nickel, to plants. Plant Cell Physiol 46:1809–1818

Kim DY, Bovet L, Maeshima M, Martinoia E, Lee Y (2007) The ABC transporter AtPDR8 is a cadmium extrusion pump conferring heavy metal resistance. Plant J 50:207–218

Koch M, Mummenhoff K, Hurka H (1998) Systematics and evolutionary history of heavy metal tolerant *Thlaspi caerulescens* in Western Europe: evidence from genetic studies based on isozyme analysis. Biochem Syst Ecol 26:823–838

Korenkov V, Hirschi K, Crutchfield JD, Wagner GJ (2007) Enhancing tonoplast Cd/H antiport activity increases Cd, Zn, and Mn tolerance, and impacts root/shoot Cd partitioning in *Nicotiana tabacum* L. Planta 226:1379–1387

Kotrba P, Najmanova J, Macek T, Ruml T, Mackova M (2009) Genetically modified plants in phytoremediation of heavy metal and metalloid soil and sediment pollution. Biotechnol Adv 27:799–810

Kraemer U (2010) Metal hyperaccumulation in plants. Annu Rev Plant Biol 61:517–534

Kraemer U, Cotter-Howells JD, Charnock JM, Baker AJM, Smith JAC (1996) Free histidine as a metal chelator in plants that accumulate nickel. Nature 379:635–638

LeDuc DL, Tarun AS, Montes-Bayon M, Meija J, Malit MF, Wu CP, AbdelSamie M, Chiang C-Y, Tagmount A, deSouza M, Neuhierl B, Boeck A, Caruso J, Terry N (2004) Overexpression of selenocysteine methyltransferase in Arabidopsis and indian mustard increases selenium tolerance and accumulation. Plant Physiol 135:377–383

Macnair MR (2002) Within and between population genetic variation for zinc accumulation in *Arabidopsis halleri*. New Phytol 155:59–66

Macnair MR, Bert V, Huitson SB, Saumitou-Laprade P, Petit D (1999) Zinc tolerance and hyperaccumulation are genetically independent characters. Proc Biol Sci 266:2175–2179

Maestri E, Marmiroli N (2011) Transgenic plants for phytoremediation. Int J Phytorem 1 First published on 4 April 2011 (First). doi:10.1080/15226514.2011.568549

Maestri E, Marmiroli M, Visioli G, Marmiroli N (2010) Metal tolerance and hyperaccumulation: costs and trade-offs between traits and environment. Environ Exp Bot 68:1–13

Maksymiec W (2007) Signaling responses in plants to heavy metal stress. Acta Physiol Plant 29:177–187

Marmiroli N, Maestri E (2008) Health implications of trace elements in the environment and the food chain. In: Prasad MNV (ed) Trace elements as contaminants and nutrients: consequences in ecosystems and human health. Wiley, New York

Marmiroli N, Restivo FM, Smith CJ, Di Cola G, Maestri E, Tassi F (1997) Induction of heat shock response and acquisition of thermotolerance in callus cultures of *Gerbera jamesonii*. In Vitro Cell Dev Biol Plant 33:49–55

Marmiroli M, Visioli G, Antonioli G, Maestri E, Marmiroli N (2009) Integration of XAS techniques and genetic methodologies to explore Cs-tolerance in *Arabidopsis*. Biochimie 91:180–191

McDiarmid CW, Milanick MA, Eide DJ (2003) Induction of the ZRC1 metal tolerance gene in zinc-limited yeast confers resistance to zinc shock. J Biol Chem 278:15065–15072

Meerts P, Van Isacker N (1997) Heavy metal tolerance and accumulation in metallicolous and non-metallicolous populations of *Thlaspi caerulescens* from continental Europe. Plant Ecol 133:221–231

Mengoni A, Gonnelli C, Galardi F, Gabbrielli R, Bazzicalupo M (2000) Genetic diversity and heavy metal tolerance in populations of *Silene paradoxa* L. (Caryophyllaceae): a random amplified polymorphic DNA analysis. Mol Ecol 9:1319–1324

Milner MJ, Kochian LV (2008) Investigating heavy-metal hyperaccumulation using *Thlaspi caerulescens* as a model system. Ann Bot 102:3–13

Mirouze M, Sels J, Richard O, Czernic P, Loubet S, Jacquier A, François IEJA, Cammue BPA, Lebrun M, Berthomieu P, Marques L (2006) A putative novel role for plant defensins: a defensin from the zinc hyper-accumulating plant, *Arabidopsis halleri*, confers zinc tolerance. Plant J 47:329–342

Miyadate H, Adachi S, Hiraizumi A, Tezuka K, Nakazawa N, Kawamoto T, Katou K, Kodama I, Sakurai K, Takahashi H, Satoh-Nagasawa N, Watanabe A, Fujimura T, Akagi H (2011) OsHMA3, a P1B-type of ATPase affects root-to-shoot cadmium translocation in rice by mediating efflux into vacuoles. New Phytol 189:190–199

O Lochlainn SO, Bowen HC, Fray RG, Hammond JP, King GJ, White PJ, Graham NS, Broadley MR (2011) Tandem quadruplication of HMA4 in the zinc (Zn) and casmium (Cd) hyperaccumulator Noccaea caerulescens. PLoS One 6(3):e17814

Papoyan A, Kochian LV (2004) Identification of Thlaspi caerulescens genes that may be involved in heavy metal hyperaccumulation and tolerance. Characterization of a novel heavy metal transporting ATPase. Plant Physiol 136:3814–3823

Pauwels M, Roosens N, Frérot H, Saumitou-Laprade P (2008) When population genetics serves genomics: putting adaptation back in a spatial and historical context. Curr Opin Plant Biol 11:129–134

Pence NS, Larsen PB, Ebbs SD, Letham DLD, Lasat MM, Garvin DF, Eide D, Kochian LV (2000) The molecular physiology of heavy metal transport in the Zn/Cd hyperaccumulator *Thlaspi caerulescens*. Proc Natl Acad Sci USA 97:4956–4960

Persans MW, Nieman K, Salt DE (2001) Functional activity and role of cation-efflux family members in Ni hyperaccumulation in *Thlaspi goesingense*. Proc Natl Acad Sci USA 98:9995–10000

Pianelli K, Mari S, Marques L, Lebrun M, Czernic P (2005) Nicotianamine over-accumulation confers resistance to nickel in *Arabidopsis thaliana*. Transgenic Res 14:739–748

Pilon-Smits EAH, Hwang SB, Lytle CM, Zhu YL, Tai JC, Bravo RC, Leustek T, Terry N (1999) Overexpression of ATP sulfurylase in *Brassica juncea* leads to increased selenate uptake, reduction and tolerance. Plant Physiol 119:123–132

Plaza S, Tearall KL, Zhao F-J, Buchner P, McGrath SP, Hawkesford MJ (2007) Expression and functional analysis of metal transporter genes in two contrasting ecotypes of the hyperaccumulator *Thlaspi caerulescens*. J Exp Bot 58:1717–1728

Plessl M, Rigola D, Hassinen V, Aarts MGM, Schat H, Ernst D (2005) Transcription profiling of the metal-hyperaccumulator *Thlaspi caerulescens* (J & C PRESL). Z Naturforsch C 60:216–223

Rausch T, Gromes R, Liedschulte V, Mueller I, Bogs J, Galovic V, Wachter A (2007) Novel insight into the regulation of GSH biosynthesis in higher plants. Plant Biol 9:565–572

Richau KH, Schat H (2009) Intraspecific variation of nickel and zinc accumulation and tolerance in the hyperaccumulator *Thlaspi caerulescens*. Plant Soil 314:253–262

Richau KH, Kozhevnikova AD, Seregin IV, Vooijs R, Koevoets PLM, Smith JAC, Ivanov VB, Schat H (2009) Chelation by histidine inhibits the vacuolar sequestration of nickel in roots of the hyperaccumulator *Thlaspi caerulescens*. New Phytol 183:106–116

Roosens NHCJ, Willems G, Saumitou-Laprade P (2008a) Using *Arabidopsis* to explore zinc tolerance and hyperaccumulation. Trends Plant Sci 13:208–215

Roosens NHCJ, Willems G, Godé C, Courseaux A, Saumitou-Laprade P (2008b) The use of comparative genome analysis and syntenic relationships allows extrapolating the position of Zn tolerance QTL regions from *Arabidopsis halleri* into *Arabidopsis thaliana*. Plant Soil 306:105–116

Sanjaya HPY, Su RC, Ko SS, Tong CG, Yang RY, Chan MT (2008) Overexpression of *Arabidopsis thaliana* tryptophan synthase beta 1 (AtTSB1) in *Arabidopsis* and tomato confers tolerance to cadmium stress. Plant Cell Environ 31:1074–1085

Schuetzenduebel A, Polle A (2002) Plant responses to abiotic stresses: heavy metal-induced oxidative stress and protection by mycorrhization. J Exp Bot 53:1351–1365

Shahzad Z, Gosti F, Frérot H, Lacombe E, Roosens N, Saumitou-Laprade P, Berthomieu P (2010) The five *AhMTP1* zinc transporters undergo different evolutionary fates towards adaptive evolution to zinc tolerance in *Arabidopsis halleri*. PLoS Genet 6(4):e1000911

Singla-Pareek SL, Yadav SK, Pareek A, Reddy MK, Sopory SK (2006) Transgenic tobacco overexpressing glyoxalase pathway enzymes grow and set viable seeds in zinc-spiked soils. Plant Physiol 140:613–623

Sinibaldi RM, Mettler IJ (1992) Intron splicing and intron-mediated enhanced expression in monocots. In: Cohn WE, Moldave K (eds) Progress in nucleic acid research and molecular biology. Academic, New York, pp 227–259

Song WY, Martinoia E, Lee J, Kim D, Kim DY, Vogt E, Shim D, Choi KS, Hwang I, Lee Y (2004) A novel family of Cys-rich membrane proteins mediates cadmium resistance in *Arabidopsis*. Plant Physiol 135:1027–1039

Suzuki N, Koizumi N, Sano H (2001) Screening of cadmium-responsive genes in *Arabidopsis thaliana*. Plant Cell Environ 24:1177–1188

Talke IN, Hanikenne M, Kraemer U (2006) Zinc-dependent global transcriptional control, transcriptional deregulation, and higher gene copy number for genes in metal homeostasis of the hyperaccumulator *Arabidopsis halleri*. Plant Physiol 142:148–167

Taylor SI, Macnair MR (2006) Within and between population variation for zinc and nickel accumulation in two species of *Thlaspi* (Brassicaceae). New Phytol 169:505–514

Tolrà R, Pongrac P, Poschenrieder C, Vogel-Mikus K, Regvar M, Barcelo J (2006) Distinctive effects of cadmium on glucosinolate profiles in Cd hyperaccumulator *Thlaspi praecox* and non-hyperaccumulator *Thlaspi arvense*. Plant Soil 288:333–341

Urano K, Kurihara Y, Seki M, Shinozaki K (2010) 'Omics' analyses of regulatory networks in plant abiotic stress responses. Curr Opin Plant Biol 13:132–138

van de Mortel JE, Villanueva LA, Schat H, Kwekkeboom J, Coughlan S, Moerland PD, ver Loren van Themaat E, Koornneef M, Aarts MGM (2006) Large expression differences in genes for iron and zinc homeostasis, stress response, and lignin biosynthesis distinguish roots of Arabidopsis thaliana and the related metal hyperaccumulator Thlaspi caerulescens. Plant Physiol 142:1127–1147

van de Mortel JE, Schat H, Moerland PD, ver Loren van Themaat E, van der Ent S, Blankestijn H, Ghandilyan A, Tsiatsiani S, Aarts MGM (2008) Expression differences for genes involved in lignin, glutathione and sulphate metabolism in response to cadmium in Arabidopsis thaliana and the related Zn/Cd-hyperaccumulator Thlaspi caerulescens. Plant Cell Environ 31:301–324

van der Zaal BJ, Neuteboom LW, Pinas JE, Chardonnens AN, Schat H, Verkleij JAC, Hooykaas PJJ (1999) Overexpression of a novel *Arabidopsis* gene related to putative zinc-transporter genes from animals can lead to enhanced zinc resistance and accumulation. Plant Physiol 119:1047–1056

van Hoewyk D, Garifullina GF, Ackley AR, Abdel-Ghany SE, Marcus MA, Fakra S, Ishiyama K, Inoue E, Pilon M, Takahashi H, Pilon-Smits EAH (2005) Overexpression of AtCpNifS enhances selenium tolerance and accumulation in *Arabidopsis*. Plant Physiol 139:1518–1528

van Huysen T, Abdel-Ghany S, Hale KL, LeDuc D, Terry N, Pilon-Smits EAH (2003) Overexpression of cystathionine-γ-synthase in Indian mustard enhances selenium volatilization. Planta 218:71–78

Verbruggen N, Hermans C, Schat H (2009) Molecular mechanisms of metal hyperaccumulation in plants. New Phytol 181:759–776

Verkleij JAC, Golan-Goldhirsh A, Antosiewisz DM, Schwitzguébel J-P, Schröder P (2009) Dualities in plant tolerance to pollutants and their uptake and translocation to the upper plant parts. Environ Exp Bot 67:10–22

Verret F, Gravot A, Auroy P, Leonhardt N, David P, Nussaume L, Vavasseur A, Richaud P (2004) Overexpression of AtHMA4 enhances root-to-shoot translocation of zinc and cadmium and plant metal tolerance. FEBS Lett 576:306–312

Vielle-Calzada J-P, Martínez de la Vega O, Hernández-Guzmán G, Ibarra-Laclette E, Alvarez-Mejía C, Vega-Arreguín JC, Jiménez-Moraila B, Fernández-Cortés A, Corona-Armenta G, Herrera-Estrella L, Herrera-Estrella A (2009) The Palomero genome suggests metal effects on domestication. Science 326:1078

Vierling E (1991) The roles of heat shock proteins in plants. Annu Rev Plant Phys 42:579–620

Weber M, Harada E, Vess C, Roepenack-Lahaye E, Clemens S (2004) Comparative microarray analysis of Arabidopsis thaliana and Arabidopsis halleri roots identifies nicotianamine synthase, a ZIP transporter and other genes as potential metal hyperaccumulation factors. Plant J 37:269–281

Weber M, Trampczynska A, Clemens S (2006) Comparative transcriptome analysis of toxic metal responses in *Arabidopsis thaliana* and the Cd2 + −hypertolerant facultative metallophyte *Arabidopsis halleri*. Plant Cell Environ 29:950–963

Wei W, Chai T, Zhang Y, Han L, Xu J, Guan Z (2009) The *Thlaspi caerulescens* NRAMP homologue TcNRAMP3 is capable of divalent cation transport. Mol Biotechnol 41:15–21

Willems G, Draeger DB, Courbot M, Godé C, Verbruggen N, Saumitou-Laprade P (2007) The genetic basis of zinc tolerance in the metallophyte *Arabidopsis halleri* ssp. *halleri* (Brassicaceae): an analysis of quantitative trait loci. Genetics 176:659–674

Wojas S, Hennin J, Plaza S, Geisler M, Siemianowski O, Sklodowska A, Ruszczynska A, Bulska E, Antosiewicz DM (2009) Ectopic expression of *Arabidopsis* ABC transporter MRP7 modifies cadmium root-to-shoot transport and accumulation. Environ Poll 157:2781–2789

Wojas S, Clemens S, Sklodowska A, Antosiewicz DM (2010) Arsenic response of AtPCS1- and CePCS-expressing plants–effects of external as(V) concentration on as-accumulation pattern and NPT metabolism. J Plant Physiol 167:169–175

Wong CKW, Cobbett CS (2009) HMA P-type ATPases are the major mechanism for root-to-shoot Cd translocation in *Arabidopsis thaliana*. New Phytol 181:71–78

Xiong J, Fu G, Tao L, Zhu C (2010) Roles of nitric oxide in alleviating heavy metal toxicity in plants. Arch Biochem Biophys 497:13–20

Zha HG, Jiang RF, Zhao FJ, Vooijs R, Schat H, Barker JHA, McGrath SP (2004) Co-segregation analysis of cadmium and zinc accumulation in *Thlaspi caerulescens* interecotypic crosses. New Phytol 163:299–312

Zhang Y, Wang Z, Chai T, Wen Z, Zhang H (2008) Indian mustard aquaporin improves drought and heavy-metal resistance in tobacco. Mol Biotechnol 40:280–292

Zhao C, Qiao M, Yu Y, Xia G, Xiang F (2010) The effect of the heterologous expression of *Phragmites australis* γ-glutamylcysteine synthetase on the Cd^{2+} accumulation of *Agrostis palustris*. Plant Cell Environ 22:877–887

Zhigang A, Cuijie L, Yuangang Z, Yejie D, Wachter A, Gromes R, Rausch T (2006) Expression of BjMT2, a metallothionein 2 from *Brassica juncea*, increases copper and cadmium tolerance in *Escherichia coli* and *Arabidopsis thaliana*, but inhibits root elongation in *Arabidopsis thaliana* seedlings. J Exp Bot 57:3575–3582

Zientara K, Wawrzynska A, Lukomska J, Lopez-Moya JR, Liszewska F, Assunçao AGL, Aarts MGM, Sirko A (2009) Activity of the AtMRP3 promoter in transgenic *Arabidopsis thaliana* and *Nicotiana tabacum* plants is increased by cadmium, nickel, arsenic, cobalt and lead but not by zinc and iron. J Biotechnol 139:258–263

Cadmium and Copper Stress Induce a Cellular Oxidative Challenge Leading to Damage Versus Signalling

Ann Cuypers, Els Keunen, Sacha Bohler, Marijke Jozefczak, Kelly Opdenakker, Heidi Gielen, Hanne Vercampt, An Bielen, Kerim Schellingen, Jaco Vangronsveld, and Tony Remans

Abstract Contamination of soils with the potentially toxic elements cadmium (Cd) and copper (Cu) affects plant growth and crop production, and bioaccumulation in the food chain poses a threat to human health. Toxic levels of Cd or Cu both impose an oxidative challenge on plants, even though these trace elements have a different chemical (non-redox active versus redox-active) and biological (non-essential versus essential element) behaviour. Through (in)direct mechanisms, Cd and Cu cause an increased production of reactive oxygen species (ROS) as well as interference with redox-regulated compounds in different cellular compartments. This chapter highlights general and/or specific mechanisms of interference with the cellular redox homeostasis by Cd and Cu, which may be part of the sensing mechanism to these stresses. Furthermore, it emphasises the metal-induced oxidative challenge and its involvement in either cellular damage and/or downstream signalling responses.

1 Introduction

Together with the industrial revolution and the use of fertilizers and sewage effluents in agriculture, metals are spread into the environment at concentrations significantly exceeding those originating from natural sources (Nriagu and Pacyna 1988). Metals can be subdivided in essential micronutrients like copper (Cu), critical for normal development and plant growth (Marschner 1995) and

A. Cuypers (✉) • E. Keunen • S. Bohler • M. Jozefczak • K. Opdenakker • H. Gielen • H. Vercampt • A. Bielen • K. Schellingen • J. Vangronsveld • T. Remans
Centre for Environmental Sciences, Research Group Environmental Biology, Hasselt University–Campus Diepenbeek, Diepenbeek, Belgium
e-mail: ann.cuypers@uhasselt.be; els.keunen@uhasselt.be; sacha.bohler@uhasselt.be; marijke.jozefczak@uhasselt.be; kelly.opdenakker@uhasselt.be; heidi.gielen@uhasselt.be; hanne.vercampt@uhasselt.be; an.bielen@uhasselt.be; kerim.schellingen@uhasselt.be; jaco.vangronsveld@uhasselt.be; tony.remans@uhasselt.be

D.K. Gupta and L.M. Sandalio (eds.), *Metal Toxicity in Plants: Perception, Signaling and Remediation*, DOI 10.1007/978-3-642-22081-4_4,

non-essential elements such as cadmium (Cd). Whereas deficiencies of micronutrients can seriously disturb normal development, excess of metals in general adversely affects biochemical reactions and physiological processes in organisms causing a major risk to the environment and, more specifically, human health (Cuypers et al. 2009). Therefore, increasing our knowledge on plant metal stress responses is pivotal for making progress in applied technologies such as metal phytoextraction of contaminated soils to improve and secure the health of the environment (Seth et al. 2011). After a general introduction on Cd and Cu uptake and homeostasis, the present chapter will focus especially on the mechanisms of Cd or Cu-induced oxidative challenge (ROS production and antioxidant defence mechanisms) that leads either to oxidative damage or signalling (Fig. 1). The latter

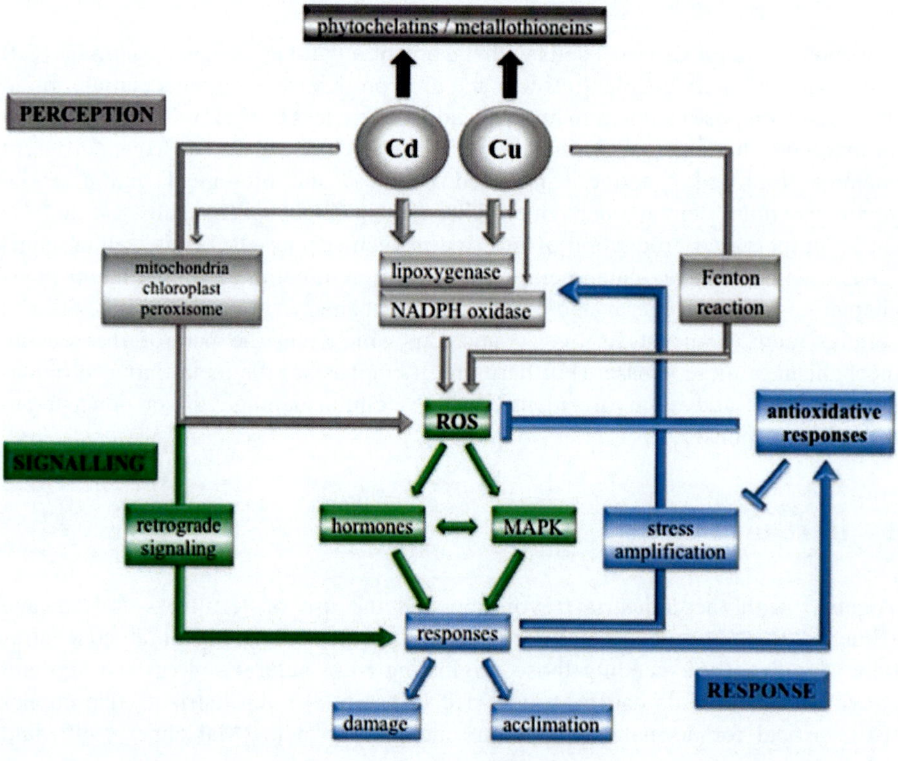

Fig. 1 Generalised model for perception and signalling responses in plants exposed to Cd or excess Cu. Although plant cells immediately react by chelating free metal ions via phytochelatins (*Cd*-specific) and metallothioneins (*Cu*-specific), free Cu can directly cause ROS production via the Fenton reaction. Since Cd is not redoxactive, indirect enzymatic and organellar mechanisms, that could also play a role during exposure to excess Cu, result in a disturbed cellular redox state. This triggers signalling responses via a complex but integrated network of hormones, MAPK and retrograde mechanisms, leading to cellular responses that can influence the new cellular redox state via stress amplification and/or antioxidative responses. Depending on the metal stress intensity and plant species, the outcome will either be damage or acclimation to Cd or excess Cu

is becoming a scientific subject area of intense investigation and will provide essential information to comprehend the cellular responses to metal toxicity.

1.1 Cadmium and Copper Uptake and Homeostasis

As important suppliers of dietary minerals for humans and animals, plants form a bridge between the soil elemental composition and the food chain. Consequently, contaminated soils with potentially toxic elements such as Cd and Cu may affect crop production and the food chain and, hence, human health (Cuypers et al. 2011a).

1.1.1 Uptake of Excess Cu and Cd by the Plant Is Unavoidable

To avoid Cu deficiency, plants have high-affinity Cu transporters belonging to the COPT family to ensure Cu^{+}-uptake under low Cu availability, after reduction of Cu^{2+} by ferric reductase (FRO). The uptake systems when plants are exposed to excess Cu have not yet been revealed. These could be COPT3–5, which are not affected by Cu levels, unlike COPT1 and COPT2 that are down regulated by excess Cu. Otherwise, Cu^{2+}, which is the most abundant form present in soil, could be taken up by ZIP (zinc-regulated transporter/iron-regulated transporter-like or ZRT/IRT-like) transporters (Burkhead et al. 2009; Palmer and Guerinot 2009). However, these bivalent cation transporters are also important uptake systems for non-essential elements such as Cd. This is due to the chemical similarity of Cd^{2+} ions with the ions of these essential elements (Clemens 2006; Perfus-Barbeoch et al. 2002; Verbruggen et al. 2009). Therefore, when growing on soils contaminated with excess metals, unspecific metal uptake seems unavoidable and plants need to have systems in place for chelation and sequestration of these excess metals.

1.1.2 Chelation and Sequestration of Excess Metals

Due to their redox-active properties, free Cu-ions in the cell are very toxic. Therefore, Cu is bound to chaperones that deliver it directly to the particular protein where its redox-active properties will be used in cellular functioning. For example, CCS chaperones deliver Cu to CuZnSOD in the chloroplast, COX chaperones deliver Cu to cytochrome c oxidase in mitochondria, and even for recycling of Cu from senescing tissues, CCH is used as a Cu chaperone (Palmer and Guerinot 2009; Robinson and Winge 2010). Besides these chaperones that are essential in normal functioning, detoxification of metals is achieved by specific chelators, which form non-toxic complexes with metals and facilitate their sequestration away from sensitive sites in cells. In this way, chelation prevents interaction of free metal ions with physiologically important proteins. The favoured ligands of the bivalent cations Cd^{2+} and Cu^{2+} are thiols (SH-groups), which are present on cysteine residues of glutathione (GSH), phytochelatins (PCs) and metallothioneins (MTs).

Glutathione (γ-glu-cys-gly) is the major low molecular mass peptide in plants and is present at the millimolar level within cells. It protects potentially susceptible protein thiol groups from binding free metal ions and consecutively hampering their function (Foyer and Noctor 2011; Herbette et al. 2006; Verbruggen et al. 2009). Phytochelatins (PCs) (γ-(glu-cys)$_n$-gly) are a polymerisation of 2–11 GSH molecules catalysed by phytochelatin synthase (PCS). These polymerised forms are more efficient in chelating several metal ions, because of multiple thiol-binding sites. Whereas Cd bound to GSH can be transported into the vacuole by an ABC transporter, PCs facilitate vacuolar sequestration of Cd (Noctor and Foyer 1998; Verbruggen et al. 2009). *Arabidopsis* mutants, deficient in PCs, are hypersensitive to Cd because they lack this high affinity metal chelator (Howden et al. 1995). Moreover, it has also been shown that PCs are involved in root-to-shoot transport of Cd^{2+} in order to alleviate metal accumulation in root cells (Gong et al. 2003; Van Belleghem et al. 2007).

Metallothioneins (MTs) are small gene-encoded proteins with many thiol groups due to their high cysteine content (Cobbett and Goldsbrough 2002; Guo et al. 2008). Their gene expression is strongly induced by Cu exposure in a number of plant species, and a correlation between MT gene expression and Cu tolerance has been inferred from studies of *Arabidopsis* ecotypes and *Silene* populations (Guo et al. 2008, and references therein). Metallothioneins have been mainly associated with Cu homeostasis. However, expression of a *Brassica juncea* MT gene in *Arabidopsis* increased not only Cu but also Cd tolerance. Zhigang et al. (2006) and Guo et al. (2008) provided evidence that MTs and PCs may have overlapping functions as they contribute both to Cu and Cd tolerance.

1.2 The Perception of Cd and Cu Stress and the Generation of Excess Reactive Oxygen Species

Plants exposed to excess metals generate excessive amounts of reactive oxygen species (ROS) leading to oxidative stress (Sharma and Dietz 2008). Under normal physiological conditions, levels of ROS are tightly controlled by a large antioxidant network (Mittler et al. 2004). This section describes how exposure of plants to Cd or excess Cu can lead to increased production of ROS in different cellular compartments, and how the altered redox-status is part of the perception of Cd and Cu stress by the plant.

1.2.1 Direct and Indirect Mechanisms of ROS Generation

Redox-active metals such as Cu can be directly involved in ROS generation through the formation of hydroxyl radicals (•OH) from H_2O_2 via the Fenton reaction.

Hydroxyl radicals can also be formed from reaction of superoxide ($O_2^{\bullet -}$) with H_2O_2 in the Haber-Weiss reaction. Hydroxyl radicals initiate radical chain reactions that damage various cellular components. For example, together with the protonated form of $O_2^{\bullet -}$, •OH cause non-enzymatic peroxidation of polyunsaturated fatty acids in membranes, leading to membrane damage (Mithöfer et al. 2004).

Cadmium is not redox-active and generates ROS mainly by indirect mechanisms. It has a high affinity for thiyl-, histidyl- and carboxyl-groups of proteins, and can also displace essential ions from their functional site in proteins, thereby inhibiting their proper functioning in electron transport chains or antioxidative defence (Sharma and Dietz 2008; DalCorso et al. 2008). This malfunctioning in its turn results in either elevated electron leakage and hence ROS production or a diminished ROS scavenging, with a net result in enhanced cellular ROS levels. Besides the direct effects of both metals, Cd and Cu also stimulate enzymatic ROS production and elevated ROS levels originating from subcellular organelles as will be discussed in the next sections.

1.2.2 Enzymatic ROS Generation

NADPH oxidases are membrane-localised and catalyze the reduction of molecular oxygen ($^3O_2$) to extracellular $O_2^{\bullet -}$ using intracellular NADPH-derived electrons. These enzymes are also called RBOHs (respiratory burst oxidase homologues) due to their homology with the mammalian respiratory burst oxidase gp91^{phox}. Plant NADPH oxidases are involved in development, in plant responses to biotic and abiotic stress and in signal transduction (reviewed in Glyan'ko and Ischenko 2010; Sagi and Fluhr 2006). Studies with inhibitors suggested that ROS were generated by NADPH oxidase in leaves from pea plants and rice leaves exposed to Cd (Romero-Puertas et al. 2004; Hsu and Kao 2007) and that increased NADPH oxidase activity may be involved in growth inhibition of roots and leaves of plants exposed to Cu (Maksymiec and Krupa 2007). Gene expression analysis of all members of the NADPH oxidase gene family in roots of *Arabidopsis thaliana* seedlings revealed that those genes responsive to the treatment were upregulated by Cd exposure and downregulated by Cu, which may indicate that NADPH oxidase generated ROS production is stimulated by Cd exposure and suppressed under Cu exposure (Remans et al. 2010). Also Yeh et al. (2007) found that NADPH oxidase activity was important for Cd^{2+}-, but not Cu^{2+}-, induced MAP kinase activities in rice roots. However, the involvement of NADPH-oxidases under Cu stress cannot be ruled out as gene expression is induced in leaves of Cu-exposed *Arabidopsis* seedlings (Cuypers et al. 2011b) and Navari-Izzo et al. (2006) also observed an early, but transient induction of NADPH-oxidase activities in Cu-exposed wheat roots.

The mechanism by which excess metals influence NADPH oxidase activity remains to be determined. This influence may occur at any of the steps in the induction and regulation of ROS production by NADPH oxidases. In general, (1)

Ca^{2+} directly binds EF-hand motifs in the cytoplasmic *N*-terminal domain, and (2) stimulates a CDPK (calcium-dependent kinase) that phosphorylates the *N*-terminal domain (Sagi and Fluhr 2006; Kobayashi et al. 2007; Ogasawara et al. 2008). This facilitates binding of Rop GTPase that leads to the activation of NADPH oxidase and ROS production (Wong et al. 2007; Jones et al. 2007). This ROS may then activate Ca^{2+} channels, generating a second Ca^{2+} influx, and a negative feedback loop is created by decreased Rop GTPase-NADPH oxidase interaction at high Ca^{2+} levels. Alternatively, a self-enhancing stimulation of ROS production may be created by direct, Rop GTPase-independent stimulation by Ca^{2+} (Sagi and Fluhr 2001). Whereas no indications are known for the direct activation or inhibition of NADPH oxidase activities or gene expression by Cu, Cd^{2+} ions may directly stimulate NADPH oxidase activity through mimicking Ca^{2+}. Alternatively, excess metals may cause increased intracellular Ca^{2+} levels similar to other stress stimuli that induce ROS generation by NADPH oxidases.

Lipoxygenases (LOXs) catalyse the dioxygenation of polyunsaturated fatty acids producing hydroperoxy fatty acids (Feng et al. 2010). LOX activities are highest in the cytoplasm, but have also been observed in chloroplasts, mitochondria, vacuoles, lipid bodies and membranes (Liavonchanka and Feussner 2006). The main substrates of plant LOXs are linoleic acid (C18:2) and linolenic acid (C18:3), which are converted into either 9- or 13-hydroxyperoxides, depending on the enzyme isoform. These can be turned into a large number of structurally different oxylipins by a range of enzymatic modifications (Mithofer et al. 2004; Montillet et al. 2004). Lipoxygenase activities are involved in normal physiological processes that regulate, e.g., growth and fertility, but are also often induced during stress conditions (Caldelari et al. 2011; Holkova et al. 2009; Vellosillo et al. 2007). For example, controlled LOX activities are involved in root growth under normal physiological conditions through the synthesis of oxylipin signalling molecules that regulate development (Vellosillo et al. 2007), but excessive LOX activity in barley root tips exposed to Cd led to enhanced lipid peroxidation and root growth inhibition (Tamas et al. 2009). Increased LOX activity is observed in *Arabidopsis thaliana* plants under Cd and Cu stress, even though a lower amount of protein was found (Skorzynska-Polit et al. 2006), and in algae exposed to Cd^{2+} the rise of LOX activity was concomitant with the induction of two new isoforms (Kumar et al. 2010). In roots of *Arabidopsis thaliana*, expression of specific isoforms was induced by Cu but not by Cd, whereas other LOX genes were induced by both metals in roots and leaves (Smeets et al. 2008, 2009; Remans et al. 2010; Cuypers et al. 2011b). The significance of the differential induction of different isoforms remains to be revealed, as well as the relative importance of increased LOX activities in synthesising signalling molecules or producing damaging oxidised lipids under different stress intensities. Indeed, non-enzymatic lipid peroxidation may be a more important damaging mechanism than enzymatic lipid peroxidation when plants are exposed to metals. Rahoui et al. (2010), for example, found excessive lipid peroxidation after exposure to Cd of germinating pea seeds, even though they observed decreased LOX activities.

1.2.3 Cd and Cu Disturb Redox Homeostasis in Plant Organelles

ROS are continuously produced as by-products of the normal metabolism in peroxisomes, chloroplasts and mitochondria, due to their oxidising nature and the presence of electron transfer chains in these organelles (Mittler 2002; Halliwell 2006; Cuypers et al. 2009). Both Cd and Cu enhance these organellar ROS generation by disturbing metabolic processes such as photosynthesis and (photo) respiration (Heyno et al. 2008; Bi et al. 2009; Rodríguez-Serrano et al. 2009).

Plant *mitochondria* are suggested to be an important source of abiotic stress-induced ROS, thereby implying their possible involvement in Cd- and Cu-mediated oxidative stress. The mitochondrial electron transfer chain (ETC) is a major site of ROS production in non-photosynthetic cells and photosynthetic tissues in the dark (Dutilleul et al. 2003; Foyer and Noctor 2003). Next to complexes I to IV, plants contain several non-proton-pumping bypasses such as alternative oxidase (AOX) limiting ROS production (Møller 2001). However, the ETC is considered to be a primary target of metal toxicity, and ROS production may increase during metal stress. Heyno et al. (2008) demonstrated an immediate Cd-induced stimulation of ROS generation originating from inside root cells, mainly from the mitochondrial ETC. Bi et al. (2009) observed an increase in ROS accumulation in Cd-exposed *Arabidopsis* mitochondria preceding ROS changes in chloroplasts. However, mitochondrial $O_2^{\bullet -}$ production was preceded by NAPDH oxidase activity at the plasma membrane of Cd-exposed tobacco cell cultures (Garnier et al. 2006).

Plant mitochondria appear to be highly sensitive to ROS-mediated oxidative damage, with the highest content of oxidised proteins as compared to peroxisomes and chloroplasts (Bartoli et al. 2004). Mitochondrial enzymes often require metals as cofactors such as Fe and Cu during electron transfer. However, higher concentrations of these essential and also non-essential metals such as Cd induce ROS production, which inhibits enzyme activity and thus respiration depending on the stress intensity. Exposure to Cd and Cu has been described to affect respiratory gas exchange rates in several plant species (Lösch 2004). Tan et al. (2010) have recently demonstrated great variation in the susceptibility of mitochondrial respiratory chain pathways and matrix enzymes to metal-induced loss of function. The presence of free metal ions in plant mitochondria could be crucial in the initiation and propagation of oxidative stress, resulting in oxidative damage to respiratory and other mitochondrial proteins. Protein oxidative damage could then serve as an alarm signal for initiating plant responses under Cd and Cu stress (Bartoli et al. 2004; Møller and Kristensen 2004).

It has been shown that Cd can accumulate in chloroplasts (Baryla et al. 2001), but the effects of this trace element depend strongly on the plant species and its sensitivity and on the Cd concentration and duration of exposure used during the experiment. Ying et al. (2010) showed a decrease in chloroplast size and an increase in chloroplast number in the hyperaccumulator *Picris diavaricata*, while Baryla et al. (2001) observed a decrease in chloroplast number in *Brassica napus*, and hypothesised that Cd inhibits chloroplast division. Decreases in chlorophyll content are commonly observed (Mobin and Khan 2007; Hédiji et al. 2010; Küpper et al.

1998; Baryla et al. 2001), although there is no agreement on whether it is due to perturbations in chlorophyll synthesis (Hédiji et al. 2010) a substitution of the Mg ion in the chlorophyll molecule by a Cd ion (Küpper et al. 1998) or a decrease in chloroplast number (Baryla et al. 2001). It is generally agreed that exposure to Cd decreases photosynthesis (Mobin and Khan 2007), which is probably due to a combination of events. The electron transport is impaired under Cd stress because of the peroxidation and loss of thylakoid membrane integrity (Kucera et al. 2008; DalCorso et al. 2008). Pagliano et al. (2006) showed that electron transfer through PSII was decreased with increasing Cd concentrations, possibly due to inhibition of the water-splitting complex assembly. Faller et al. (2005) observed that the inhibition of oxygen evolution by Cd decreased in the presence of Ca, and suggested that Cd had substituted the Ca ion needed for the assembly of the complex. Cadmium is a competitive inhibitor of the Ca^{2+} site in the catalytic centre of PSII and blockage of the electron flow in PSII leads to the formation of excited triplet chlorophyll that reacts with molecular oxygen ($^3O_2$) to form the highly reactive singlet oxygen ($^1O_2$) (Kucera et al. 2008; Sharma and Dietz 2008). Reduced abundance of ferredoxin-$NADP^+$-oxidoreductase (FNR) was also observed during Cd treatment (Grzyb et al. 2004, 2011; Durand et al. 2010). This enzyme produces NADPH and can also reduce thioredoxin. NADPH is the major reducing molecule of the chloroplast and is among other things used for the regeneration of ascorbate (AsA) and GSH. Thus, FNR inhibition would lead to an increase in the general oxidation state of the chloroplast and influence the redox-regulation by thioredoxin, causing a decrease in activation of enzymes of the Calvin cycle and a decreased de-activation of pentose phosphate pathway enzymes (Heldt and Heldt 2005; Quick and Neuhaus 1997; Stitt et al. 2010; Semane et al. 2010; Kruger and von Schaewen 2003). Indeed, a decrease of RuBisCO, sedoheptulose–1,7–bisphosphatase and fructose–1,5–bisphosphatase activity and protein abundance was shown in several publications, indicating that there is an impact of Cd on the Calvin cycle (Mobin and Khan 2007; Kieffer et al. 2009; Durand et al. 2010; Chugh and Sawhney 1999; Semane et al. 2010). Although the pentose phosphate pathway is capable of producing NADPH, the fixation of carbon would be inhibited, which is needed for delivering carbon skeletons to the chloroplast-localised synthesis of cysteine, which is one of the components of Cd chelating molecules like GSH and PCs (Howarth et al. 2003; Harada et al. 2002).

These assembled studies show that the effect of Cd on chloroplast processes is twofold, a direct effect of the accumulated Cd ions by substitution of Ca or binding to sulfhydryl groups, and the indirect effect on the redox-regulation of key enzymes. Both effects result in changes in the redox-regulation of key enzymes of chloroplast processes, which could be one of the major reasons for decreased carbon fixation during Cd exposure. Compared to Cd, research on the effects of Cu on chloroplasts is rare. Toxicity only appears at very high concentrations of Cu. Even though it is a naturally occurring cofactor for many proteins, increased concentrations of Cu decrease photosynthesis (Hattab et al. 2009). It has been proposed that Cu does not have a direct effect on the water-splitting complex (Pádua et al. 2010), but could have effects on other components of PSII like the secondary quinine acceptors for which it is a cofactor (Mijovilovich et al. 2009;

Patsikka et al. 1998; Mohanty et al. 1989). Even if potential direct effects of Cu on chloroplast functions are relatively unknown, the fact that Cu can produce ROS through a Fenton reaction (Sutton and Winterbourn 1989) leads to an oxidative stress that can lead to perturbations in redox regulation similar to those described for Cd above.

Peroxisomes contain various enzymes that produce ROS: Xanthine oxidase and NADPH-dependent oxidase generate $O_2^{\bullet -}$, while glycolate oxidase, flavin oxidase and β-oxidation of fatty acids yields H_2O_2. These ROS are removed mainly by catalase (CAT) and superoxide dismutase (SOD) activity (Rodríguez-Serrano et al. 2009 and references therein; Sharma and Dietz 2008). Cadmium causes increased H_2O_2 levels in peroxisomes, induces senescence symptoms and transition into glyoxysomes, and causes an increase in the number and mobility of the peroxisomes (Rodríguez-Serrano et al. 2009; Sharma and Dietz 2008, and references herein), but the biological significance remains to be revealed.

1.2.4 Perception of the Stress Signal

From the above paragraphs, it is clear that Cd and excess Cu influence the redox homeostasis in the cell, due to the accumulation of ROS and changes in the thiol-disulphide status of proteins like thioredoxin. This changed redox status is part of the stress sensing leading to changes in gene expression (Kacperska 2004; Foyer and Noctor 2009; Potters et al. 2010). The cellular redox state comprises the collective status of many interactive redox-active compounds, such as AsA, GSH, NAD(P)H and thioredoxins. Furthermore, redox states can be defined for subcellular sites like mitochondria, chloroplasts and cell walls (Potters et al. 2010). The sites in the cell where ROS are produced or changes in redox-related compounds are generated are likely sensors of stress (Kacperska 2004). Thus, perception of stress is likely occurring through monitoring by several molecules of the redox status and production of ROS species in different cellular compartments (Jaspers and Kangasjärvi 2010). In addition, a number of possible abiotic stress sensing mechanisms at the cell wall–plasmamembrane interface, mediated by receptor-like kinases, mechanosensors, Ca-channels, or phospholipid signalling, has been reviewed by Kacperska (2004).

2 The Oxidative Stress Signature Consists of Altered Redox-Related Gene Expression, Enzyme Activities and Metabolites, and Is Informative for the Oxidative Challenge Induced by Metals

The oxidative challenge induced by metal exposure causes ROS levels to increase via stimulation of pro-oxidative mechanisms or through inhibition of antioxidant defence mechanisms. Antioxidative components can also be stimulated to counteract these increased ROS levels to prevent oxidative damage or to direct oxidative signalling. As described in the previous section, pro-oxidative components can be affected in a metal-specific way. In the next paragraphs, you will notice that antioxidative gene expression, enzyme activities and metabolites can be influenced in a similar way. Depending on the metal stress intensity, i.e. exposure level and duration, and plant species, Cd and Cu can either increase or inhibit antioxidative enzyme activities (Smeets et al. 2009; Cuypers et al. 2011b).

2.1 Superoxide Scavenging by Superoxide Dismutases

Superoxide radicals are the primary ROS formed after electron transfer and superoxide dismutase (SOD) converts these molecules to H_2O_2 and O_2. In plants, three different groups of isoforms exist, each containing a redox-active metal in the active site to perform the reaction: CuZnSOD, FeSOD and MnSOD. In multiple studies, a diverse outcome on SOD activity in plants exposed to Cd was noticed, an increase in wheat, a decrease in pea plants (DalCorso et al. 2008) and no significant effect was seen on total SOD activity in *Arabidopsis* seedlings exposed to either Cu or Cd (Cuypers et al. 2011b). It should be mentioned that different SOD isoforms are expressed in cytosol, chloroplasts/plastids and mitochondria, which are differentially influenced under Cd or Cu stress, and even between roots and shoots. Cuypers et al. (2011b) have shown that exposure to Cd leads to an upregulation of *FSD1* (FeSOD) in roots in contrast to a down-regulation in leaves. The *CSD* (CuZnSOD) isoforms, however, were down-regulated under Cd exposure, whereas they were up-regulated under Cu stress. Mitochondrial *MSD* (MnSOD) transcripts were not so influenced, and nor by Cd or Cu exposure. Posttranscriptional regulation of *CSD1*–2 by microRNA 398 has been described by Sunkar et al. (2006) and also plays a role in metal-specific antioxidant responses (Cuypers et al. 2011b).

MicroRNAs (miRNAs) together with small interfering RNAs (siRNAs) belong to a class of endogenous non-coding small RNAs and are important regulators of plant development and stress responses (Vazquez et al. 2010). Via homologous base pairing, these small RNAs regulate gene transcription by targeting epigenetic modifications at the DNA level, or act posttranscriptionally via RNA cleavage or translational repression (Mallory and Bouché 2008; Jamalkandi and Masoudi-Nejad 2009; Khraiwesh et al. 2010). Several miRNAs could play a role in environmental stress responses, as was shown via microarrays, realtime PCR and sequence analysis of small RNA libraries (Sunkar and Zhu 2004; Hsieh et al. 2009; Huang et al. 2009, 2010; Pant et al. 2009; Wei et al. 2009). Zhou et al. (2008) noticed up- and downregulated miRNAs (miR171, miR319, miR393, miR529, miR166, miR398, miR160 and miR395) after Hg, Cd or Al exposure in *Medicago truncatula*, which were also found in Cd-exposed *Oryza sativa* (Huang et al. 2009). In *Arabidopsis,* miR398 expression is downregulated by Cu excess but induced by Cu deficiency (Sunkar et al. 2006; Abdel-Ghany and Pilon 2008; Cuypers et al. 2011b). In the case of Cu deficiency, CuZn-SOD (*CSD1* and

CSD2) transcripts decrease to enhance Cu levels for vital functions, which causes FeSOD (*FSD1*) to take over $O_2^{\bullet-}$ scavenging (Abdel-Ghany and Pilon 2008; Cohu and Pilon 2007). On the other hand, Cu excess leads to induced *CSD1* and *CSD2* transcription regulated by miR398 reduction (Sunkar et al. 2006; Cuypers et al. 2011b). Yamasaki et al. (2009) found that, according to the Cu availability, miR398 is regulated by SPL7 (SQUAMOSA promoter binding protein-like7) that binds onto GTAC motifs in the miR398 promoter. The expression of miR398 greatly increases during Cd stress, resulting in lower *CSD* transcripts (Cuypers et al. 2011b). The opposite effects on miR398 and CSDs under Cu and Cd stress indicate metal-specific regulation.

2.2 H_2O_2 Scavenging: Catalases and Ascorbate Peroxidases

Hydrogen peroxide scavenging can be accomplished by both catalases (CAT) and ascorbate peroxidases (APX). In the regulation of H_2O_2 contents, CATs have a high reaction rate, but lower affinity to H_2O_2 as compared to APX. So CATs are more involved in H_2O_2 detoxification rather than the regulation of H_2O_2 as a signalling molecule. *Catalases* are heme-containing enzymes catalysing the reduction of H_2O_2 to H_2O without any cellular reducing equivalents. A delicate balance between the CAT isoforms at transcriptional level, downregulation of *CAT2* and upregulation of *CAT1*/3 was noticed in *Arabidopsis* plants exposed to Cd and Cu (Cuypers et al. 2011b). This correlates with the regulatory effects of senescence on CAT activities, indicating that *CAT2* downregulation appears as the initial step in producing an elevated H_2O_2 level, which then leads to the induction of *CAT3* expression and activity (Zimmermann et al. 2006).

Ascorbate peroxidases are class I heme peroxidases important in H_2O_2 scavenging, of which eight types are expressed in *Arabidopsis thaliana*: three cytosolic (APX1, APX2 and APX6), two chloroplastic (stromal APX and thylakoid APX) and three microsomal types (APX3, APX4 and APX5) (Panchuk et al. 2002). This enables a crucial cross-compartment protection since H_2O_2 is transported across biological membranes (Davletova et al. 2005a). APXs catalyse the reduction of H_2O_2 to H_2O and O_2 using AsA as a reducing agent in the AsA–GSH cycle (Fig. 2, cfr. infra), connecting enzymatic antioxidant systems with antioxidant metabolites.

Cytosolic APX1 and APX2 are the best studied isoforms because of their location. Under normal conditions, plants without functional APX1 accumulate H_2O_2, show retarded growth and development and altered stomatal responses (Pnueli et al. 2003). Davletova et al. (2005a) demonstrated oxidation of chloroplastic proteins to occur in *ko-apx1* plants exposed to mild or moderate light stress. Thus, APX1 preserves the cellular reactive oxygen network by protecting the organelles against oxidative stress under normal circumstances and after exposure to various stress factors including metals (Smeets et al. 2008, 2009; Cuypers et al. 2011b). In contrast, *APX2* expression is only induced during stress conditions (Davletova et al. 2005a) and was clearly induced in roots of Cu and Cd

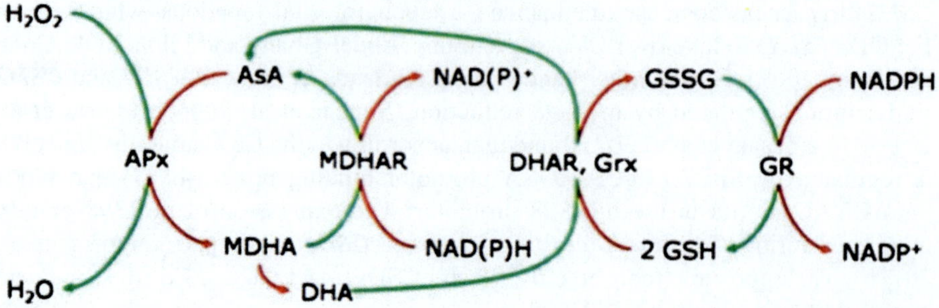

Fig. 2 AsA–GSH cycle. The *red arrows* represent oxidation and the *green arrows* reduction (Adapted from Asada 1999)

exposed *Arabidopsis* seedlings (Cuypers et al. 2011b). At the level of enzyme activities, H_2O_2 accumulation and decreased APX activity was observed in different plants under excess Cu, which appears to be differentially regulated in roots and leaves (Drazkiewicz et al. 2010; Zhang et al. 2010; Cuypers et al. 2011b). Exposure to Cd does evoke H_2O_2 generation (Cuypers et al. 2011b), but to a lesser extent as compared to Cu exposure (Drazkiewicz et al. 2010). However, APX activity increased with increasing Cd concentrations, suggesting the involvement of APX in Cd-induced oxidative stress and tolerance (Cho and Seo 2005; Cuypers et al. 2011b).

2.3 *Detoxification of ROS Via the Ascorbate-Glutathione Cycle*

Distinct from its direct metal-chelating capacities, GSH can indirectly neutralise metal-induced ROS, creating the oxidised glutathione disulphide (GSSG). The antioxidative mechanism, in which GSH participates, is known as the AsA–GSH cycle that is located in various subcellular compartments. In Fig. 2, the successive oxidation and reduction of AsA and GSH in order to create a cyclic transfer of reducing equivalents is shown. This ultimately results in the reduction of H_2O_2 to water using electrons derived from NAD(P)H (Noctor et al. 1998; Mittler et al. 2004; Foyer and Noctor 2005). Metal stress however, causes a highly oxidised redox balance, especially in plant roots. Therefore, AsA and GSH act as important redox buffers, with their oxidation/reduction ratios reflecting cellular toxicity (Cuypers et al. 2011b). Since the GSH–GSSG redox pair can only function with adequate NADPH supplies, GSH is also suggested to function as a cellular NADPH sensor (May et al. 1998). Apart from its role in the AsA–GSH cycle, GSH is also the substrate for GSH peroxidase (GPx), an antioxidative enzyme detoxifying peroxides such as H_2O_2 and lipid peroxides (Ghezzi and Bonetto 2003).

Exposure to Cd and Cu affects the AsA–GSH cycle in plants in both metabolites and antioxidant enzymes. In multiple studies, an immediate change in enzyme activities and/or a strong oxidation of the metabolites was observed under Cu stress (Cuypers et al. 2000; Drazkiewicz et al. 2010). In comparison, Cd exposure also affected the different components of the AsA–GSH cycle but more delayed and/or with a smaller oxidizing potential (Smeets et al. 2005; Drazkiewicz et al. 2010; Cuypers et al. 2011b).

2.4 *Antioxidant Metabolites*

Next to *GSH*, several other metabolites participate in the cellular antioxidant network to combat an increased ROS accumulation during metal stress. In the following paragraphs, these metabolites and their potential involvement in the oxidative challenge mediated by Cd and Cu are discussed.

In plants, *AsA* has a regulatory role during cell growth, cell wall biosynthesis, photosynthesis and cell differentiation. This metabolite is present in all subcellular compartments including the apoplast, chloroplast, cytosol, vacuoles, mitochondria and peroxisomes. In chloroplasts and mitochondria, AsA plays a key protective role by decreasing the damage caused by ROS formed during photosynthetic and respiratory processes (Potters et al. 2002). Ascorbic acid directly reacts with all different ROS ($^1O_2$, $O_2^{\bullet -}$, H_2O_2, •OH) in which AsA is being oxidised (Fig. 2). As mentioned above, the oxidation/reduction ratio of AsA reflects the cellular toxicity level in Cd- and Cu-exposed plants. Besides its role as primary antioxidant, AsA also functions as a secondary antioxidant in the regeneration of other membrane antioxidants such as vitamin E (Hahnel et al. 1999).

While *vitamin E* is synthesised in photosynthetic organisms, it is an essential vitamin in animal and human diet (Maeda and DellaPenna 2007). Vitamin E comprises a class of membrane-located metabolites consisting of four *tocopherols* (α-, β-, δ-, γ-tocopherol) and tocotrienol (Smirnoff 2005). All tocopherols are incorporated in plastid membranes, with their hydrophobic tails associated with membrane lipids and their polar chromanol head situated at the membrane surface (Smirnoff 2005). As electron donor, the chromanol ring lies at the heart of the antioxidative capacities of tocopherols. These metabolites protect unstable polyunsaturated fatty acids (PUFAs) from ROS-mediated lipid peroxidation. α-tocopherol is the most active form in conditions of oxidative stress due to its high abundance in plant leaves (Munne-Bosch and Falk 2004; Yusuf et al. 2010). Tocopherols are suggested to play a significant role in the plant cell antioxidant network. Growing evidence indicates a significant contribution of tocopherols in the metal-induced oxidative challenge at multiple biological organisation levels (Artetxe et al. 2002; Zengin and Munzuroglu 2005; Gajewska and Sklodowska 2007; Collin et al. 2008; Sun et al. 2010; Yusuf et al. 2010). Plants exposed to Cu or Cd show increased tocopherol levels, a shift in tocopherol composition and altered transcription of enzymes involved in the tocopherol biosynthetic pathway (Zengin and Munzuroglu 2005; Collin et al. 2008; Sun et al. 2010). Overexpression or downregulation of transcripts encoding for biosynthetic components of the tocopherol pathway in mutant plants also indicates the importance of tocopherols in antioxidative responses. In *Brassica juncea* mutants deficient in γ-tocopherol methyltransferase (γ-TMT), which is the last enzyme in the α-tocopherol biosynthesis pathway converting γ- into α-tocopherol, tocopherol composition shifted from α- to γ-tocopherol in response to Cd stress (Yusuf et al. 2010). Collin et al. (2008) investigated the role of tocopherol by use of tocopherol deficient (*vte1*) and ascorbate deficient (*vtc*) mutants. *vte1* mutants, showed enhanced lipid peroxidation and oxidative stress in the presence of both Cd and Cu in comparison to wild-type plants. *vtc2* mutants, however, showed wild-type responses to metal exposure, suggesting that vitamin E plays a crucial role in the tolerance of *Arabidopsis thaliana* to metal-induced oxidative stress.

2.5 Description of the Oxidative Stress Signature

Changes in pro-and antioxidative components together are regarded as an oxidative stress signature, and the more components at different biological organisation levels of this stress signature can be measured, the more likely metal-specific differences can be found. Metal-specific differences can be informative of or can be used to unravel the underlying mechanisms of stress perception, signalling and responses.

3 The Oxidative Challenge Can Cause Damage and Trigger Signalling Pathways Leading to Acclimation Responses

Exposure to Cd and excess Cu causes an oxidative challenge that may result in cellular damage. On the other hand, to be able to cope with this oxidative challenge, plants use the produced ROS for the onset of diverse signalling processes that regulate cellular responses leading to cellular protection and/or acclimation. The complete sequence of events between the perception of Cd or Cu stress and the onset of cellular responses has not been revealed, but a number of parameters involved in response signalling to Cd and Cu-induced oxidative challenge have been described. In this section, we will discuss ROS-induced oxidative damage, as well as cellular components that participate in ROS-induced signalling pathways during Cd and Cu stress.

3.1 Metal-Induced Oxidative Damage

Free •OH, produced due to Cu excess via Fenton reactions, or triggered by Cd and Cu via increased production of H_2O_2 and $\bullet O_2^-$ that can take part in Haber-Weiss reactions, can lead to oxidative damage to virtually all biomolecules including lipids, proteins and DNA (Kucera et al. 2008; reviewed in Møller et al. 2007). Lipid peroxidation causes an increase in plasma membrane permeability and leakage of ions from the roots. High rates of lipid peroxidation are observed in Cu exposed *Arabidopsis* seedlings, leading to K^+-leakage in roots (Smeets et al. 2009; Cuypers et al. 2011b). Increased ROS production when *Arabidopsis thaliana* plants are exposed to Cd lead to increased lipid peroxidation (Collin et al. 2008) and protein denaturation (Roth et al. 2006).

3.2 The Cd- and Cu-Induced Oxidative Challenge Activates and Interferes with Signalling Pathways

Although excessive ROS can react with biomolecules, cause cellular damage and may lead to necrosis and cell death, controlled levels of ROS play an important role in modulating signalling networks that control physiological processes and stress responses (Maksymiec 2007; Miller et al. 2010). Oxidative stress is a common factor in many stress situations, but the underlying oxidative signalling may be very specific for each stress response. An important specification may lay in the subcellular localisation, the type, the source, the quantity and the perception of the ROS produced (Miller et al. 2008). Several reviews have been written about the involvement of ROS in abiotic stress signalling (Jaspers and Kangasjärvi 2010; Miller et al. 2008; Foyer and Noctor 2009; Maksymiec 2007).

Mitogen activated protein kinase (*MAPK*) *pathways* are responsive to even small changes in levels of ROS and are important signalling modules that convert signals generated from the receptors/sensors to cellular responses. Several authors have reported the involvement of MAPK-signalling in Cd and Cu stress. In *Medicago sativa* roots as well as in rice roots, the orthologues of *Arabidopsis* MPK3 and MPK6, SAMK/SIMK and OsMPK3/MPK6, respectively, were activated after exposure to excess Cd or Cu ions (Jonak et al. 2004; Yeh et al. 2007). These authors reported that Cd and Cu both induce MAPKs via ROS generation, but that these metals make use of distinct upstream signalling pathways, which is probably related to direct (Cu) or indirect (Cd) ROS generation. Cadmium- but not Cu-induced MAPkinase activity was dependent on NADPH oxidase activity and functional mitochondria (Yeh et al. 2007). Also, in *Arabidopsis thaliana*, MPK3 and MPK6 were activated by Cd in a ROS-dependent manner (Liu et al. 2010). Further, Wang et al. (2010) demonstrated in *Zea mays* that *ZmMPK3* transcript levels are induced after exposure to 500 μM $CdCl_2$ for 6 h.

ROS-stimulated MAPkinase signalling pathways have been extensively examined and described, especially in the model organism *Arabidopsis thaliana*, and involve the activation of, for example, the H_2O_2-responsive serine/threonine kinase *Oxidative signal-inducible 1* (OXI1) (reviewed by Colcombet and Hirt 2008). However, little is known about the exact upstream signalling pathways and the downstream targets of these pathways when plants are exposed to metal stress. Yet, defined end points of some MAPK signalling pathways are also involved in anti-oxidative defence against metal stress. The zinc-finger protein ZAT12 is a MAPK-regulated transcription factor in *Arabidopsis* that in response to H_2O_2 results in the enhanced expression of *APX1* (Davletova et al. 2005b), and also *CAT1* expression is regulated via a MAPK-cascade in *Arabidopsis* (Xing et al. 2007, 2008). Davletova et al. (2005a) suggested that MAPK dependent regulation of NAPDH oxidase (RBOHD) expression may be involved in amplification of the ROS signal and, indeed, Miller et al. (2009) showed that RBOHD mediates systemic signalling in response to abiotic stress. Whereas different parts of the mentioned components are well analysed in plants under Cd and/or Cu exposure (cfr. supra), an integration of these data in combination with a specific MAPK-related research question are essential to reveal the sequence of events in the metal-induced signalling cascade from perception to response.

Cd and Cu-induced MAPK signalling pathways cross-talk with *hormonal pathways*. MPK3 and MPK6, which are activated by Cd and Cu (Jonak et al. 2004; Liu et al. 2010), can also phosphorylate 1-aminocyclopropane-1-carboxylate synthase (ACS) isoforms, resulting in an increase in ethylene production (Kim et al. 2003; Liu and Zhang 2004; Fiil et al. 2009; Xu et al. 2008). Furthermore, Yoo et al. (2008) identified a MKK9–MPK3/6 cascade acting positively in ethylene signal transduction by phosphorylating the nuclear transcription factor EIN3. In addition, the production of other plant hormones like auxins (e.g., IAA), cytokinins, jasmonic acid (JA), salicylic acid (SA) and abscisic acid (ABA) can be influenced by toxic metals such as Cd and Cu, indirectly affecting the ethylene biosynthesis or signal transduction (Maksymiec 2007; Lequeux et al. 2010; Stepanova and Alonso 2005; Argueso et al. 2007; Overmyer et al. 2003; Lin et al. 2009; Chae et al. 2003; Wang et al. 2005; Stepanova et al. 2007).

A large number of *oxylipins* exist in plants that originate from LOX activity and subsequent enzymatic modifications, and they represent a pool of signalling molecules that contribute to the plasticity of responses in plants (Mithofer et al. 2004). Induction of LOX gene expression (Smeets et al. 2008; Remans et al. 2010; Cuypers et al. 2011b) and increased LOX activity have been observed under Cd and Cu stress (Skorzynska-Polit et al. 2006), and JA are shown to be involved in signalling under Cd and Cu stress (Maksymiec and Krupa 2006; Maksymiec 2007). It remains, however, to be determined whether differential expression of LOX genes under Cd and Cu stress underlies the production of specific oxylipin signals (Remans et al. 2010).

All together, it is clear that Cd- and Cu-induced oxidative challenge involves a multitude of interacting signalling components. For example, studies with pea plants exposed to Cd revealed that response signalling is likely to involve a multitude of components like ROS, NO, cGMP, SA, Ca^{2+}, JA and ethylene (Romero-Puertas et al. 2007; Rodríguez-Serrano et al. 2009). Nevertheless, other regulating systems such as microRNAs (cfr. supra) can also take part in the signalling cascade and need to be taken into account in future studies.

3.3 Retrograde Signalling by Cellular Organelles

Mitochondria and chloroplasts are part of an integrated network controlling cellular energy and redox metabolism, which influences the impact of environmental challenges on plant development (Noctor et al. 2007). As such, the responses in both organelles during Cd and Cu stress are pivotal in the signalling cascade from perception to downstream responses. As discussed in Sect. 2.3, inhibition of photosynthetic reactions in chloroplasts can cause excessive ROS production, changes in the redox status of regulatory proteins like thioredoxins, and decreased NADPH formation that influences levels of reduced GSH. Stable ROS (H_2O_2 and lipid peroxidation-derived products) have been implied in retrograde signalling and less stable ROS (e.g. singlet oxygen $^1O_2$) in triggering these pathways. Also redox changes due to changes in GSH content and reduction state are associated with retrograde signalling. When plants are subjected to mild stress, the retrograde signalling causes an acclimation response, but under severe stress, ROS-induced retrograde signalling can trigger programmed cell death. The involvement of ROS in chloroplast–nucleus retrograde signalling has been reviewed by Galvez-Valdivieso et al. (2010).

Abiotic stress effects are also related to oxidative processes inhibiting mitochondrial functioning in plants (Schwarzländer et al. 2009). Metals such as Cd and Cu impair mitochondrial function partly by altering the redox regulation at the ETC level. In germinating pea seeds, Cd induces differential redox responses, which may not be improved by glutaredoxin and thioredoxin protein redox systems present in mitochondria (Smiri et al. 2010). Excess Cu arrests mitochondrial biogenesis and causes a marked decline in cell respiration of sycamore suspension cells. It also enhances AOX at transcriptional and protein levels (Pádua et al. 1999), which could also indicate a disturbed redox balance. In addition, AOX is often regarded as a model to study mitochondrial retrograde regulation, where changes at the organelle level influence nuclear gene expression (Rhoads and Subbaiah 2007). However, van Aken et al. (2009) offer new model systems for mitochondrial retrograde regulation next to AOX. Since retrograde mechanisms are known to be involved in stress responses evoked by other metals (Yamamoto et al. 2002, Rhoads and Subbaiah 2007), a role during Cd- and Cu-mediated oxidative stress is plausible.

Cross-talk and acclimation between mitochondria and other organelles appears vital to maintain whole cell redox balance as demonstrated by Dutilleul et al. (2003). As shown by Schwarzländer et al. (2009), mitochondria are highly sensitive to redox perturbation evoked by Cd, with their redox state recovering more slowly from the oxidative insult. Plant mitochondria are thus suggested to be involved in both perception and response signalling during Cd- and Cu-mediated oxidative challenge in plants.

4 Conclusion

Uptake of Cd or excess Cu by the plant is unavoidable, and beyond the capacity of chelation and sequestration mechanisms, these metals will impose an oxidative challenge to the plant. Cd and Cu can induce the oxidative challenge in a metal-specific way, through different mechanisms of ROS production and interference with specific redox-related compounds in different cellular compartments. Whether the oxidative challenge can be overcome depends on damage versus acclimation responses that may depend on the stress intensity. ROS can cause cellular damage but are also involved in signalling pathways triggering cellular responses. Measurements of changes in pro-and antioxidative components establish an oxidative stress signature that is different for Cd and Cu, due to metal-specific induction of signalling pathways that lead to, at least partly, differential responses. A more detailed characterisation of the oxidative stress signature will help us unravel the general and metal-specific underlying mechanisms of stress perception, signalling and responses caused by Cd and Cu.

References

Abdel-Ghany SE, Pilon M (2008) MicroRNA-mediated systemic down-regulation of copper protein expression in response to low copper availability in *Arabidopsis*. J Biol Chem 283:15932–15945

Argueso CT, Hansen M, Kieber JJ (2007) Regulation of ethylene biosynthesis. J Plant Growth Regul 26:92–105

Artetxe U, García-Plazoala JI, Hernández A, Becerril JM (2002) Low light grown duckweed plants are more protected against the toxicity induced by Zn and Cd. Plant Physiol Biochem 40:859–863

Asada K (1999) The water-water cycle in chloroplasts: scavenging of active oxygen and dissipation of excess photons. Annu Rev Plant Physiol Plant Mol Biol 50:601–639

Bartoli CG, Gómez F, Martínez DA, Guiamet JJ (2004) Mitochondria are the main target for oxidative damage in leaves of wheat (*Triticum aestivum L*). J Exp Bot 55:1663–1669

Baryla A, Carrier P, Franck F, Coulomb C, Sahut C, Havaux M (2001) Leaf chlorosis in oilseed rape plants (*Brassica napus*) grown on cadmium-polluted soil: causes and consequences for photosynthesis and growth. Planta 212:696–709

Bi YH, Chen WL, Zhang WN, Zhou Q, Yun LJ, Xing D (2009) Production of reactive oxygen species, impairment of photosynthetic function and dynamic changes in mitochondria are early events in cadmium-induced cell death in *Arabidopsis thaliana*. Biol Cell 101:629–643

Burkhead JL, Reynolds KA, Abdel-Ghany SE, Cohu CM, Pilon M (2009) Copper homeostasis. New Phytol 182:799–816

Caldelari D, Wang GG, Farmer EE, Dong XN (2011) *Arabidopsis* lox3 and lox4 double mutants are male sterile and defective in global proliferative arrest. Plant Mol Biol 75:25–33

Chae HS, Faure F, Kieber JJ (2003) The eto1, eto2 and eto3 mutations and cytokinin treatment increase ethylene biosynthesis in *Arabidopsis* by increasing the stability of ACS protein. Plant Cell 15:545–559

Cho UH, Seo NH (2005) Oxidative stress in *Arabidopsis thaliana* exposed to cadmium is due to hydrogen peroxide accumulation. Plant Sci 168:113–120

Chugh LK, Sawhney SK (1999) Photosynthetic activities of *Pisum sativum* seedlings grown in presence of cadmium. Plant Physiol Biochem 37:297–303

Clemens S (2006) Toxic metal accumulation, responses to exposure and mechanisms of tolerance in plants. Biochimie 88:1707–1719

Cobbett C, Goldsbrough P (2002) Phytochelatins and metallothioneins: roles in heavy metal detoxification and homeostasis. Annu Rev Plant Biol 53:159–182

Cohu CM, Pilon M (2007) Regulation of superoxide dismutase expression by copper availability. Physiol Plant 129:747–755

Colcombet J, Hirt H (2008) *Arabidopsis* MAPKs: a complex signalling network involved in multiple biological processes. Biochem J 413:217–226

Collin VC, Eymery F, Genty B, Rey P, Havaux M (2008) Vitamin E is essential for the tolerance of *Arabidopsis thaliana* to metal-induced oxidative stress. Plant Cell Environ 31:244–257

Cuypers A, Vangronsveld J, Clijsters H (2000) Biphasic effect of copper on the ascorbate glutathione pathway in primary leaves of *Phaseolus vulgaris* seedlings during the early stages of metal assimilation. Physiol Plant 110:512–517

Cuypers A, Smeets K, Vangronsveld J (2009) Heavy metal stress in plants. In: Hirt H (ed) Plant stress biology: from genomics to systems biology. Wiley, Weinheim

Cuypers A, Remans T, Weyens N, Colpaert J, Vassilev A, Vangronsveld J (2011a) Soil-plant relationships of metals and metalloids. In: Alloway BJ (ed) Heavy metals in soils. Environmental pollution series. Springer, Berlin

Cuypers A, Smeets K, Ruytinx J, Opdenakker K, Keunen E, Remans T, Horemans N, Vanhoudt N, Van Sanden S, Van Belleghem F, Guisez Y, Colpaert J, Vangronsveld J (2011b) The cellular redox state as a modulator in cadmium and copper responses in *Arabidopsis thaliana* seedlings. J Plant Physiol 168:309–316

DalCorso G, Farinati S, Maistri S, Furini A (2008) How plants cope with cadmium: staking on all metabolism and gene expression. J Integr Plant Biol 50:1268–1280

Davletova S, Rizhsky L, Liang HJ, Zhong SQ, Oliver DJ, Coutu J, Shulaev V, Schlauch K, Mittler R (2005a) Cytosolic ascorbate peroxidase 1 is a central component of the reactive oxygen gene network of *Arabidopsis*. Plant Cell 17:268–281

Davletova S, Schlauch K, Coutu J, Mittler R (2005b) The zinc-finger protein Zat12 plays a central role in reactive oxygen and abiotic stress signaling in *Arabidopsis*. Plant Physiol 139:847–856

Drazkiewicz M, Skorzynska-Polit E, Krupa Z (2010) Effect of BSO-supplemented heavy metals on antioxidant enzymes in *Arabidopsis thaliana*. Ecotox Environ Safety 73:1362–1369

Durand TC, Sergeant K, Planchon S, Carpin S, Label P, Morabito D, Hausman JF, Renaut J (2010) Acute metal stress in *Populus tremula* × *P. alba* (717-1B4 genotype): leaf and cambial proteome changes induced by cadmium^{2+}. Proteomics 10:349–368

Dutilleul C, Garmier M, Noctor G, Mathieu C, Chétrit P, Foyer CH, de Paepe R (2003) Leaf mitochondria modulate whole cell redox homeostasis, set antioxidant capacity, and determine stress resistance through altered signaling and diurnal regulation. Plant Cell 15:1212–1226

Faller P, Kienzler K, Krieger-Liszkay A (2005) Mechanism of Cd^{2+} toxicity: Cd^{2+} inhibits photoactivation of photosystem II by competitive binding to the essential Ca^{2+} site. Biochim Biophys Acta 1706:158–164

Feng B, Dong ZY, Xu ZB, An XL, Qin HJ, Wu N, Wang DW, Wang T (2010) Molecular analysis of lipoxygenase (LOX) genes in common wheat and phylogenetic investigation of LOX proteins from model and crop plants. J Cereal Sci 52:387–394

Fiil BK, Petersen K, Petersen M, Mundy J (2009) Gene regulation by MAP kinase cascades. Curr Opin Plant Biol 12:615–621

Foyer CH, Noctor G (2003) Redox sensing and signalling associated with reactive oxygen species in chloroplasts, peroxisomes and mitochondria. Physiol Plant 119:355–364

Foyer CH, Noctor G (2005) Oxidant and antioxidant signalling in plants: a re-evaluation of the concept of oxidative stress in a physiological context. Plant Cell Environ 28:1056–1071

Foyer CH, Noctor G (2009) Redox regulation in photosynthetic organisms: signaling, acclimation, and practical implications. Antioxid Redox Signal 11:861–905

Foyer CH, Noctor G (2011) Ascorbate and glutathione: the heart of the redox hub. Plant Physiol 155:2–18

Gajewska E, Sklodowska M (2007) Relations between tocopherol, chlorophyll and lipid peroxides contents in shoots of Ni-treated wheat. J Plant Physiol 164:364–366

Galvez-Valdivieso G, Mullineaux PM (2010) The role of reactive oxygen species in signalling from chloroplasts to the nucleus. Physiol Plant 138:430–439

Garnier L, Simon-Plas F, Thuleau P, Agnel JP, Blein JP, Ranjeva R, Montillet JL (2006) Cadmium affects tobacco cells by a series of three waves of reactive oxygen species that contribute to cytotoxicity. Plant Cell Environ 29:1956–1969

Ghezzi P, Bonetto V (2003) Redox proteomics: identification of oxidatively modified proteins. Proteomics 3:1145–1153

Glyan'ko AK, Ischenko AA (2010) Structural and functional characteristics of plant NADPH oxidase: a review. Appl Biochem Micro 46:463–471

Gong JM, Lee DA, Schroeder JI (2003) Long-distance root-to-shoot transport of phytochelatins and cadmium in *Arabidopsis*. Proc Natl Acad Sci U S A 100:10118–10123

Grzyb J, Waloszek A, Latowski D, Wieckowski S (2004) Effect of cadmium on ferredoxin: $NADP^+$ oxidoreductase activity. J Inorg Biochem 98:1338–1346

Grzyb J, Bojko M, Waloszek A, Strzalka K (2011) Ferredoxin: $NADP^+$ oxidoreductase as a target of Cd^{2+} inhibitory action-biochemical studies. Phytochemistry 72:14–20

Guo WJ, Meetam M, Goldsbrough PB (2008) Examining the specific contributions of individual *Arabidopsis* metallothioneins to copper distribution and metal tolerance. Plant Physiol 146:1697–1706

Hahnel D, Beyer K, Engelmann B (1999) Inhibition of peroxyl radical-mediated lipid oxidation by plasmalogen phospholipids and alpha-tocopherol. Free Radical Bio Med 27:1087–1094

Halliwell B (2006) Reactive species and antioxidants: redox biology is a fundamental theme of aerobic life. Plant Physiol 141:312–322

Harada E, Yamaguchi Y, Koizumi N, Hiroshi S (2002) Cadmium stress induces production of thiol compounds and transcripts for enzymes involved in sulfur assimilation pathways in *Arabidopsis*. J Plant Physiol 159:445–448

Hattab S, Dridi B, Chouba L, Ben Kheder M, Bousetta H (2009) Photosynthesis and growth responses of pea *Pisum sativum* L. under heavy metals stress. J Environ Sci (China) 21:1552–1556

Hédiji H, Djebali W, Cabasson C, Maucourt M, Baldet P, Bertrand A, Boulila Zoghlami L, Deborde C, Moing A, Brouquisse R, Chaïbi W, Gallusci P (2010) Effects of long-term cadmium exposure on growth and metabolomic profile of tomato plants. Ecotox Environ Safety 73:1965–1974

Heldt HW, Heldt F (2005) The Calvin cycle catalyzes photosynthetic CO_2 assimilation. In: Plant biochemistry, 3rd edn. Academic Press, Burlington

Herbette S, Taconnat L, Hugouvieux V, Piette L, Magniette ML, Cuine S, Auroy P, Richaud P, Forestier C, Bourguignon J, Renou JP, Vavasseur A, Leonhardt N (2006) Genome-wide transcriptome profiling of the early cadmium response of *Arabidopsis* roots and shoots. Biochimie 88:1751–1765

Heyno E, Klose C, Krieger-Liszkay A (2008) Origin of cadmium-induced reactive oxygen species production: mitochondrial electron transfer versus plasma membrane NADPH oxidase. New Phytol 179:687–699

Holkova I, Bezakova L, Vanko M, Bilka F, Oblozinsky M (2009) Lipoxygenases and their significance in biochemical processes in plant organisms. Chem Listy 103:487–595

Howarth JR, Domínguez-Solís JR, Gutiérrez-Alcalá G, Wray JL, Romero LC, Gotor C (2003) The serine acetyltransferase gene family in *Arabidopsis thaliana* and the regulation of its expression by cadmium. Plant Mol Biol 51:589–598

Howden R, Goldsbrough PB, Andersen CR, Cobbett CS (1995) Cadmium-sensitive, cad1 mutants of *Arabidopsis thaliana* are phytochelatin deficient. Plant Physiol 107:1059–1066

Hsieh LC, Lin SI, Shih ACC, Chen JW, Lin WY, Tseng CY, Li WH, Chiou TJ (2009) Uncovering small RNA-mediated responses to phosphate deficiency in *Arabidopsis* by deep sequencing. Plant Physiol 151:2120–2132

Hsu YT, Kao CH (2007) Toxicity in leaves of rice exposed to cadmium is due to hydrogen peroxide accumulation. Plant Soil 298:231–241

Huang SQ, Peng J, Qiu CX, Yang ZM (2009) Heavy metal-regulated new microRNAs from rice. J Inorg Biochem 103:282–287

Huang SQ, Xiang AL, Che LL, Chen S, Li H, Song JB, Yang ZM (2010) A set of miRNAs from *Brassica napus* in response to sulphate deficiency and cadmium stress. Plant Biotechnol J 8: 887–899

Jamalkandi SA, Masoudi-Nejad A (2009) Reconstruction of *Arabidopsis thaliana* fully integrated small RNA pathway. Funct Integrated Genomics 9:419–432

Jaspers P, Kangasjärvi J (2010) Reactive oxygen species in abiotic stress signaling. Physiol Plant 138:405–413

Jonak C, Nakagami H, Hirt H (2004) Heavy metal stress: activation of distinct mitogen-activated protein kinase pathways by copper and cadmium. Plant Physiol 136:3276–3283

Jones MA, Raymond MJ, Yang Z, Smirnoff N (2007) NADPH oxidase-dependent reactive oxygen species formation required for root hair growth depends on ROP GTPase. J Exp Bot 58: 1261–1270

Kacperska A (2004) Sensor types in signal transduction pathways in plant cells responding to abiotic stressors: do they depend on stress intensity? Physiol Plantarum 122:159–168

Khraiwesh B, Arif MA, Seumel GI, Ossowski S, Weigel D, Reski R (2010) Transcriptional control of gene expression by microRNAs. Cell 140:111–122

Kieffer P, Schroder P, Dommes J, Hoffmann L, Renaut J, Hausman JF (2009) Proteomic and enzymatic response of *poplar* to cadmium stress. J Proteomics 72:379–396

Kim CY, Liu Y, Thorne ET, Yang H, Fukushig H, Gassmann W, Hildebrand D, Sharp RE, Zhang S (2003) Activation of a stress-responsive mitogen-activated protein kinase cascade induces the biosynthesis of ethylene in plants. Plant Cell 15:2707–2718

Kobayashi M, Ohura I, Kawakita K, Yokota N, Fujiwara M, Shimamoto K, Doke N, Yoshioka H (2007) Calcium-dependent protein kinases regulate the production of reactive oxygen species by potato NADPH oxidase. Plant Cell 19:1065–1080

Kruger NJ, von Schaewen A (2003) The oxidative pentose phosphate pathway: structure and organisation. Curr Opin Plant Biol 6:236–246

Kucera T, Horakova H, Sonska A (2008) Toxic metal ions in photoautotrophic organisms. Photosynthetica 46:481–489

Kumar M, Kumari P, Gupta V, Anisha PA, Reddy CR, Jha B (2010) Differential responses to cadmium induced oxidative stress in marine macroalga *Ulva lactuca* (Ulvales, Chlorophyta). Biometals 23:315–325

Küpper H, Küpper F, Spiller M (1998) In situ detection of heavy metal substituted chlorophylls in water plants. Photosynth Res 58:123–133

Lequeux H, Hermans C, Lutts S, Verbruggen N (2010) Response to copper excess in *Arabidopsis thaliana*: impact on root system architecture, hormone distribution, lignin accumulation and mineral profile. Plant Physiol Biochem 48:673–682

Liavonchanka A, Feussner N (2006) Lipoxygenases: occurrence, functions and catalysis. J Plant Physiol 163:348–357

Lin Z, Zhong S, Grierson D (2009) Recent advances in ethylene research. J Exp Bot 60:3311–3336

Liu Y, Zhang S (2004) Phosphorylation of 1-aminocyclopropane-1-carboxylic acid synthase by MPK6, a stress-responsive mitogen-activated protein kinase, induces ethylene biosynthesis in *Arabidopsis*. Plant Cell 16:3386–3399

Liu XM, Kim KE, Kim KC, Nguyen XC, Han HJ, Jung MS, Kim HS, Kim SH, Park HC, Yun DJ, Chung WS (2010) Cadmium activates *Arabidopsis* MPK3 and MPK6 via accumulation of reactive oxygen species. Phytochemistry 71:614–618

Lösch R (2004) Plant mitochondrial respiration under the influence of heavy metals. In: Prasad MNV (ed) Heavy metal stress in plants: from biomolecules to ecosystems, 2nd edn. Springer, Berlin/Heidelberg

Maeda H, DellaPenna D (2007) Tocopherol functions in photosynthetic organisms. Curr Opin Plant Biol 10:260–265

Maksymiec W (2007) Signaling responses in plants to heavy metal stress. Acta Physiol Plant 29: 177–187

Maksymiec W, Krupa Z (2006) The effects of short-term exposition to Cd, excess Cu ions and jasmonate on oxidative stress appearing in *Arabidopsis thaliana*. Environ Exp Bot 57:187–194

Maksymiec W, Krupa Z (2007) Effects of methyl jasmonate and excess copper on root and leaf growth. Biol Plantarum 51:322–326

Mallory AC, Bouché N (2008) MicroRNA-directed regulation: to cleave or not to cleave. Trends Plant Sci 13:359–367

Marschner H (1995) Functions of mineral nutrients: micronutrients. In: Marschner H (ed) Mineral nutrition of higher plants. Academic, London

May M, Vernoux T, Leaver C, Montagu MV, Inzé D (1998) Glutathione homeostasis in plants: implications for environmental sensing and plant development. J Exp Bot 49:649–667

Mijovilovich A, Leitenmaier B, Meyer-Klaucke W, Kroneck PMH, Gotz B, Kupper H (2009) Complexation and toxicity of copper in higher plants. II: different mechanisms for copper versus cadmium detoxification in the copper-sensitive cadmium/zinc hyperaccumulator *Thlaspi caerulescens* (Ganges ecotype). Plant Physiol 151:715–731

Miller G, Shulaev V, Mittler R (2008) Reactive oxygen signaling and abiotic stress. Physiol Plantarum 133:481–489

Miller G, Schlauch K, Tam R, Cortes D, Torres MA, Shulaev V, Dangl JL, Mittler R (2009) The plant NADPH oxidase RBOHD mediates rapid systemic signaling in response to diverse stimuli. Sci Signal 2:A26–A35

Miller G, Suzuki N, Ciftci-Yilmaz S, Mittler R (2010) Reactive oxygen species homeostasis and signalling during drought and salinity stresses. Plant Cell Environ 33:453–467

Mithofer A, Schulze B, Boland W (2004) Biotic and heavy metal stress response in plants: evidence for common signals. FEBS Lett 566:1–5

Mittler R (2002) Oxidative stress, antioxidants and stress tolerance. Trends Plant Sci 7:405–410

Mittler R, Vanderauwera S, Gollery M, Van Breusegem F (2004) Reactive oxygen gene network of plants. Trends Plant Sci 9:490–498

Mobin M, Khan NA (2007) Photosynthetic activity, pigment composition and antioxidative response of two mustard (*Brassica juncea*) cultivars differing in photosynthetic capacity subjected to cadmium stress. J Plant Physiol 164:601–610

Mohanty N, Vass I, Demeter S (1989) Copper toxicity affects photosystem II electron transport at the secondary quinone acceptor, QB. Plant Physiol 90:175–179

Møller IM (2001) Plant mitochondria and oxidative stress: electron transport, NADPH turnover, and metabolism of reactive oxygen species. Annu Rev Plant Physiol Plant Mol Biol 52:561–591

Møller IM, Kristensen BK (2004) Protein oxidation in plant mitochondria as a stress indicator. Photochem Photobiol Sci 3:730–735

Møller IM, Jensen PE, Hansson A (2007) Oxidative modifications to cellular components in plants. Ann Rev Plant Biol 58:459–481

Montillet JL, Cacas JL, Garnier L, Montané MH, Douki T, Bessoule JJ, Polkowska-Kowalczyk L, Maciejewska U, Agnel JP, Vial A, Triantaphylidès C (2004) The upstream oxylipin profile of *Arabidopsis thaliana*: a tool to scan for oxidative stresses. Plant J 40:439–451

Munne-Bosch S, Falk J (2004) New insights into the function of tocopherols in plants. Planta 218:323–326

Navari-Izzo F, Cestone B, Cavallini A, Natali L, Giordani T, Quartacci MF (2006) Copper excess triggers phospholipase D activity in wheat roots. Phytochemistry 67:1232–1242

Noctor G, Foyer CH (1998) Ascorbate and glutathione: keeping active oxygen under control. Annu Rev Plant Physiol Plant Mol Biol 49:249–279

Noctor G, Arisi AM, Jouanin L, Kunert KJ, Rennenberg H, Foyer CH (1998) Glutathione: biosynthesis, metabolism and relationship to stress tolerance explored in transformed plants. J Exp Bot 49:623–647

Noctor G, De Paepe R, Foyer CH (2007) Mitochondrial redox biology and homeostasis in plants. Trends Plant Sci 12:125–134

Nriagu J, Pacyna J (1988) Quantitative assessment of worldwide contamination of air, water and soils by trace metals. Nature 333:134–139

Ogasawara Y, Kaya H, Hiraoka G, Yumoto F, Kimura S, Kadota Y, Hishinuma H, Senzaki E, Yamagoe S, Nagata K, Nara M, Suzuki K, Tanokura M, Kuchitsu K (2008) Synergistic activation of the *Arabidopsis* NADPH oxidase Atrbodh by Ca^{2+} and phosphorylation. J Biol Chem 283:8885–8892

Overmyer K, Brosché M, Kangasjarvi J (2003) Reactive oxygen species and hormonal control of cell death. Trends Plant Sci 8:335–342

Pádua M, Aubert S, Casimiro A, Bligny R, Millar AH, Day DA (1999) Induction of alternative oxidase by excess copper in sycamore cell suspensions. Plant Physiol Biochem 37:131–137

Pádua M, Cavaco AM, Aubert S, Bligny R, Casimiro A (2010) Effects of copper on the photosynthesis of intact chloroplasts: interaction with manganese. Physiol Plant 138:301–311

Pagliano C, Raviolo M, Dalla Vecchia F, Gabbrielli R, Gonnelli C, Rascio N, Barbato R, La Rocca N (2006) Evidence for PSII donor-side damage and photoinhibition induced by cadmium treatment on rice (*Oryza sativa L.*). J Photochem Photobiol B 84:70–78

Palmer CM, Guerinot ML (2009) Facing the challenges of Cu, Fe and Zn homeostasis in plants. Nat Chem Biol 5:333–340

Panchuk II, Volkov RA, Schoffl F (2002) Heat stress- and heat shock transcription factor dependent expression and activity of ascorbate peroxidase in *Arabidopsis*. Plant Physiol 129:838–853

Pant BD, Musialak-Lange M, Nuc P, May P, Buhtz A, Kehr J, Walther D, Scheible WR (2009) Identification of nutrient-responsive *Arabidopsis* and rapeseed microRNAs by comprehensive real-time polymerase chain reaction profiling and small RNA sequencing. Plant Physiol 150:1541–1555

Patsikka E, Aro EM, Tyystjarvi E (1998) Increase in the quantum yield of photoinhibition contributes to copper toxicity in vivo. Plant Physiol 117:619–627

Perfus-Barbeoch L, Leonhardt N, Vavasseur A, Forestier C (2002) Heavy metal toxicity: cadmium permeates through calcium channels and disturbs the plant water status. Plant J 32:539–548

Pnueli L, Liang H, Rozenberg M, Mittler R (2003) Growth suppression, altered stomatal responses, and augemented induction of heatshock proteins in cytosolic ascorbate peroxidase (Apx1)-deficient *Arabidopsis* plants. Plant J 34:187–203

Potters G, De Gara L, Asard H, Horemans N (2002) Ascorbate and glutathione: guardians of the cell cycle, partners in crime? Plant Physiol Biochem 40:537–548

Potters G, Horemans N, Jansen MAK (2010) The cellular redox state in plant stress biology–a charging concept. Plant Physiol Biochem 5:292–300

Quick WP, Neuhaus HE (1997) The regulation and control of photosynthetic carbon assimilation. In: Foyer CH, Quick WP (eds) A molecular approach to primary metabolism in higher plants. Taylor & Francis, London

Rahoui S, Chaoui A, El Ferjani E (2010) Membrane damage and solute leakage from germinating pea seed under cadmium stress. J Hazard Mater 178:1128–1131

Remans T, Opdenakker K, Smeets K, Mathijsen D, Vangronsveld J, Cuypers A (2010) Metal-specific and NADPH oxidase dependent changes in lipoxygenase and NADPH oxidase gene expression in *Arabidopsis thaliana* exposed to cadmium or excess copper. Funct Plant Biol 37:532–544

Rhoads DM, Subbaiah CC (2007) Mitochondrial retrograde regulation in plants. Mitochondrion 7:177–194

Robinson NJ, Winge DR (2010) Cu metallochaperones. Annu Rev Biochem 79:537–562

Rodríguez-Serrano M, Romero-Puertas MC, Sparkes I, Hawes C, del Río LA, Sandalio LM (2009) Peroxisome dynamics in *Arabidopsis* plants under oxidative stress induced by cadmium. Free Radic Biol Med 47:1632–1639

Romero-Puertas MC, Rodriguez-Serrano M, Corpas FJ, Gomez M, Del Rio LA, Sandalio LM (2004) Cadmium-induced subcellular accumulation of $O^{-2}(^{-})$ and H_2O_2 in pea leaves. Plant Cell Environ 27:1122–1134

Romero-Puertas MC, Corpas FJ, Rodriguez-Serrano M, Gomez M, del Rio LA, Sandalio LM (2007) Differential expression and regulation of antioxidative enzymes by cadmium in pea plants. J Plant Physiol 164:1346–1357

Roth U, van Roepenack-Lahaye E, Clemens S (2006) Proteome changes in *Arabidopsis thaliana* roots upon exposure to Cd^{2+}. J Exp Bot 57:4003–4013

Sagi M, Fluhr R (2001) Superoxide production by plant homologues of the gp91(phox) NADPH oxidase: modulation of activity by calcium and by tobacco mosaic virus infection. Plant Physiol 126:1281–1290

Sagi M, Fluhr R (2006) Production of reactive oxygen species by plant NADPH oxidases. Plant Physiol 141:336–340

Schwarzländer M, Fricker MD, Sweetlove LJ (2009) Monitoring the in vivo redox state of plant mitochondria: effect of respiratory inhibitors, abiotic stress and assessment of recovery from oxidative challenge. Biochim Biophys Acta 1787:468–475

Semane B, Dupae J, Cuypers A, Noben JP, Tuomainen M, Tervahauta A, Kärenlampi S, Van Belleghem F, Smeets K, Vangronsveld J (2010) Leaf proteome responses of *Arabidopsis thaliana* exposed to mild cadmium stress. J Plant Physiol 167:247–254

Seth CS, Remans T, Keunen E, Jozefczak M, Gielen H, Opdenakker K, Weyens N, Vangronsveld J, Cuypers A (2011) Phytoextraction of toxic metals: a central role for glutathione. Plant Cell Environ. doi:10.1111/j.1365-3040.2011.02338.x

Sharma SS, Dietz KJ (2008) The relationship between metal toxicity and cellular redox imbalance. Trends Plant Sci 14:43–50

Skorzynska-Polit E, Pawlikowska-Pawlega B, Szczuka E, Drazkiewicz M, Krupa Z (2006) The activity and localization of lipoxygenases in *Arabidopsis thaliana* under cadmium and copper stresses. Plant Growth Regul 48:29–39

Smeets K, Cuypers A, Lambrechts A, Semane B, Hoet P, Van Laere A, Vangronsveld J (2005) Induction of oxidative stress and antioxidative mechanisms in *Phaseolus vulgaris* after Cd application. Plant Physiol Biochem 43:437–444

Smeets K, Ruytinx J, Semane B, Van Belleghem F, Remans T, Van Sanden S, Vangronsveld J, Cuypers A (2008) Cadmium-induced transcriptional and enzymatic alterations related to oxidative stress. Environ Exp Bot 63:1–8

Smeets K, Opdenakker K, Remans T, Van Sanden S, Van Belleghem F, Semane B, Horemans N, Guisez Y, Vangronsveld J, Cuypers A (2009) Oxidative stress-related responses at transcriptional and enzymatic levels after exposure to Cd or Cu in a multipollution context. J Plant Physiol 166:1982–1992

Smiri M, Chaoui A, Rouhier N, Kamel C, Gelhaye E, Jacquot JP, El Ferjani E (2010) Cadmium induced mitochondrial redox changes in germinating pea seed. Biometals 23:973–984

Smirnoff N (2005) Ascorbate, tocopherol and carotenoids: metabolism, pathway engineering and functions. In: Smirnoff N (ed) Antioxidants and reactive oxygen species in plants. Blackwell, Oxford

Stepanova AN, Alonso JM (2005) Ethylene signaling and response pathway: unique signaling cascade with a multitude of inputs and outputs. Physiol Plant 123:195–206

Stepanova AN, Yun J, Likhacheva AV, Alonso JM (2007) Multilevel interactions between ethylene and auxin in *Arabidopsis* roots. Plant Cell 19:2169–2185

Stitt M, Lunn J, Usadel B (2010) *Arabidopsis* and primary photosynthetic metabolism – more than the icing on the cake. Plant J 61:1067–1091

Sun XM, Zhang JX, Zhang HJ, Ni YW, Zhang Q, Chen JP, Guan YF (2010) The responses of *Arabidopsis thaliana* to cadmium exposure explored via metabolite profiling. Chemosphere 78:840–845

Sunkar R, Zhu JK (2004) Novel and stress-regulated microRNAs and other small RNAs from *Arabidopsis*. Plant Cell 16:2001–2019

Sunkar R, Kapoor A, Zhu JK (2006) Posttranscriptional induction of two Cu/Zn superoxide dismutase genes in *Arabidopsis* is mediated by downregulation of miR398 and important for oxidative stress tolerance. Plant Cell 18:2051–2065

Sutton HC, Winterbourn CC (1989) On the participation of higher oxidation states of iron and copper in Fenton reactions. Free Radic Biol Med 6:53–60

Tamas L, Dudikova J, Durcekova K, Haluskova L, Huttova J, Mistrik I (2009) Effect of cadmium and temperature on the lipoxygenase activity in barley root tip. Protoplasma 235:17–25

Tan YF, O'Toole N, Taylor NL, Millar AH (2010) Divalent metal ions in plant mitochondria and their role in interactions with proteins and oxidative stress-induced damage to respiratory function. Plant Physiol 152:747–761

van Aken O, Zhang B, Carrie C, Uggalla V, Paynter E, Giraud E, Whelan J (2009) Defining the mitochondrial stress response in *Arabidopsis thaliana*. Mol Plant 2:1310–1324

van Belleghem F, Cuypers A, Semane B, Smeets K, Vangronsveld J, d'Haen J, Valcke R (2007) Subcellular localization of cadmium in roots and leaves of *Arabidopsis thaliana*. New Phytol 173:495–508

Vazquez F, Legrand S, Windels D (2010) The biosynthetic pathways and biological scopes of plant small RNAs. Trends Plant Sci 15:337–345

Vellosillo T, Martinez M, Lopez MA, Vicente J, Cascon T, Dolan L, Hamberg M, Castresana C (2007) Oxylipins produced by the 9-lipoxygenase pathway in *Arabidopsis* regulate lateral root development and defense responses through a specific signaling cascade. Plant Cell 19: 831–846

Verbruggen N, Hermans C, Schat H (2009) Mechanisms to cope with arsenic or cadmium excess in plants. Curr Opin Plant Biol 12:364–372

Wang NN, Shih MC, Li N (2005) The GUS reporter-aided analysis of the promoter activities of *Arabidopsis* ACC synthase genes AtACS4, AtACS5, and AtACS7 induced hormones and stresses. J Exp Bot 56:909–920

Wang JX, Ding HD, Zhang A, Ma FF, Cao JM, Jiang MY (2010) A novel mitogen-activated protein kinase gene in maize (*Zea mays*), ZmMPK3, is involved in response to diverse environmental cues. J Integr Plant Biol 52:442–452

Wei LY, Zhang DF, Xiang F, Zhang ZX (2009) Differentially expressed miRNAs potentially involved in the regulation of defense mechanism to drought stress in maize seedlings. Int J Plant Sci 170:979–989

Wong HL, Pinontoan R, Hayashi K, Tabata R, Yaeno T, Hasegawa K, Kojima C, Yoshioka H, Iba K, Kawasaki T, Shimamoto K (2007) Regulation of rice NADPH oxidase by binding of Rac GTPase to its N-terminal extension. Plant Cell 19:4022–4034

Xing Y, Jia WS, Zhang JH (2008) AtMKK1 mediates ABA-induced CAT1 expression and H_2O_2 production via AtMPK6-coupled signaling in *Arabidopsis*. Plant J 54:440–451

Xing Y, Jia WS, Zhang JH (2007) AtMEK1 mediates stress-induced gene expression of CAT1 catalase by triggering H_2O_2 production in Arabidopsis. J Exp Bot 58:2969–2981

Xu J, Li Y, Wang Y, Liu H, Lei L, Yang H, Liu G, Ren D (2008) Activation of MAPK kinase 9 induces ethylene and camalexin biosynthesis and enhances sensitivity to salt stress in *Arabidopsis*. J Biol Chem 283:26996–27006

Yamamoto Y, Kobayashi Y, Rama Devi S, Rikiishi S, Matsumoto H (2002) Aluminum toxicity is associated with mitochondrial dysfunction and the production of reactive oxygen species in plant cells. Plant Physiol 128:63–72

Yamasaki H, Hayashi M, Fukazawa M, Kobayashi Y, Shikanai T (2009) SQUAMOSA promoter binding protein–like7 is a central regulator for copper homeostasis in *Arabidopsis*. Plant Cell 21:347–361

Yeh CM, Chien PS, Huang HJ (2007) Distinct signalling pathways for induction of MAP kinase activities by cadmium and copper in rice roots. J Exp Bot 58:659–671

Ying RR, Qiu RL, Tang YT, Hu PJ, Qiu H, Chen HR, Shi TH, Morel JL (2010) Cadmium tolerance of carbon assimilation enzymes and chloroplast in Zn/Cd hyperaccumulator *Picris divaricata*. J Plant Physiol 167:81–87

Yoo SD, Cho YH, Tena G, Xiong Y, Sheen J (2008) Dual control of nuclear EIN3 by bifurcate MAPK cascades in C_2H_4 signalling. Nature 451:789–795

Yusuf MA, Kumar D, Rajwanshi R, Strasser RJ, Tsimilli-Michael M, Govindjee SNB (2010) Overexpression of gamma-tocopherol methyl transferase gene in transgenic *Brassica juncea* plants alleviates abiotic stress: physiological and chlorophyll a fluorescence measurements. Biochim Biophys Acta 1797:1428–1438

Zengin FK, Munzuroglu O (2005) Effects of some heavy metals on content of chlorophyll, proline and some antioxidant chemicals in bean (*Phaseolus vulgaris* L.) seedlings. Acta Biol Cracov Ser Bot 47:157–164

Zhang H, Zhang F, Xia Y, Wang G, Shen Z (2010) Excess copper induces production of hydrogen peroxide in the leaf of *Elsholtzia haichowensis* through apoplastic and symplastic CuZn-superoxide dismutase. J Hazard Mater 178:834–843

Zhigang A, Cuijie L, Yuangang Z, Yejie D, Wachter A, Gromes R, Rausch T (2006) Expression of BjMT2, a metallothionein 2 from *Brassica juncea*, increases copper and cadmium tolerance in *Escherichia coli* and *Arabidopsis thaliana* seedlings. J Exp Bot 57:3575–3582

Zhou ZS, Huang SQ, Yang ZM (2008) Bioinformatic identification and expression analysis of new microRNAs from *Medicago truncatula*. Biochem Biophys Res Commun 374:538–542

Zimmermann P, Heinlein C, Orendi G, Zentgraf U (2006) Senescence-specific regulation of catalases in *Arabidopsis thaliana* (L.) heynh. Plant Cell Environ 29:1049–1060

Insights into Cadmium Toxicity: Reactive Oxygen and Nitrogen Species Function

María C. Romero-Puertas, Ana P. Ortega-Galisteo, María Rodríguez-Serrano, and Luisa M. Sandalio

Abstract Cadmium (Cd) is a heavy metal that enters the environment mainly due to phosphate fertilizers and processes derived from industry and mining. This metal is a toxic element and the main problem of its accumulation is the rapid transference into the food chain through plants that take up the metal by their roots from where it can be loaded into the xylem for its transport into the leaves and fruits. Cd inhibits plant growth producing alterations in the photosynthesis rate, water use efficiency and the uptake and distribution of micro- and macro-nutrients. Additionally, Cd disturbs the plant antioxidant system and induces the production of reactive oxygen species (ROS) leading to an oxidative stress. Because Cd does not participate in Fenton-type reactions, the mechanisms by which Cd induces ROS production are not well understood. Recently, nitric oxide (NO), a well-known messenger in plants, has been involved in the plant Cd response although its function and sources are still largely unknown. In this chapter, we will discuss the effect of Cd on plants, the responses of the plant to the metal showing the proteomic and transcriptomic analysis that have been done in recent years and the role of ROS and NO in this response.

1 Introduction

There are almost 100 natural chemical elements on Earth including many metals. While some of them are essential for life, others can be toxic or even lethal and heavy metals such as Cd, Hg, Pb and Al, constitute some of the major environmental contaminants (Chen et al. 2006). Atmospheric emissions caused by metal mines,

M.C. Romero-Puertas (✉) • A.P. Ortega-Galisteo • M. Rodríguez-Serrano • L.M. Sandalio
Departamento de Bioquímica, Biología Celular y Molecular de Plantas, Estación Experimental del Zaidín, Consejo Superior de Investigaciones Científicas (CSIC), Granada, Spain
e-mail: maria.romero@eez.csic.es; ana.ortega@eez.csic.es; maria.rodriguez@eez.csic.es; luisamaria.sandalio@eez.csic.es

D.K. Gupta and L.M. Sandalio (eds.), *Metal Toxicity in Plants: Perception, Signaling and Remediation*, DOI 10.1007/978-3-642-22081-4_5,

industrial emissions, solid waste incineration or coal combustion, together with accidental pollution and direct deposits like fertilizers, especially phosphates, represent the major anthropogenic sources causing cadmium pollution (McLaughlin and Singh 1999; Alloway and Steinnes 1999) that severely limit plant productivity (Sharma and Dietz 2009). Cd bioavailability depends on pH, soil structure, soil organic matter and chemical speciation. In most environmental conditions Cd enters plants first by the roots through the cortical tissue and is then translocated to the above-ground tissues and in this way into the human food chain (Sanitá di Toppi and Gabrielle 1999; Yang et al. 1998). Despite the fact that Cd ions are mainly retained in the roots (Cataldo et al. 1983), it can be rapidly loaded into the xylem and transported into the leaves. There are differences across plant species in the partitioning of Cd between tissues (Wagner 1993). In general, the content of Cd in plants decreases in the order: roots > stems > leaves > fruits > seeds (Blum 1997).

Cadmium (Cd) has no demonstrated biological function in animal or plants, except for the use of Cd as a co-factor by the enzyme carbonic anhydrase of a marine diatom in conditions of micronutrient scarcity (Lane and Morel 2000; Morel and Malcolm 2005). In plants so, it is highly toxic at low concentrations and due to its high mobility and its solubility properties in water, a small concentration in soil can cause visible effects (Barceló and Pöschenrieder 1990; Pinto et al. 2004). Cd is considered as a potential threat for human health and the environment due to its accumulation in the soil, in the food chain and in drinking water (Nawrot et al. 2006; Zhao et al. 2008). Cd has chemical similarity with essential elements such as Zn, Ca and Fe, and for this reason, Cd is able to imbalance their homeostasis or substitute them in proteins (Verbruggen et al. 2009) increasing its toxicity.

It has been shown recently that reactive oxygen and nitrogen species (ROS/ RNS) are produced in response to Cd stress (Romero-Puertas et al. 2004; Rodríguez-Serrano et al. 2006, 2009a; Besson-Bard et al. 2009; Bi et al. 2009). Their function however is still not clear due to its own double-faced function inducing: (1) oxidative/nitrosative stress imposed by an accumulation of ROS/ RNS and (2) signalling to regulate gene network involved in the plant response to Cd stress when low doses of ROS/RNS are present. The main objective of this chapter will be the discussion of the plant responses to Cd stress, and especially the involvement of ROS and RNS metabolism in these responses.

2 Cadmium Toxicity in Plants

Although cadmium toxicity in many plant species has been reported by several authors (Clemens 2006; Sandalio et al. 2009) the mechanisms of its toxicity are not yet well understood. Recent experiments reveal, however, that there is a relationship between metal stress, redox homeostasis and antioxidant capacity (Sharma and Dietz 2009) and that the cellular redox state is important in the determination of metal phytotoxicity (Cuypers et al. 2011). Cadmium is able to induce different effects on plants. Some of them are visible to the naked eye such as inhibition of seed germination, chlorosis and reduced plant size, while others are not, like functionality of membranes by inducing changes in lipid composition and by affecting the enzymatic activities (Ouariti et al. 1997; Fodor et al. 1995). Metal content in leaves and roots has a linear relationship with metal concentration in the nutrient solution (Sandalio et al. 2001). Cadmium can modify not only the uptake of minerals by plants (Moreno et al. 1999; Das et al. 1997) but also the nutrient distribution of some elements in roots and leaves (Sandalio et al. 2001). The content of Zn, Fe, Mn, Cu and Ca is reduced in leaves of several plant species such as pea, wheat or sunflower (Sandalio et al. 2001; Shukla et al. 2003; Azevedo et al. 2005; Rodríguez-Serrano et al. 2009a), as a result of an excess of Cd that probably competes with those elements for the transporters promoting a reduction in both the uptake and the accumulation of those cations (Clemens 2006). In fact, iron, phosphorous or manganese deficiency could be the reason for the chlorosis observed in plants treated with Cd (Haghiri 1973; Godbold and Hutterman 1985). Cd also reduced the absorption of nitrate and its transport from roots to shoots, by inhibiting the activity of nitrate reductase in the shoot, inhibiting severely NO_3^- assimilation and inducing a reduction in the fresh tissue weight and its relative water content (Hernandez et al. 1996). The reduction of nitrate reductase activity could modify the NO production in response to cadmium.

Exposure to Cd decreased growth of aerial tissues and roots, which is characterized by a significant reduction of the primary root length and the number of lateral roots (Sandalio et al. 2001; Rodríguez-Serrano et al. 2009a; Watanabe et al. 2010). Growth inhibition is usually accompanied by a significant decrease in chlorophyll content, transpiration rate, water use efficiency and the photosynthesis rate as has been observed in pea plants, soybean and *Chlorella vulgaris* (Sandalio et al. 2001; Faller et al. 2005; Qian et al. 2009). Additionally, an induction of oxidative stress has been observed in leaves and roots of plants treated with Cd, characterized by an accumulation of lipid peroxides and oxidized proteins (Sandalio et al. 2001; Schützendübel et al. 2001; Romero-Puertas et al. 2002; Smeets et al. 2009). These results together with the probably metabolic transition of leaf peroxisomes into glyoxysomes and the increase of proteolytic activity point to a possible induction of leaf senescence by Cd (McCarthy et al. 2001). In fact, it has been shown recently that *Arabidopsis* cell cultures exposed to 100–150 μM of Cd seems to have an accelerated senescence process with an induction of the marker gene associated to senescence *SAG12* (De Michele et al. 2009). Different analysis has also shown that Cd treatment can induce plant cell death in *Arabidopsis* and tobacco cell cultures (De Michele et al. 2009; Fotjová and Kovařik 2000). This PCD could be regulated by ROS and NO although the mechanism is still not clear (Yakimova et al. 2006; De Michele et al. 2009).

3 Plant Mechanisms to Cope with Cadmium

Because plants are sessile organisms and they cannot move when they are exposed to a stress such as cadmium, they have developed different mechanisms to survive to the metal: (1) accumulate it; (2) prevent its entry or (3) exclude it once it is in the plant tissue (Watanabe et al. 2010):

1. A group of plants, called hyperaccumulators, tolerate Cd accumulation without toxicity symptoms (Verbruggen et al. 2009). Cd hyperaccumulation is present only in some populations of *Sedum alfredii, Thlaspi caerulescens, Thlaspi praecox* and *Arabidopsis halleri* which are able to hyperaccumulate Cd which is taken up in part through Zn transporters (Zhao et al. 2006). It has been described that peroxidases played an important role during Cd hyperaccumulation, and the accumulation of ROS induced by Cd treatment might be involved in the metal hyperaccumulation (Zhang and Qiu 2007). In *Arabidopsis thaliana* a cadmium-tolerant mutant (MRC-32) has been identified that accumulates more Cd than wild-type (WT) plants although the mechanism involved in this hyperaccumulation has not yet been identified (Watanabe et al. 2010).
2. At root level, plants have developed extracellular strategies to avoid Cd toxicity and especially interesting is the relationship with mycorrhizal fungus (Jentschke et al. 1999; Courbot et al. 2004; Janouskova et al. 2006) or with some bacterial strains from the rhizosphere which can reduce Cd concentration in the shoot of the hyperaccumulators *A. halleri,* and which highlights the importance of plant–microbe interactions in Cd toxicity (Farinati et al. 2011). Additionally, the evaluation of interactions between heavy metal contamination and beneficial rhizosphere microbes adapted to the contaminated soils and their effects on plant development has shown that the microbes are not only able to grow but also to improve plant development under polluted conditions (Azcon et al. 2010). Interestingly, it seems that toxicity of heavy metal to microorganisms is due, in part, to oxidative stress, and it has been hypothesised that the metal resistance of microorganisms and their beneficial effect on the plant can be ascribed partially to the microbe antioxidative enzyme metabolism (Azcon et al. 2010). Moreover, there are other mechanisms like the immobilisation of Cd by means of the cellular and extracellular carbohydrates (Verkleij and Schat 1990; Wagner 1993). Recently, a Cd-phobic mutant (MRC-22) has been shown to arrest the growth of the primary root immediately when it encounters Cd and to produce an increased number of lateral roots located in the Cd-free zone (Watanabe et al. 2010). Some authors have suggested that the primary root could act as a sensor of the root environment and send some cues to the whole root system to modulate the direction of root growth, MRC-22 could be affected in this sensitivity sending more cues than WT on encountering Cd (Watanabe et al. 2010).
3. Plants have also developed intracellular strategies against Cd toxicity such as transport to the major storage organs or tissues, chelation and subcellular compartmentalization and the efflux from the plant (Benavides et al. 2005; Verbruggen et al. 2009; Sharma and Dietz 2009). As described previously, Cd

enters first by the roots through the cortical tissue. Roots accumulate Cd during its exposure and part of the metal is then translocated to leaves (Ogawa et al. 2009). Cd can be loaded rapidly into the xylem by transport to the above-ground tissues. Once Cd has entered into the cytosol, it can bind to phytochelatins or their precursor glutathione, generating conjugates that can be transported into the vacuoles, preventing the free circulation of Cd ions in the cytosol (Cobbett 2000; Verbruggen et al. 2009). Cd can also be complexated by metallothioneins and nicotianamine (Cobbett and Goldsbrough 2002; Sharma and Dietz 2006). Proline, histidine and polyamines are also involved in the defence against metal stress because they may be involved in osmoregulation and metal chelation, or they can act as antioxidants (Sharma and Dietz 2006). Referring to compartmentalisation of Cd, it has been shown that some of the genes regulated by Cd are involved in its own transport, like AtPcr1 (Song et al. 2004). It seems that soluble phenols may be involved in Cd shoot-to-root translocation as a result of SA treatment (Kovacik et al. 2009). Translocation of Cd from root to shoots is mainly done through inorganic forms (Ueno et al. 2008), although Cd^{2+} can be directly transported into the vacuoles by Cd^{2+}/proton antiporter (Korenkov et al. 2007; Berezin et al. 2008) and the metal transporters NRAMP3 and NRAMP4 are responsible for Cd^{2+} efflux from vacuoles (Thomine et al. 2000, 2003). For the translocation of Cd, plants partially share the processes with Zn and/or Fe transport, like the Zn transporter ZNT1 that can also transport Cd, but with a lower affinity than with Zn (Pence et al. 2000; Ueno et al. 2008). The MRC-26 gene could be involved in the Cd transport from root to aerial tissues (Watanabe et al. 2010) and originally it was a defective mutant in Fe absorption or Fe transport from roots to the aboveground tissues. Very recently, a novel molecular mechanism of heavy metal tolerance in plants has been ascribed to microRNAs (miRNAs) that are small non-coding RNAs that negatively regulate specific target mRNAs at the post-transcriptional level. In fact, a total of 19 Cd-responsive miRNAs have been identified in rice which encoded transcription factors (TFs) and proteins associated with metabolic processes or stress responses (Ding et al. 2011). Many of the predicted target genes of the Cd-responsive miRNAs encode TFs and as most of the miRNAs are down-regulated, the target of these TFs will be up-regulated leading to enhanced Cd tolerance (Ding et al. 2011). Plants have also developed other strategies to cope with Cd like activation of genes involved in defence responses, production of ROS and NO and modifications of antioxidant systems that will be discussed in the following paragraphs.

4 Transcriptomic and Proteomic Analyses Under Cadmium Stress

Recently there have been an increasing number of studies related to changes in the expression of Cd-responsive genes and proteins in several plants species (Tables 1 and 2). The results obtained suggest that the regulation of gene expression in

Table 1 A summary of transcriptomic studies of plant response to Cd treatment in the last 10 years;*

Plant system/treatment	Technique	Genes regulated by cadmiun	Reference
A. thaliana plants 500 µM Cd: 2 h	Fluorescent differential display (FDD) 8000cDNAs	31 cDNAs identified differentially expressed involved in signal transduction, general and oxidative stress.	Suzuki et al. (2001)
Datura innoxia tissue culture 250 µM Cd: 24 h	Two differential screening steps in a Cd-induced cDNA library.	4 cDNAs induced: related to a sulfur transferase-family protein	Louie et al. (2003)
A. thaliana roots and leaves 50 µM Cd.: 24 h	Microchip technology and semi-quantitative PCR	4 AtMRPs were up-regulated in roots. AtMRP3 transcript levels increased in both root and shoot of young plants.	Bovet et al. (2003)
A. thaliana leaves 30–100 µM Cd: 4 day	Screening for *A. thaliana* genes that confer resistance to *ycf1*	At least 4 AtPcr genes (Cys-Rich Membrane Proteins) mediate Cd resistance.	Song et al. (2004)
A. thaliana (line 651)* 50 µM Cd: 21 day	Microchip analysis	65 genes up-regulated and 338 genes down-regulated: general and oxidative stress, transport, sugar and lipid metabolism, protein and aa metabolism.	Kovalchuk et al. (2005)
B. juncea plants. 10 µMCd: 6, 24 h and 6 weeks	cDNA-AFLP analysis	73 transcript changed: related to transcriptional factors, expression regulators, stress responding and transport facilitation genes, cellular metabolism and organization and photosynthetic process.	Fusco et al. (2005)
B. juncea leaves and roots 10 µM Cd: 48 h	Fluorescent mRNA differential display (DD)	Up-regulated: auxin and iron metabolism, oxidative and general stress, TFs. Down-regulated: Nop family nucleolar protein, ARF-like small GTPases	Minglin et al. (2005)
A.thaliana and *A. halleri* roots 25–125 and 10–50 µM Cd, 2 h	*Arabidopsis* genome array (Affymetrix, 8,000 genes)	*A. thaliana*: 23 genes induced related to general stress, abiotic and biotic stress, TFs, ethylene related, sulphate assimilation. *A. halleri*: 5 genes induced, cyt P450 and 4 in common with *A. thaliana*: ATP sulfurylase, EF-hand containing putative Ca binding, 2 TFs.	Weber et al. (2006)
A. halleri ssp halleri x *A. lyrata ssp petraea* plants. 10 µM Cd: 72 h	cDNA-AFLP analysis (±19,000 transcript tags)	188 genes overexpressed in tolerant genotypes related to cellular detoxification and repair, metal sequestration, hydric balance, signal transduction, transcriptional regulation, protein degradation.	Craciun et al. (2006)
A. thaliana roots and leaves 5–50 µM Cd: 2,6 and 30 h.	CATMA microarray (24,576 probes)	5 µM: 752 up- 340 down-regulated. 50 µM: 1,166 up- 1,089 down-regulated. Related to photosynthesis, oxidative stress, cell wall metabolism, phenylpropanoid metabolism, sulfate assimilation	Herbette et al. (2006)
O. sativa L. roots. 0.1, 1, 10, 100 µM and 1 mM Cd: 1 day	cDNA-AFLP analysis	Induced changes in metabolic enzymes, transporters and proteins involved in the degradation of oxidatively modified proteins	Aina et al. (2007)

A. thaliana and *T. caerulescens* roots 15 and 0,5, 50 μM Cd, resp. : 1 week	Agilent Arabidopsis2 60-mer oligonucleotide microarray (21,500 genes)	*A. thaliana*: 4 genes down-regulated: UPM1, phosphate-response, nodulin-like and MA3 domain-containing protein. 26 genes up-regulated: transport, transcription and stress response genes. *T. caerulescens*: 171 genes modulated related to genes involved in lignin, glutathione and sulphate metabolism.	van de Mortel et al. (2008)
H. vulgare. roots. 1 mM Cd: 24 h	Semi-quantitative RT-PCR	Genes involved in scavenging AOS, HvC2 and Prx8, water stress related genes.	Tamas et al. (2008)
G. Americana L. roots 0.5,1,2,4,8 and 16 mg/L: 72–96 h.	Construction of a cDNA library	81 ESTs identified which can be classified as primary metabolism and defense genes. The most abundance was metallothionein and PRs genes.	Souza et al. (2008)
A. thaliana roots 15 μM $CdCl_2$: 24 h.	Microarrays analyses (Agilent Technologies)	41 genes induced by Cd and others ions: Ca-binding proteins, disease resistance proteins, ROS-responsive genes. 48 genes induced only by Cd: related to primary and secondary metabolism, stress-response genes and pathogen related proteins.	Zhao et al. (2009)
O. sativa roots and shoots 10 μM $CdCl_2$: 3 h 1 μM $CdCl_2$: 24, 48, and 72 h, 8 day	Oligo-DNA microarray (Agilent Technologies)	Roots: 1,207 genes up-regulated, especially Cytochrome P450 family proteins, heat shock proteins, glutathione S-transferase and genes involved in signal transduction (DREB,NAC, and protein kinases)	Ogawa et al. (2009)
V. baoshanensis. roots. 300 μM Cd: 48 h.	Construction of a cDNA library	Half of the identified genes were involved in general stress defense, such as antioxidative enzymes, protein degradation and stress signal transduction.	Zhang et al. (2009)
S. torvum Sw. roots 0.1 mM $CdCl_2$: 3 h, 1 day, and 3 days.	Construction of a cDNA library	2,049 genes up- regulated: chaperone proteins, antioxidative and sulphur-assimilating enzymes 2,022 down-regulated: xylem-loading citrate transporter, dehydrationrelated transcription factors and aquaporin isoforms	Yamaguchi et al. (2009)
Arabidopsis seedlings 0–2 mg/L Cd: 60 h.	RT-PCR analysis	0.5 mg/L: 4 genes induced: DNA mismatch repair (MMR) system and cell division genes. 1 mg/L: inhibition in expression of three (MMR) genes	Liu et al. (2009)
O. sativa roots. 60 μM CdCl2: 0, 3, 6, 12, 24 h.	Microarray assay.	Metal stress-responsive cis-elements tended to be up-regulated.	Ding et al. (2011)

*: see Swoboda et al. 1994

Table 2 A summary of proteomic studies of plant response to Cd treatment in the last 5 years

Plant system/treatment	Technique	Proteins regulated by cadmiun	Reference
Arabidopsis thaliana cells 0–200 μM Cd: 24 h	Proteomic and metabolomic differential display	Proteins upregulated: related to carbon, nitrogen and sulphur metabolic pathways and six families of PCs.	Sarry et al. (2006)
A.thaliana roots 10 μM Cd: 24 h	2D gels and matrix-assisted laser desorption/ionization time-of-flight mass spectrometry	41 Proteins up-regulated: related to metabolic, antioxidant system and stress response.	Roth et al. (2006)
Populus tremula leaves 20 μM Cd: 3, 7 and 14 days	2D gels and MALDI-TOF-TOF	52 proteins up-regulated: related to pathogenesis, remobilization of carbon 73 proteins down-regulated: related to oxidative stress regulation and carbon metabolism.	Kieffer et al. (2008)
Brassica juncea (L.) roots 250 μM Cd: 3 days	2-D DIGE and multiplexed isobaric tagging technology (iTRAQ)	Several proteins upregulated: involved in sulfur assimilation, redox homeostasis, and xenobiotic detoxification.	Alvarez et al. (2009)
Spinacia oleracea (L.) apical and basal leaves 20–100 μM Cd: 0,5 and 18 days	2D IEF-SDSPAGE and Nano-RP-HPLC-ESI-MS/MS	*Apical leaves*: proteins upregulated involved in photosynthesis, carbohydrate metabolism and CO_2 fixation. *Basal leaves*: proteins upregulated involved in oxidative stress response, carbohydrate metabolism, pathogen related proteins and cysteine biosynthesis. Down regulated: related to photosynthesis, carbohydrate metabolism, O_2 production and protein folding.	Fagioni and Zolla (2009)
Oryza sativa roots and leaves 100 μM Cd: 24 h	MALDI-TOF MS	36 proteins up or down-regulated: related to oxidative stress, carbohydrate metabolism, protein degradation, amino acid synthesis and photosyntesis.	(Lee et al. 2010)
Lycopersicon sculentum roots 0, 10, 100 μM Cd: 10 days	2D gels and MALDI–TOF–MS, LIFT TOF–TOF	Up-regulation in stress and detoxification proteins. Down-regulation in carbon metabolism.	Rodríguez-Celma et al. (2010)
Populus tremula leaves and cambial zone. 360 mg Cd/Kg SDW: 61 days	2D gels and EST database on an inhouse MASCOT platform	Reduced in photosynthesis, resulting in a profound reorganisation of carbon and carbohydrate metabolisms.	Durand et al. (2010)
A. thaliana leaves 1, 10 μM Cd: 1 week	2D gels and LC–ESI-MS/MS	21 proteins up-regulated: related to oxidative stress response, photosynthesis energy production, protein metabolism and gene expression.	Semane et al. (2010)

response to Cd is more dependent on time than dose exposure, and it is also regulated differentially depending on the plant tissue (Herbette et al. 2006; Ogawa et al. 2009; Harada et al. 2010). Additionally, the cDNA library constructions carried out in different plant species have allowed the identification of several genes related to the toxicity of Cd that are involved in photosynthesis, antioxidative stress, defence response, cellular detoxification, transporters and signal transduction genes (Fusco et al. 2005; Souza et al. 2008; Yamaguchi et al. 2009; Harada et al. 2010). Most transcriptomic and proteomic studies have in common that genes related to photosynthesis and carbon metabolism are down-regulated (Fagioni and Zolla 2009; Durand et al. 2010; Rodríguez-Celma et al. 2010), whereas genes related to pathogens or oxidative stress, detoxification of stress, transporters and stress signal transduction are up-regulated (Tables 1 and 2; Zhao et al. 2009; Aloui et al. 2009; Lee et al. 2010).

Photosynthesis: While only a very few photosynthesis-related genes and proteins were up-regulated (Fagioni and Zolla 2009; Semane et al. 2010), an important number of these, such as enzymes of the chlorophyll pathway, the proteins of PSI and PSII, electron transporters, H^+-ATPases and enzymes involved in the Calvin cycle, were down-regulated by Cd exposure and, in some of them, there is also a decrease in the protein content (Fusco et al. 2005; Herbette et al. 2006; Ogawa et al. 2009; Qian et al. 2009; Fagioni and Zolla 2009; Durand et al. 2010). Due to that, the presence of Cd causes an overall inhibition of photosynthesis in different plant species, which could be the first response in plants against different stress conditions to avoid oxidative damage.

Pathogenesis-related proteins (PRs): they were initially defined as proteins produced by the host plant, whose induction was thought to be specifically in pathological or related situations following the development of a local or systemic resistance (van Loon and van Strien 1999; Muthukrishnan et al. 2001). However, many PR proteins were found to be expressed without pathogen/wounding stress. In fact, many PR genes and proteins can be induced by Cd exposure (Weber et al. 2006; Kieffer et al. 2008; Souza et al. 2008; Rodríguez-Serrano et al. 2009a; Fagioni and Zolla 2009; Harada et al. 2010). The effect of Cd in plant species mimics the response of plant to pathogen attack, in which an increase in reactive oxygen species is observed, followed by reactions of secondary defences (Schützendübel and Polle 2002).

Oxidative stress response: Different transcriptomic and proteomics analyses have shown that oxidative stress-related genes and proteins could be up- or down-regulated in several plants species in response to Cd exposure depending on the concentration and timing of the metal treatment (Minglin et al. 2005; Roth et al. 2006; Kieffer et al. 2008; Fagioni and Zolla 2009; Semane et al. 2010). Up-regulation of producing enzymes of ROS and down-regulation of antioxidative enzymes by Cd could facilitate the accumulation of ROS and oxidative stress (Lee et al. 2010; Rodríguez-Celma et al. 2010). In *Arabidopsis* plants, a transitory increase in the expression of *rboh*F has been shown in response to Cd (Horemans et al. 2007), and recently, a differential regulation of the NADPH oxidase genes has also been involved in *Arabidopsis* response to Cd (Remans et al. 2010). Protein oxidation has been described in different plant species under Cd stress. In pea plants, different oxidised proteins were identified such as Rubisco, GR, SOD and CAT (Romero-Puertas et al. 2002). These proteins were more efficiently degraded and the proteolytic activity increased due to the metal treatment. Additionally, an increase of lipid peroxidation has been detected after Cd exposure in different plant species (Romero-Puertas et al. 1999; Sandalio et al. 2001; Tamas et al. 2008; Semane et al. 2010).

Chelation of Cd: as described before, one of the mechanisms that plants have developed to cope with Cd toxicity is the chelation of the metal with molecules such as GSH, phytochelatins or metallothioneins. In fact, many genes and proteins related to GSH and phytochelatins were up-regulated under Cd stress in different plant species (Sarry et al. 2006; Herbette et al. 2006; Fagioni and Zolla 2009; Ogawa et al. 2009). In particular, enzymes involved in GSH synthesis are up-regulated in *A. thaliana* and *O. sativa* roots and in basal leaves of *Spinacia oleracea* (Herbette et al. 2006; Ogawa et al. 2009; Fagioni and Zolla 2009), which could be related to the increased level of GSH observed in *A. thaliana* roots (Herbette et al. 2006). In contrast, GSH levels decreased in *Arabidopsis* leaves, which was accompanied by an increased level of phytochelatins probably due to the induction of PC synthases such as AtPCS1 (Herbette et al. 2006). This result fits with previous studies showing that GSH biosynthesis/level is a mechanism to modulate the PC production and accumulation (Zhu et al. 1999; Gong et al. 2003). In addition, metallothionein-related genes, another chelating molecule, were up-regulated by Cd exposure (Souza et al. 2008).

Transporters: Some of the genes that are regulated by Cd are members of different transporter families (Kovalchuk et al. 2005; Fusco et al. 2005; Aina et al. 2007). Most of them, such as members of the ABC, PDR, MATE or ZIP family transporters, are up-regulated under Cd exposure in *Arabidopsis* and rice (Bovet et al. 2005; Ogawa et al. 2009), which supports the idea that Cd enters into the plant through different transporters (Clemens 2006).

Signal transduction: The expression of many genes involved in signal transduction has been shown to be up-regulated in response to Cd in different tissues and plant species (Suzuki et al. 2001; Weber et al. 2006; Zhao et al. 2009). While several ROS responsive or Ca-binding proteins are induced by different ions, there are some genes such as specific protein kinases, heat shock and defence response proteins that are induced exclusively by Cd (Zhao et al. 2009). Most of them encode calcium-dependent protein kinases (CDPKs), mitogen-activated protein kinases (MAPKs), transcription factors and genes involved in ethylene and jasmonic acid signalling pathways (Herbette et al. 2006; Ogawa et al. 2009). Ca^{2+}/calmodulin-dependent protein kinases are serine/threonine-specific protein kinases that are primarily regulated by the Ca^{2+}/calmodulin complex. Up-regulation of some calmodulins and CDPKs by Cd suggests that the metal interferes with calcium signalling pathways as has been described elsewhere (Rodríguez-Serrano et al. 2009a), and it has been shown that Cd produces a decrease of Ca content in pea plant (Sandalio et al. 2001). MAPKs are serine/threonine-specific protein kinases that respond to extracellular stimuli mitogens, osmotic stress, heat shock and proinflammatory cytokines and regulate various cellular activities, such as gene expression, mitosis, differentiation, proliferation, and cell survival/apoptosis. In *Arabidopsis*, Cd activates the mitogen-activated protein kinases, MPK3 and MPK6, in a dose-dependent manner (Liu et al. 2010) and a MAP kinase cascade may function in the Cd-signalling pathway in rice (Yeh et al. 2004). Both CDPKs and MAPKs pathways are thought to engage with ROS production (Kobayashi et al. 2007; Ren et al. 2002). Several transcription factors are also regulated by Cd. In fact, members of the families of DREB, WRKY, NAC, MYB and AP2 are up-regulated after a cadmium exposure (Fusco et al. 2005; Minglin et al. 2005; Weber et al. 2006).

Plant hormones: An increase in JA and ET levels under Cd treatment has been observed (Rodríguez-Serrano et al. 2009a). In fact, the ethylene signalling pathway seems to be involved in the early phases of response to Cd (Weber et al. 2006; Herbette et al. 2006) and the genes encoding for ERF2 and ERF5 ethylene responsive factors were up-regulated (Herbette et al. 2006). It has also been shown that Cd induced PCD in tomato suspension cells requires ethylene-signalling pathways (Yakimova et al. 2006). The induction of ET was reversed by supplying Ca to the plant, which could be due to an indirect effect of Ca on ROS and NO production (Rodríguez-Serrano et al. 2009a). In this context, it was observed the induction of some PRs genes, which could be a response to the overproduction of JA, ET and ROS in order to protect proteins from damage associated with Cd toxicity (Rodríguez-Serrano et al. 2009a). Genes involved in the biosynthesis of jasmonic acid were up-regulated in rice and *Arabidopsis* in response to Cd which indicates that this pathway is also important in signal transduction pathways in the response to the metal (Ogawa et al. 2009). It was found additionally that JA regulates genes involved in glutathione and phytochelatin synthesis under Cd treatment (Xiang and Oliver 1998). The increase of JA could also contribute to metal toxicity through the activation of lipoxygenase activity, H_2O_2 production and lipid peroxidation (Wang and Wu 2005; Maksymiec et al. 2007). Salicylic acid (SA) which is also an important signalling element in plants, has been observed to alleviate Cd-induced growth inhibition and oxidative damage in barley roots although the mechanism is not well understood (Metwally et al. 2003). It has also been reported that salicylic acid could prevent Cd-induced photosynthetic damage and cell death, probably due to the inhibition of ROS overproduction (Zhang and Chen 2011). SA enhanced total soluble phenols, particularly in the roots and it has been reported that soluble phenols may be involved in Cd shoot-to-root translocation (Kovacik et al. 2009). SA decreases and MeSA increases in pea roots treated with Cd while in pea leaves none of them changed its level which suggests a differential response of roots and leaves (Rodríguez-Serrano et al. 2009a).

5 ROS Metabolism in Response to Cadmium

5.1 ROS Production Under Cd Stress

Reactive oxygen species (ROS) such as $O_2^{\bullet -}$, $^{\bullet}OH$ or H_2O_2 are produced as a result of normal aerobic metabolism (Gutteridge and Halliwell 2000). An excess of ROS, however, is dangerous for the plant, mainly due to the reaction with lipids, proteins and nucleic acids giving rise to lipid peroxidation, membrane leakage, enzyme inactivation and DNA break or mutations, which can induce severe damage to the cell (Halliwell and Gutteridge 1999). Although Cd^{2+} is not able to directly generate ROS by a Fenton reaction, it might inhibit antioxidant enzymes, impair the respiratory chain, or displace copper and iron ions in metalloproteins, which eventually trigger a Fenton reaction (Valko et al. 2005). It has been reported that Cd produces concentration-dependent imbalances in the antioxidant defence of plants and induces oxidative stress (Romero-Puertas et al. 1999; Dixit et al. 2001; Sandalio et al. 2001). Additionally, ROS production at subcellular levels was demonstrated in plants grown under Cd exposure (Olmos et al. 2003; Romero-Puertas et al. 2004; Garnier et al. 2006; Rodríguez-Serrano et al. 2006; Ortega-Villasante et al. 2007). Different techniques such as absorption, fluorescence and spin-trapping electron paramagnetic resonance (EPR) spectroscopy have been used to study ROS production under Cd stress and its localization in different plant species (Aravind et al. 2009; Bi et al. 2009; Heyno et al. 2008). In peroxisomes purified from pea leaves, Cd exposition increases the H_2O_2 content measured by fluorometry, mainly due to the induction of glycolate oxidase activity (Romero-Puertas et al. 1999). By cytochemical or spectrophotometric methods, H_2O_2 was found as starting point at the cell plasma membrane in Cd treated tobacco culture cells (Olmos et al. 2003; Garnier et al. 2006), followed by the accumulation of $O_2^{\bullet -}$ in mitochondria (Garnier et al. 2006). It has also been shown that, in isolated mitochondria from potato tuber, there is an increase of both H_2O_2 and $O_2^{\bullet -}$; while a reduction of $O_2^{\bullet -}$ production, measured by EPR, was observed in isolated plasma membranes from soybean under Cd stress (Heyno et al. 2008). Localization of ROS production has been done also by electron microscopy showing that the accumulation of H_2O_2 in leaves from pea plants, was observed mainly in the plasma membrane of transfer, mesophyll and epidermal cells, as well as in the tonoplast of bundle sheath cells (Romero-Puertas et al. 2004). In mesophyll cells, a small accumulation of H_2O_2 was observed in mitochondria and peroxisomes. Moreover, the subcellular localization of $O_2^{\bullet -}$ production was demonstrated in the tonoplast of bundle sheath cells, and plasma membrane from mesophyll cells (Romero-Puertas et al. 2004). Confocal laser microscopy has been used to detect ROS production in the different organs of the plant such as intact roots where Cd induced the over-accumulation of $O_2^{\bullet -}$ and H_2O_2. In lateral roots, $O_2^{\bullet -}$ is located mainly in phloem and cortex cells, whereas H_2O_2 is located in cell wall of epidermis, phloem xylem and cortex cells (Rodríguez-Serrano et al. 2006). In principal roots, both $O_2^{\bullet -}$ as H_2O_2, were mainly detected in xylem cells. These studies reveal that ROS production was regulated by different enzymes including NADPH oxidases, peroxidases and different sources from organelles and the responsibility for the metal toxicity/signalling depends on the tissue, timing and plant conditions (Romero-Puertas et al. 2004; Garnier et al. 2006; Horemans et al. 2007; Cuypers et al. 2011). The Cd-induced oxidative burst follows a signalling cascade that is dependent on calmodulin, intracellular Ca^{2+} mobilization, entry and competition with Cd. H_2O_2 production is also dependent on kinases and protein phosphatases and, in part, on ADPR cyclase and guanylate cyclase (Olmos et al. 2003; Romero-Puertas et al. 2004; Garnier et al. 2006; Rodríguez-Serrano et al. 2009a). Additionally, in *Arabidopsis* suspension cells exposed to 100 and 150 μM of Cd that undergo PCD, H_2O_2 production is dependent on a previous NO release (De Michele et al. 2009).

5.2 *Antioxidant Systems Under Cd Stress*

Plants possess homeostatic cellular mechanisms to regulate the concentration of the metal inside the cell to minimize the potential damage that could result from the exposure to nonessential metal ions (Benavides et al. 2005). Recent experiments revealed a clear relationship between metal stress, redox homeostasis and antioxidant capacity (Sharma and Dietz 2009). Actually, Cd-dependent increase of cellular ROS levels is probably due to the inhibition of antioxidative enzymes (Sandalio et al. 2001; Schützendübel and Polle 2002; Benavides et al. 2005).

5.2.1 Enzymatic Antioxidants Systems

A variety of proteins function as scavengers of ROS such as superoxide dismutase (SOD), catalase (CAT), enzymes of the ascorbate–glutathione cycle, peroxiredoxin and thioredoxin families. The main response for the removal of ROS appears to be via the induction of SOD and CAT activities, and by the induction of GR to ensure the availability of reduced glutathione for the synthesis of Cd-binding peptides. This fact could be related to the inhibition of APX activity probably due to glutathione and ascorbate depletion (Gomes-Junior et al. 2006). Different effects were observed in antioxidant enzymes depending on Cd concentration, period of treatment, the plant tissue studied and the plant itself. SOD plays an important role as a defence mechanism carrying out the catalytic dismutation or disproportionation of $O_2^{\bullet -}$ with production of water and hydrogen peroxide. While Cd-dependent reduction of SOD activity has been reported in wheat (Milone et al. 2003) or pea (Sandalio et al. 2001; Romero-Puertas et al. 2007), in other species like sunflower (Laspina et al. 2005) or coffee cells (Gomes-Junior et al. 2006), an increase of this activity was observed.

CAT activity, that catalyses the decomposition of H_2O_2, decreases with increasing concentrations of Cd in pea plants (Sandalio et al. 2001; Rodríguez-Serrano et al. 2006; Romero-Puertas et al. 2007), pine (Schützendübel et al. 2001), pepper

(León et al. 2002) and sunflower (Laspina et al. 2005), while the opposite effect was observed in radish roots (Vitoria et al. 2001). Especially interesting is that light may interfere with the Cd-dependent catalase activity, as it has been shown that Cd^{2+} treatment of leaf discs under light decreased CAT activity and increased carbonyl groups content, which suggested that CAT inactivation could be due in part, to the oxidation of the protein under this condition (Azpilicueta et al. 2007), although in pea plants treated with Cd, no changes in the pattern of CAT oxidation were observed (Romero-Puertas et al. 2002). In addition to CAT, the enzymes of the ascorbate-glutathione cycle also remove H_2O_2, especially in cellular compartments where H_2O_2 exits and there is no catalase (Gutteridge and Halliwell 2000). This cycle is composed of four enzymes: ascorbate peroxidase (APX), monodehydroascorbate reductase (MDHAR), dehydroascorbate reductase (DHAR) and glutathione reductase (GR) which results in the elimination of H_2O_2 using ascorbate (ASC), and involving reduced glutathione (GSH) and reducing power (NADPH). APX activity was increased in *Phaseolus vulgaris* roots and leaves (Chaoui et al. 1997) and in *Ceratophyllum demersum* (Aravind and Prasad 2005), while it was decreased in wheat (Milone et al. 2003) and no changes were observed in pea leaves (Romero-Puertas et al. 2007). MDHAR activity was induced in onions (Fatima and Ahmad 2005) while its activity was reduced in *C. Demersum* (Aravind and Prasad 2005) in which was observed the same effect in DHAR activity. In contrast, no significant change was observed in DHAR activity in onions (Fatima and Ahmad 2005). Changes in GR activity also depend on the plant species, so GR did not show statistically significant changes with Cd in pea leaves (Sandalio et al. 2001), whereas it was reduced in onions and pea seeds (Fatima and Ahmad 2005; Smiri et al. 2010) and increased in different cultivars of pepper plants (León et al. 2002). Interestingly, it has been observed in different cultivars of pepper plants that the tolerance to Cd was more dependent on the availability of NADPH than on its antioxidant capacity (León et al. 2002). Finally, there are two family of proteins, peroxiredoxins that control peroxide levels and play an important function in cellular detoxification of Cd (Finkemeier et al. 2005; Dietz et al. 2006), and thioredoxins that act as antioxidants, making the reduction of other proteins through thiol-disulfide exchange easier, which may also act in response to Cd stress. In fact, it has been shown a pronounced stimulation of glutaredoxin (Gpx) expression, a thioredoxin-dependent enzyme in plants, under Cd stress (Smiri et al. 2010).

5.2.2 Non-enzymatic Antioxidants Systems

The glutathione (GSH)–ascorbate (ASC) couple and emerging antioxidative players like α-tocopherol, proline and carbon monoxide (Sharma and Dietz 2009; Zhang and Chen 2011; Aravind et al. 2009) are non-enzymatic antioxidant systems and some of them have been shown to have an important role in plant responses to Cd stress. Actually, a Cd-dependent reduction of glutathione (GSH) has been described in different tissues and plant species (Rodríguez-Serrano et al. 2006; Gomes-Junior et al. 2006; Romero-Puertas et al. 2007; Chen et al. 2010). Cadmium caused a reduction of the total glutathione content (GSH + GSSG), with the reduced form of glutathione (GSH) being the most affected (Romero-Puertas et al. 2007), and this may cause oxidative stress and in turn, short-term toxicity (Schützendübel and Polle 2002; Nocito et al. 2006). Like glutathione, ascorbate content decreases in different plant species under Cd exposure (Rodríguez-Serrano et al. 2006; Gomes-Junior et al. 2006; Romero-Puertas et al. 2007; Chen et al. 2010). Interestingly, exogenous application of proline significantly restored the membrane integrity in tobacco and increased the activities of ASC-GSH cycle enzymes under Cd stress. Therefore, proline offered an efficient protection to membranes against Cd stress (Islam et al. 2009). Also, experiments with *Medicago sativa* root tissues suggest a role for CO release as a signal element for the alleviation of Cd-induced oxidative damage by modulating glutathione metabolism (Han et al. 2008). Although Zn is an essential micronutrient for plants and it could be toxic at high concentrations, it has been observed that Zn can act as an antioxidant against Cd exposure mainly due to competition with the heavy metal. In *Ceratophyllum demersum*, a Zn supplementation inhibits the oxidative products of proteins measured as carbonyls caused by Cd-enhanced ROS production (Aravind et al. 2009).

6 NO Metabolism in Response to Cadmium

Nitric oxide (NO) is a hydrophobic gaseosus molecule and a diffusible free radical. It is considered as a biological mediator involved in key physiological processes which includes germination, root growth, gravitropic binding, stomatal closure and growth regulation of pollen tubes (Wilson et al. 2008; Besson-Bard et al. 2008). Moreover, it has also been implicated in the plant adaptive response to biotic and abiotic stresses, acting as a signalling compound in the molecular cascade leading to changes in gene expression (Gould et al. 2003; Delledonne 2005). In cells, NO can react with oxygen radicals giving rise to reactive nitrogen species (RNS) as happens with $O_2^{\bullet-}$ that generates peroxinitrite, a potent oxidant molecule.

6.1 NO Production Under Cd Stress

In plants, NO could be generated by nitrate reductase (NR) that catalyses the reduction of nitrite to NO, or by a putative NOS-like protein which generate NO from L-arginine (Neill et al. 2008; Gupta and Kaiser 2010). Despite the fact that the NOS protein has not been identified up to now, NOS activities have been measured in plant tissue extracts as well as in several purified organelles, and at least 11 different plant species showed this activity (del Rio 2011). Recently, a nitric oxide synthase was characterized from the green alga *Ostreococcus tauri* (Foresi et al. 2010). Another enzyme that has been shown to produce NO is a plasma membrane-bound enzyme of tobacco roots, nitrite NO oxidoreductase (Stohr et al. 2001; Stohr and Stremlau 2006), and NO can be produced in plants by non-enzymatic reactions from nitrite under acidic pH or through nitrification/denitrification cycles (del Rio et al. 2004). The production of NO after Cd exposure has a time- and dose-dependent pattern. Treatments with high doses of Cd (100–200 μM of $CdCl_2$) show an increase in NO content in cells in different plant species (De Michele et al. 2009; Bartha et al. 2005; Groppa et al. 2008; Besson-Bard et al. 2008). Treatment with 50 μM of $CdCl_2$ showed an increase in NO content in cells with short time treatments (24 h) (De Michele et al. 2009) while higher doses of $CdCl_2$ (50–100 μM) and a longer time of treatment (>7 days) showed a decrease in NO production (Rodríguez-Serrano et al. 2006, 2009a; Xiong et al. 2009) which could be due in part to an inactivation of NOS-like activity as a consequence of the Cd-induced Ca deficiency in leaves (Rodríguez-Serrano et al. 2009a). Production of NO in plants under Cd stress can be detected both in roots and leaves which suggests that Cd transporters from roots to leaves could trigger NO synthesis in leaves (Besson-Bard et al. 2009). In leaves, NO was found in trichomes and stomata of the leaf surface (Besson-Bard et al. 2009) and in xylem vessels, sclerenchyma and epidermal cells (Rodríguez-Serrano et al. 2009a). In principal roots, NO was detected in the cortex, xylem and, to a minor extent, in phloem, while in lateral roots it was mainly observed in the epidermis (Rodríguez-Serrano et al. 2006). NO was also detected in crown roots (CR) in rice seedlings. In this tissue, NO plays a crucial role in CR primordium initiation in plants exposed to Cd (Xiong et al. 2009).

6.2 NO Function and Protection Under Cd Stress

Nitric oxide is an intracellular and intercellular messenger with a broad spectrum of regulatory functions in many physiological processes and responses to biotic and abiotic stresses (del Rio et al. 2006; Delledonne 2005; Wilson et al. 2008). It has also been reported that NO can act as pro-oxidant and antioxidant in plants. Actually, it can scavenge directly some ROS like the superoxide anion or indirectly changing both enzymatic or non-enzymatic antioxidant systems (Kopyra and Gwóźdź 2003). It has been described that NO increases SOD activity in *Lupinus luteus* (Kopyra and Gwóźdź 2003), delays the loss of SOD activity in barley aleurone cells (Beligni et al. 2002) or prevents Cd-induced increase of SOD from *Helianthus annus* (Laspina et al. 2005). Additionally, NO can restore catalase (CAT) activity, which was reduced in Cd-treated sunflowers plants (Laspina et al. 2005), also delays the loss of CAT activity in barley aleurone cells (Beligni et al. 2002) and negatively affects the CAT and APX capacity of the cadmium-treated *Arabidopsis* cells (De Michele et al. 2009). An NO donor reversed the increment of ASC under Cd stress in *Helianthus annuus* (Laspina et al. 2005) and induced APX and especially GR expression under Cd stress in pea plants (Romero-Puertas et al. 2007). Altogether, these results suggest that NO could protect the plant against Cd toxicity by increasing the antioxidant capacity, and confer tolerance to heavy metal stress in plants.

Recently, a study in which CATMA arrays were used, has shown the effects of NO in roots of Cd^{2+} treated plants (Besson-Bard et al. 2009). Forty-three genes related to iron homeostasis, proteolysis, nitrogen assimilation/metabolism and root growth, were identified to as being regulated by nitric oxide. NO can also regulate cellular responses through posttranscriptional modifications such as S-nitrosylation of proteins (Lindermayr et al. 2006; Romero-Puertas et al. 2008). A decrease in S-nitrosylation of CAT protein under Cd stress was observed (Romero-Puertas et al. submitted). CAT, a key enzyme involved in H_2O_2 detoxification in peroxisomes is S-nitrosylated during physiological conditions (Romero-Puertas et al. submitted) and the reduction in S-nitrosylation could be related to the slight increase in CAT activity previously observed in isolated pea leaf peroxisomes (Romero-Puertas et al. 1999). Additionally, it has been shown that in plants treated with Cd there is an increase in the level of PCs nitrosylated, that are probably less effective in chelating Cd^{2+} (De Michele et al. 2009).

7 Organelles Involvement in Cd Stress

ROS production after cadmium stress differs in the nature of the source and localization in the cell. It seems that NADPH oxidase from plasma membrane is one of the first enzymes involved in ROS production in response to heavy metal (Olmos et al. 2003; Garnier et al. 2006). Other organelles, however, such as mitochondria (Garnier et al. 2006), peroxisomes (Romero-Puertas et al. 1999) and chloroplasts (Bi et al. 2009), are also involved and play a key role in ROS release and/or detoxification in response to Cd. Several studies have shown that the vacuole is a site of accumulation of different heavy metals (Cobbett 2000; Verbruggen et al. 2009). Actually, chelating complexes formed by Cd can be transported into the vacuoles, preventing the free circulation of Cd ions and placing them into a limited area (Sanitá di Toppi and Gabrielle 1999). On the other hand, it has been reported that Cd^{2+} increase H_2O_2 and $O_2^{\bullet-}$ accumulation in mitochondria (Heyno et al. 2008; Garnier et al. 2006). Additionally, it has been shown that after Cd treatment, mitochondria can change their distribution and mobility that together with ROS production suggests that these organelles play an important role in Cd-induced cell death (Bi et al. 2009). In this work, it was observed that, in *Arabidopsis* protoplast treated with Cd, an accumulation of H_2O_2 occurred first in mitochondria, just after 1 h treatment and subsequently H_2O_2 also appeared in chloroplasts, preceding cell death (Bi et al. 2009). The chloroplast is one of the major targets of Cd stress due to the down-regulation of genes involved in photosynthesis that causes an overall inhibition of photosynthetic rate and the accumulation of ROS in this organelle (Azpilicueta et al. 2007), which causes a change in the distribution and mobility of chloroplast similar to mitochondria (Bi et al. 2009). Peroxisomes are probably the major sites of intracellular H_2O_2 production although it also generates superoxide radicals $O_2^{\bullet-}$ (del Río and Puppo 2009). Different analyses showed H_2O_2 and $O_2^{\bullet-}$ accumulation in peroxisomes from pea plants treated with

Cd (Romero-Puertas et al. 2004; Rodríguez-Serrano et al. 2009a). Cadmium treatment also produced a proliferation of peroxisomes (Romero-Puertas et al. 1999) probably due to an induction of peroxisome biogenesis genes like it was observed in both plant and animal cells as a result of the induction of H_2O_2 production (Lopez-Huertas et al. 2000). Cd also causes an increase in peroxisome movement rate which is independent on changes in the actin cytoskeleton and dependent of calcium and peroxisomal ROS production (Rodríguez-Serrano et al. 2009b). Peroxisomes possess a battery of antioxidants that could play an important role in controlling ROS production under Cd stress. Thus, CAT can protect *Arabidopsis* protoplast against Cd-induced cell death (Bi et al. 2009).

8 Conclusion

Given the dramatic increase in data about plant response to Cd obtained by the numerous proteomic, metabolomic and transcriptomic analyses that have been done in the last ten years (see Tables 1 and 2), we have clearly only just begun to tap an immense well of knowledge that should provide a deeper understanding of the biology of plants related to this abiotic stress. It is difficult, however, to discriminate specific responses to this heavy metal from general stress responses and there is a need to analyse this data in depth. A point in common in almost all studies about plant responses to Cd is the induction of ROS production. Cd causes an oxidative stress and protein degradation that are important components of its toxicity, even giving rise to cell death (Fig. 1). It seems to be clear from proteomic and transcriptomic data, that induction of antioxidative defences and protein chaperones (like HSP) that remove oxidized proteins are observed. The role of these ROS is not only to provoke oxidative stress but also to take part in the signalling cascade triggering the stress response. Different messengers have been shown be involved in ROS production such as Ca, calmodulin, kinases and phosphatases, and different stress-responsive proteins or messengers have been shown to be downstream ROS production, such as MAPKinases, PRs or HSP, involved in the transcriptional responses to Cd. Recently, NO has also been involved as a component in the signalling pathway in response to Cd leading together with ROS to PCD. Further attention should be paid however to the sources of ROS and NO, the compartments, cross-talk and spatial distribution of both and the timing of the cascade that is still not clear. The cross-talk between the different compartments of the cell needs to be investigated because it is probably an important cue to decide cell fate. The complete ROS/RNS-dependent signalling cascade will help to understand the TFs and mechanisms involved specifically in the response to Cd stress.

Acknowledgments This study was supported by Marie Curie Actions-European re-integration Grants (FP7-PEOPLE-2007-2-2-ERG; Project nº 208146), by ERDF-cofinanced grant BIO2008-04067 from MICINN and *Junta de Andalucía* (project P06-CVI-01820), Spain.

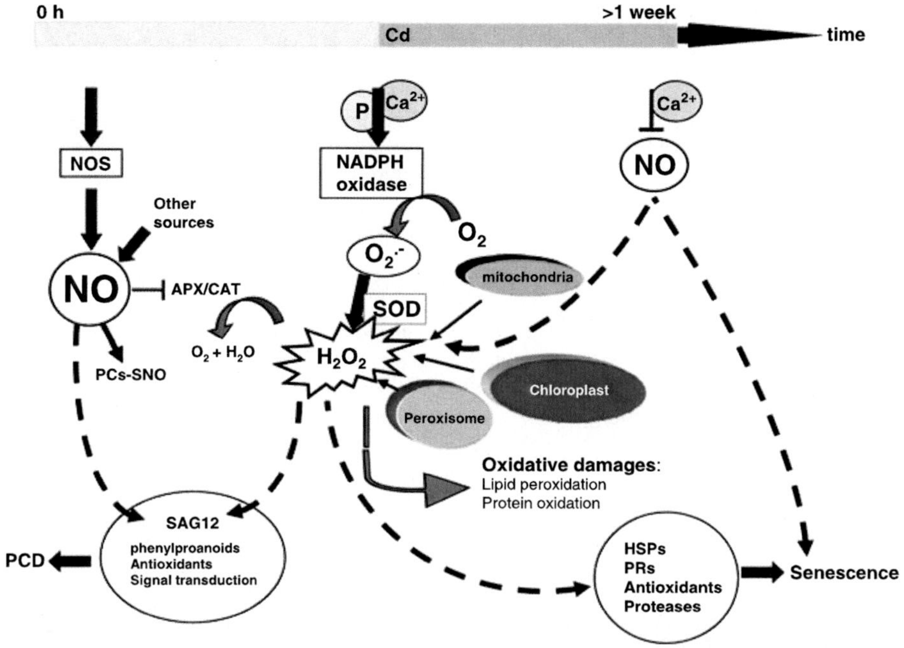

Fig. 1 Plants response to Cd treatment. Short-period treatment produces NO and ROS which induces gene expression to prevent oxidative damages caused by the metal although PCD is observed in culture cells. Long term treatment produces a reduction of NO and over-accumulation of ROS, giving rise to severe damage to proteins and lipids. Gene regulation in long-term treatment is focused on repairing oxidative damages and senescence. *HSPs* heat shock proteins, *PCs* phytochelatins, *PRs* pathogenesis-related proteins

References

Aina R, Labra M, Fumagalli P, Vannini C, Marsoni M, Cucchi U, Bracale M, Sgorbati S, Citterio S (2007) Thiol-peptide level and proteomic changes in response to cadmium toxicity in *Oryza sativa* L. roots. Environ Exp Bot 59:381–392

Alloway BJ, Steinnes E (1999) Antropogenic additions of cadmium to soils. In: Cadmium in soils and plants. Kluwer, Dordrecht

Aloui A, Recorbet G, Gollotte A, Robert F, Valot B, Gianinazzi-Pearson V, Aschi-Smiti S, Dumas-Gaudot E (2009) On the mechanisms of cadmium stress alleviation in *Medicago truncatula* by arbuscular mycorrhizal symbiosis: a root proteomic study. Proteomics 9:420–433

Alvarez S, Berla BM, Sheffield J, Cahoon RE, Jez JM, Hicks LM (2009) Comprehensive analysis of the *Brassica juncea* root proteome in response to cadmium exposure by complementary proteomic approaches. Proteomics 9:2419–2431

Aravind P, Prasad MN (2005) Modulation of cadmium-induced oxidative stress in *Ceratophyllum demersum* by zinc involves ascorbate-glutathione cycle and glutathione metabolism. Plant Physiol Biochem 43:107–116

Aravind P, Prasad MNV, Malec P, Waloszek A, Strzalka K (2009) Zinc protects *Ceratophyllum demersum* L. (free-floating hydrophyte) against reactive oxygen species induced by cadmium. J Trace Elem Med Biol 23:50–60

Azcón R, Peralvarez MC, Roldan A, Barea JM (2010) Arbuscular mycorrhizal fungi, *Bacillus cereus*, and *Candida parapsilosis* from a multicontaminated soil alleviate metal toxicity in plants. Microb Ecol 59:668–677

Azevedo H, Pinto C, Santos C (2005) Cd effects in sunflower: nutritional imbalances in plants and calluses. J Plant Nutr 28:2221–2231

Azpilicueta CE, Benavides MP, Tomaro ML, Gallego SM (2007) Mechanism of CATA3 induction by cadmium in sunflower leaves. Plant Physiol Biochem 45:589–595

Barceló J, Pöschenrieder C (1990) Plant water relations as affected by heavy metal stress: a review. J Plant Nutr 13(1):1–37

Bartha B, Kolbert Z, Erdei L (2005) Nitric oxide production induced by heavy metals in *Brassica juncea* L.Cern. and *Pisum sativum* L. In: Proceedings of the 8th Hungarian Congress of Plant Physiology and the 6th Hungarian Conference of Photosyntesis. Hungary

Beligni MV, Fath A, Bethke PC, Lamattina L, Jones RL (2002) Nitric oxide acts as an antioxidant and delays programmed cell death in barley aleurone layers. Plant Physiol 129:1642–50

Benavides MP, Gallego MS, Tomaro ML (2005) Cadmium toxicity in plants. Braz J Plant Physiol 17:21–34

Berezin I, Mizrachy-Dagry T, Brook E, Mizrahi K, Elazar M, Zhuo S, Saul-Tcherkas V, Shaul O (2008) Overexpression of AtMHX in tobacco causes increased sensitivity to Mg^{2+}, Zn^{2+}, and Cd^{2+} ions, induction of V-ATPase expression, and a reduction in plant size. Plant Cell Rep 27:939–949

Besson-Bard A, Pugin A, Wendehenne D (2008) New insights into nitric oxide signaling in plants. Annu Rev Plant Biol 59:21–39

Besson-Bard A, Gravot A, Richaud P, Auroy P, Duc C, Gaymard F, Taconnat L, Renou JP, Pugin A, Wendehenne D (2009) Nitric oxide contributes to cadmium toxicity in *Arabidopsis* by promoting cadmium accumulation in roots and by up-regulating genes related to iron uptake. Plant Physiol 149:1302–1315

Bi Y, Chen W, Zhang W, Zhou Q, Yun L, Xing D (2009) Production of reactive oxygen species, impairment of photosynthetic function and dynamic changes in mitochondria are early events in cadmium-induced cell death in *Arabidopsis thaliana*. Biol Cell 101:629–643

Blum W (1997) Cadmium uptakeby higher plants. In: Proceedings of extendedabstracts from the Fourth International Conference on the Biogeochemistry of Trace Elements. University of California, Berkrley, USA, pp 109–110

Bovet L, Eggmann T, Meylan-Bettex M, Polier J, Kammer P, Marin E, Feller U, Martinoia E (2003) Transcript levels of AtMRPs after cadmium treatment: induction of AtMRP3. Plant Cell Environ 26:371–381

Bovet L, Feller U, Martinoia E (2005) Possible involvement of plant ABC transporters in cadmium detoxification: a cDNA sub-microarray approach. Environ Int 31:263–267

Cataldo DA, Garland TR, Wildung RE (1983) Cadmium uptake kinetics in intact soybean plants. Plant Physiol 73:844–848

Chaoui A, Mazhoudi S, Ghorbal M (1997) Cadmiun and zinc induction of lipid peroxidation and effects on antioxidant enzyme activities in bean. Plant Sci 127:139–147

Chen A, Komives EA, Schroeder JI (2006) An improved grafting technique for mature *Arabidopsis* plants demonstrates long-distance shoot-to-root transport of phytochelatins in *Arabidopsis*. Plant Physiol 141:108–120

Chen F, Wang F, Wu F, Mao W, Zhang G, Zhou M (2010) Modulation of exogenous glutathione in antioxidant defense system against Cd stress in the two barley genotypes differing in Cd tolerance. Plant Physiol Biochem 48:663–672

Clemens S (2006) Toxic metal accumulation, responses to exposure and mechanisms of tolerance in plants. Biochimie 88:1707–1719

Cobbett CS (2000) Phytochelatins and their roles in heavy metal detoxification. Plant Physiol 123:825–832

Cobbett C, Goldsbrough P (2002) Phytochelatins and metallothioneins: roles in heavy metal detoxification and homeostasis. Annu Rev Plant Biol 53:159–182

Courbot M, Diez L, Ruotolo R, Chalot M, Leroy P (2004) Cadmium-responsive thiols in the ectomycorrhizal fungus *Paxillus involutus*. Appl Environ Microbiol 70:7413–7417

Craciun AR, Courbot M, Bourgis F, Salis P, Saumitou-Laprade P, Verbruggen N (2006) Comparative cDNA-AFLP analysis of Cd-tolerant and -sensitive genotypes derived from crosses between the Cd hyperaccumulator *Arabidopsis halleri* and *Arabidopsis lyrata* ssp. petraea. J Exp Bot 57:2967–2983

Cuypers A, Smeets K, Ruytinx J, Opdenakker K, Keunen E, Remans T, Horemans N, Vanhoudt N, Van Sanden S, Van Belleghem F, Guisez Y, Colpaert J, Vangronsveld J (2011) The cellular redox state as a modulator in cadmium and copper responses in *Arabidopsis thaliana* seedlings. J Plant Physiol 168:309–316

Das P, Samantaray S, Rou G (1997) Studies of cadmium toxicity in plants: a review. Environ Poll 98:29–36

De Michele R, Vurro E, Rigo C, Costa A, Elviri L, Di Valentin M, Careri M, Zottini M, di Toppi LS, Lo Schiavo F (2009) Nitric oxide is involved in cadmium-induced programmed cell death in *Arabidopsis* suspension cultures. Plant Physiol 150:217–228

del Río LA (2011) Peroxisomes as a cellular source of reactive nitrogen species signal molecules. Arch Biochem Biophys 506:1–11

del Río LA, Puppo A (2009) Reactive oxygen species in plant signaling. Springer, Germany

del Río LA, Corpas FJ, Barroso JB (2004) Nitric oxide and nitric oxide synthase activity in plants. Phytochemistry 65:783–792

del Río LA, Sandalio LM, Corpas FJ, Palma JM, Barroso JB (2006) Reactive oxygen species and reactive nitrogen species in peroxisomes. Production, scavenging, and role in cell signaling. Plant Physiol 141:330–335

Delledonne M (2005) NO news is good news for plants. Curr Opin Plant Biol 8:390–396

Dietz KJ, Jacob S, Oelze ML, Laxa M, Tognetti V, de Miranda SM, Baier M, Finkemeier I (2006) The function of peroxiredoxins in plant organelle redox metabolism. J Exp Bot 57:1697–1709

Ding Y, Chen Z, Zhu C (2011) Microarray-based analysis of cadmium-responsive microRNAs in rice (*Oryza sativa*). J Exp Bot. doi:DOI:10.1093/jxb/err046

Dixit V, Pandey V, Shyam R (2001) Differential antioxidative responses to cadmium in roots and leaves of pea (*Pisum sativum* L. cv. Azad). J Exp Bot 52:1101–1109

Durand TC, Sergeant K, Planchon S, Carpin S, Label P, Morabito D, Hausman JF, Renaut J (2010) Acute metal stress in *Populus tremula x P. alba* (717-1B4 genotype): leaf and cambial proteome changes induced by Cd^{2+}. Proteomics 10:349–368

Fagioni M, Zolla L (2009) Does the different proteomic profile found in apical and basal leaves of spinach reveal a strategy of this plant toward cadmium pollution response? J Proteome Res 8:2519–2529

Faller P, Kienzler K, Krieger-Liszkay A (2005) Mechanism of Cd^{2+} inhibits photoactivation of photosystem II by competitive binding to the essential Ca^{2+} site. Biochim Biophys Acta 1706:158–164

Farinati S, Dalcorso G, Panigati M, Furini A (2011) Interaction between selected bacterial strains and *Arabidopsis halleri* modulates shoot proteome and cadmium and zinc accumulation. J Exp Bot. doi:10.1093/jxb/err015

Fatima RA, Ahmad M (2005) Certain antioxidant enzymes of *Allium cepa* as biomarkers for the detection of toxic heavy metals in wastewater. Sci Total Environ 346:256–273

Finkemeier I, Goodman M, Lamkemeyer P, Kandlbinder A, Sweetlove LJ, Dietz KJ (2005) The mitochondrial type II peroxiredoxin F is essential for redox homeostasis and root growth of *Arabidopsis thaliana* under stress. J Biol Chem 280:12168–12180

Fodor A, Szabó-Nagy A, Erdei L (1995) The effects of cadmium on the fluidity and H^+-ATPase activity of plasma membrane from sunflower and wheat roots. J Plant Physiol 14:787–792

Foresi N, Correa-Aragunde N, Parisi G, Calóc G, Salernoc G, Lamattina L (2010) Characterization of a nitric oxide synthase from the plant kingdom: NO generation from the green alga Ostreococcus tauri is light irradiance and growth phase dependent. Plant Cell 22:3816–3830

Fotjová M, Kovařik A (2000) Genotoxic effect of cadmium is associated with apoptotic changes in tobacco cells. Plant Cell Environ 23:531–537

Fusco NM, Micheletto L, Dal Corso G, Borgato L, Furini A (2005) Identification of cadmium-regulated genes by cDNA-AFLP in the heavy metal accumulator *Brassica juncea* L. J Exp Bot 56:3017–3027

Garnier L, Simon-Plas F, Thuleau P, Agnel JP, Blein JP, Ranjeva R, Montillet JL (2006) Cadmium affects tobacco cells by a series of three waves of reactive oxygen species that contribute to cytotoxicity. Plant Cell Environ 29:1956–1969

Godbold DL, Hutterman A (1985) Effect of zinc, cadmium and mercury on root elongation on *Picea abies* (Karst.) seedlings and the significance of these methals to forest die-back. Environ Pollut 38:375–381

Gomes-Junior RA, Moldes CA, Delite FS, Pompeu GB, Gratao PL, Mazzafera P, Lea PJ, Azevedo RA (2006) Antioxidant metabolism of coffee cell suspension cultures in response to cadmium. Chemosphere 65:1330–1337

Gong JM, Lee DA, Schroeder JI (2003) Long-distance root-to-shoot transport of phytochelatins and cadmium in *Arabidopsis*. Proc Natl Acad Sci U S A 100:10118–10123

Gould K, Lamotte O, Klinguer A, Pugin A, Wendehenne D (2003) Nitric oxide production in tobacco leaf cells: a generalized stress response? Plant Cell Environ 26:1851–1862

Groppa MD, Rosales EP, Iannone MF, Benavides MP (2008) Nitric oxide, polyamines and Cd-induced phytotoxicity in wheat roots. Phytochemistry 69:2609–2615

Gupta KJ, Kaiser WM (2010) Production and scavenging of nitric oxide by barley root mitochondria. Plant Cell Physiol 51:576–584

Gutteridge JM, Halliwell B (2000) Free radicals and antioxidants in the year 2000: a historical look to the future. Ann N Y Acad Sci 899:136–147

Haghiri F (1973) Plant uptake of cadmium as influenced by cation exchange capacity, organic matter, zinc and soil temperature. J Environ Qual 2:93–96

Halliwell B, Gutteridge JMC (1999) Free radicals in biology and medicine. Oxford University Press, New York

Han Y, Zhang J, Chen X, Gao Z, Xuan W, Xu S, Ding X, Shen W (2008) Carbon monoxide alleviates cadmium-induced oxidative damage by modulating glutathione metabolism in the roots of *Medicago sativa*. New Phytol 177:155–166

Harada E, Kim JA, Meyer AJ, Hell R, Clemens S, Choi YE (2010) Expression profiling of tobacco leaf trichomes identifies genes for biotic and abiotic stresses. Plant Cell Physiol 51:1627–1637

Herbette S, Taconnat L, Hugouvieux V, Piette L, Magniette ML, Cuine S, Auroy P, Richaud P, Forestier C, Bourguignon J, Renou JP, Vavasseur A, Leonhardt N (2006) Genome-wide transcriptome profiling of the early cadmium response of *Arabidopsis* roots and shoots. Biochimie 88:1751–1765

Hernandez LE, Carpena Ruiz RO, Garate A (1996) Alterations in the mineral nutrition of pea seedlings exposed to cadmium. J Plant Nutr 19:1581–1598

Heyno E, Klose C, Krieger-Liszkay A (2008) Origin of cadmium-induced reactive oxygen species production: mitochondrial electron transfer versus plasma membrane NADPH oxidase. New Phytol 179:687–699

Horemans N, Raeymaekers T, Van Beek K, Nowocin A, Blust R, Broos K, Cuypers A, Vangronsveld J, Guisez Y (2007) Dehydroascorbate uptake is impaired in the early response of *Arabidopsis* plant cell cultures to cadmium. J Exp Bot 58:4307–4317

Islam MM, Hoque MA, Okuma E, Jannat R, Banu MN, Jahan MS, Nakamura Y, Murata Y (2009) Proline and glycinebetaine confer cadmium tolerance on tobacco bright yellow-2 cells by increasing ascorbate-glutathione cycle enzyme activities. Biosci Biotechnol Biochem 73:2320–2323

Janouskova M, Pavlikova D, Vosatka M (2006) Potential contribution of arbuscular mycorrhiza to cadmium immobilisation in soil. Chemosphere 65:1959–1965

Jentschke G, Winter S, Godbold DL (1999) Ectomycorrhizas and cadmium toxicity in Norway spruce seedlings. Tree Physiol 19:23–30

Kieffer P, Dommes J, Hoffmann L, Hausman JF, Renaut J (2008) Quantitative changes in protein expression of cadmium-exposed poplar plants. Proteomics 8:2514–2530

Kobayashi M, Ohura I, Kawakita K, Yokota N, Fujiwara M, Simamoto K (2007) Calcium dependent protein kinases regulate the production or reactive oxygen species by potato NADPH oxidase. Plant Cell 19:1065–1080

Kopyra M, Gwóźdź EA (2003) Nitric oxide stimulates seed germination and counteracts the inhibitory effect of heavy metals and salinity on root growth of *Lupinus luteus*. Plant Physiol Biochem 41:1011–1017

Korenkov V, Hirschi K, Crutchfield JD, Wagner GJ (2007) Enhancing tonoplast Cd/H antiport activity increases Cd, Zn, and Mn tolerance, and impacts root/shoot Cd partitioning in *Nicotiana tabacum L.* Planta 226:1379–1387

Kovacik J, Gruz J, Hedbavny J, Klejdus B, Strnad M (2009) Cadmium and nickel uptake are differentially modulated by salicylic acid in *Matricaria chamomilla* plants. J Agric Food Chem 57:9848–9855

Kovalchuk I, Titov V, Hohn B, Kovalchuk O (2005) Transcriptome profiling reveals similarities and differences in plant responses to cadmium and lead. Mutat Res 570:149–161

Lane TW, Morel FM (2000) A biological function for cadmium in marine diatoms. Proc Natl Acad Sci U S A 97:4627–4631

Laspina NV, Groppa MD, Tomaro M, Benavides MP (2005) Nitric oxide protects sunflower leaves against Cd-induced oxidative stress. Plant Sci 169:323–330

Lee K, Bae DW, Kim SH, Han HJ, Liu X, Park HC, Lim CO, Lee SY, Chung WS (2010) Comparative proteomic analysis of the short-term responses of rice roots and leaves to cadmium. J Plant Physiol 167:161–168

León AM, Palma JM, Corpas FJ, Gomez M, Romero-Puertas MC, Chatterjee D, Mateos RM, del Río LA, Sandalio LM (2002) Antioxidative enzymes in cultivars of pepper plants with different sensitivity to cadmium. Plant Physiol Biochem 40:813–820

Lindermayr C, Saalbach G, Bahnweg G, Durner J (2006) Differential inhibition of *Arabidopsis* methionine adenosyltransferases by protein S-nitrosylation. J Biol Chem 281:4285–4291

Liu W, Zhou Q, Li P, Gao H, Han YP, Li XJ, Yang YS, Li Y (2009) DNA mismatch repair related gene expression as potential biomarkers to assess cadmium exposure in *Arabidopsis* seedlings. J Hazard Mater 167:1007–1013

Liu XM, Kim KE, Kim KC, Nguyen XC, Han HJ, Jung MS, Kim HS, Kim SH, Park HC, Yun DJ, Chung WS (2010) Cadmium activates *Arabidopsis* MPK3 and MPK6 via accumulation of reactive oxygen species. Phytochemistry 71:614–618

López-Huertas E, Charlton WL, Johnson B, Graham IA, Baker A (2000) Stress induces peroxisome biogenesis genes. EMBO J 19:6770–6777

Louie M, Kondor N, DeWitt JG (2003) Gene expression in cadmium-tolerant *Datura innoxia*: detection and characterization of cDNAs induced in response to Cd^{2+}. Plant Mol Biol 52:81–89

Maksymiec W, Wojcik M, Krupa Z (2007) Variation in oxidative stress and photochemical activity in *Arabidopsis thaliana* leaves subjected to cadmium and excess copper in the presence or absence of jasmonate and ascorbate. Chemosphere 66:421–427

McCarthy I, Romero-Puertas MC, Palma JM, Sandalio LM, Corpas FJ, Gómez M, del Río LA (2001) Cadmium induces senescence symptoms in leaf peroxisomes of pea plants. Plant Cell Environ 24:1065–1073

McLaughlin MJ, Singh BR (1999) Cadmium in soils and plants. In: Cadmium in soils and plants. Kluwer, Dordrecht

Metwally A, Finkemeier I, Georgi M, Dietz KJ (2003) Salicylic acid alleviates the cadmium toxicity in barley seedlings. Plant Physiol 132:272–281

Milone MT, Sgherri C, Clijsters H, Navarri-Izzo F (2003) Antioxidant responses of wheat treated with realistic concentration of cadmiun. Environ Exp Bot 50:265–276
Minglin L, Yuxiu Z, Tuanyao C (2005) Identification of genes up-regulated in response to Cd exposure in *Brassica juncea L.* Gene 363:151–158
Morel FM, Malcolm EG (2005) The biogeochemistry of cadmium. Met Ions Biol Syst 43:195–219
Moreno JL, Hernandez T, Garcia C (1999) Effects of cadmium containing sewage sludge compost on dynamics of organic matter and microbial activity in an arid soils. Biol Fert Soils 28:230–237
Muthukrishnan S, Liang G, Trick H, Gill B (2001) Pathogenesis-related proteins and their genes in cereals. Plant Cell Tiss Org 64:94–113
Nawrot T, Plusquin M, Hogervorst J, Roels HA, Celis H, Thijs L, Vangronsveld J, Van Hecke E, Staessen JA (2006) Environmental exposure to cadmium and risk of cancer: a prospective population-based study. Lancet Oncol 7:119–126
Neill S, Bright J, Desikan R, Hancock J, Harrison J, Wilson I (2008) Nitric oxide evolution and perception. J Exp Bot 59:25–35
Nocito FF, Lancilli C, Crema B, Fourcroy P, Davidian JC, Sacchi GA (2006) Heavy metal stress and sulfate uptake in maize roots. Plant Physiol 141:1138–1148
Ogawa I, Nakanishi H, Mori S, Nishizawa N (2009) Time course analysis of gene regulation under cadmium stress in rice. Plant Soil 325:97–108
Olmos E, Martinez-Solano JR, Piqueras A, Hellin E (2003) Early steps in the oxidative burst induced by cadmium in cultured tobacco cells (BY-2 line). J Exp Bot 54:291–301
Ortega-Villasante C, Hernandez LE, Rellán-Alvarez R, Del Campo FF, Carpena-Ruiz RO (2007) Rapid alteration of cellular redox homeostasis upon exposure to cadmium and mercury in alfalfa seedlings. New Phytol 176:96–107
Ouariti O, Boussama N, Zarrouk M, Cherif A, Ghorbal MH (1997) Cadmium-and copper-induced changes in tomato membrane lipids. Phytochemistry 45:1343–1350
Pence NS, Larsen PB, Ebbs SD, Letham DL, Lasat MM, Garvin DF, Eide D, Kochian LV (2000) The molecular physiology of heavy metal transport in the Zn/Cd hyperaccumulator *Thlaspi caerulescens*. Proc Natl Acad Sci U S A 97:4956–4960
Pinto AP, Mota AM, de Varennes A, Pinto FC (2004) Influence of organic matter on the uptake of cadmium, zinc, copper and iron by sorghum plants. Sci Total Environ 326:239–247
Qian H, Li J, Sun L, Chen W, Sheng GD, Liu W, Fu Z (2009) Combined effect of copper and cadmium on *Chlorella vulgaris* growth and photosynthesis-related gene transcription. Aquat Toxicol 94:56–61
Remans T, Opdenakker K, Smeets K, Mathijsen D, Vangronsveld J, Cuypers A (2010) Metal-specific and NADPH oxidase dependent changes in lipoxygenase and NADPH oxidase gene expression in Arabidopsis thaliana exposed to cadmium or excess copper. Functional Plant Biol 37:532–544
Ren D, Yang H, Zhang S (2002) Cell death mediated by MAPK is associated with hydrogen peroxide production in *Arabidopsis*. J Biol Chem 277:559–565
Rodríguez-Celma J, Rellan-Alvarez R, Abadia A, Abadia J, Lopez-Millan AF (2010) Changes induced by two levels of cadmium toxicity in the 2-DE protein profile of tomato roots. J Proteomics 73:1694–1706
Rodríguez-Serrano M, Romero-Puertas MC, Zabalza A, Corpas FJ, Gomez M, del Río LA, Sandalio LM (2006) Cadmium effect on oxidative metabolism of pea (*Pisum sativum* L.) roots. Imaging of reactive oxygen species and nitric oxide accumulation in vivo. Plant Cell Environ 29:1532–1544
Rodríguez-Serrano M, Romero-Puertas M, Pazmino DM, Testillano PS, Risueno MC, del Río LA, Sandalio LM (2009a) Cellular response of pea plants to cadmium toxicity: cross talk between reactive oxygen species, nitric oxide, and calcium. Plant Physiol 150:229–243
Rodríguez-Serrano M, Romero-Puertas MC, Sparkes I, Hawes C, del Río LA, Sandalio LM (2009b) Peroxisome dynamics in *Arabidopsis* plants under oxidative stress induced by cadmium. Free Radic Biol Med 47:1632–1639

Romero-Puertas MC, McCarthy I, Sandalio LM, Palma JM, Corpas FJ, Gomez M, del Río LA (1999) Cadmium toxicity and oxidative metabolism of pea leaf peroxisomes. Free Radic Res 31:25–31

Romero-Puertas MC, Palma JM, Gómez M, del Río LA, Sandalio LM (2002) Cadmium causes the modification of proteins in pea plants. Plant Cell Environ 25:677–686

Romero-Puertas MC, Rodríguez-Serrano M, Corpas FJ, Gómez M, del Río LA, Sandalio LM (2004) Cadmium induced subcellular accumulation of $O_2^{\bullet -}$and H_2O_2 in pea plants. Plant Cell Environ 27:1122–1134

Romero-Puertas MC, Corpas FJ, Rodríguez-Serrano M, Gomez M, del Río LA, Sandalio LM (2007) Differential expression and regulation of antioxidative enzymes by cadmium in pea plants. J Plant Physiol 164:1346–1357

Romero-Puertas MC, Campostrini N, Matte A, Righetti PG, Perazzolli M, Zolla L, Roepstorff P, Delledonne M (2008) Proteomic analysis of S-nitrosylated proteins in *Arabidopsis thaliana* undergoing hypersensitive response. Proteomics 8:1459–1469

Romero-Puertas MC, Ortega-Galisteo AP, Rodríguez-Serrano M, Pazmino DM, Gupta DK, Sandalio LM (2011) S-nitrosylated proteins in pea (*Pisum sativum L.*) leaf peroxisomes: changes under abiotic stress. Submitted

Roth U, von Roepenack-Lahaye E, Clemens S (2006) Proteome changes in *Arabidopsis thaliana* roots upon exposure to Cd^{2+}. J Exp Bot 57:4003–4013

Sandalio LM, Dalurzo HC, Gomez M, Romero-Puertas MC, del Río LA (2001) Cadmium-induced changes in the growth and oxidative metabolism of pea plants. J Exp Bot 52:2115–2126

Sandalio LM, Rodríguez-Serrano M, del Río LA, Romero-Puertas MC (2009) Reactive oxygen species and signalling in cadmium toxicity. In: del Río LA, Puppo A (eds) Reactive oxygen species in plant signaling. Springer, Berlin/Heidelberg, pp 175–190

Sanitá di Toppi L, Gabrielle R (1999) Response to Cadmiun in higher plants. Environ Exp Bot 41:105–130

Sarry JE, Kuhn L, Ducruix C, Lafaye A, Junot C, Hugouvieux V, Jourdain A, Bastien O, Fievet JB, Vailhen D, Amekraz B, Moulin C, Ezan E, Garin J, Bourguignon J (2006) The early responses of *Arabidopsis thaliana* cells to cadmium exposure explored by protein and metabolite profiling analyses. Proteomics 6:2180–2198

Schützendübel A, Polle A (2002) Plant responses to abiotic stresses: heavy metal-induced oxidative stress and protection by mycorrhization. J Exp Bot 53:1351–1365

Schützendübel A, Schwanz P, Teichmann T, Gross K, Langenfeld-Heyser R, Godbold DL, Polle A (2001) Cadmium-induced changes in antioxidative systems, hydrogen peroxide content, and differentiation in Scots pine roots. Plant Physiol 127:887–898

Semane B, Dupae J, Cuypers A, Noben JP, Tuomainen M, Tervahauta A, Karenlampi S, Van Belleghem F, Smeets K, Vangronsveld J (2010) Leaf proteome responses of *Arabidopsis thaliana* exposed to mild cadmium stress. J Plant Physiol 167:247–254

Sharma SS, Dietz KJ (2006) The significance of amino acids and amino acid-derived molecules in plant responses and adaptation to heavy metal stress. J Exp Bot 57:711–726

Sharma SS, Dietz KJ (2009) The relationship between metal toxicity and cellular redox imbalance. Trends Plant Sci 14:43–50

Shukla UC, Si J, Joshi PC, Kakkar P (2003) Effect of bioaccumulation of cadmium on biomass productivity, essential trace elements, chorophyl biosyntesis, and macromolecules of wheat seedlings. Biol Trace Elem Res 92:257–274

Smeets K, Opdenakker K, Remans T, Van Sanden S, Van Belleghem F, Semane B, Horemans N, Guisez Y, Vangronsveld J, Cuypers A (2009) Oxidative stress-related responses at transcriptional and enzymatic levels after exposure to Cd or Cu in a multipollution context. J Plant Physiol 166:1982–1992

Smiri M, Chaoui A, Rouhier N, Gelhaye E, Jacquot JP, El Ferjani E (2010) Cadmium affects the glutathione/glutaredoxin system in germinating pea seeds. Biol Trace Elem Res. doi: DOI:10.1007/s12011-010-8749-3

Song WY, Martinoia E, Lee J, Kim D, Kim DY, Vogt E, Shim D, Choi KS, Hwang I, Lee Y (2004) A novel family of cys-rich membrane proteins mediates cadmium resistance in *Arabidopsis*. Plant Physiol 135:1027–1039

Souza VL, Almeida AA, Hora BT Jr, Gesteira AS, Cascardo JC (2008) Preliminary analysis of expressed sequences of genes in *Genipa americana* L. plant roots exposed to cadmium in nutrient solution. Genet Mol Res 7:1282–1288

Stohr C, Stremlau S (2006) Formation and possible roles of nitric oxide in plant roots. J Exp Bot 57:463–470

Stohr C, Strube F, Marx G, Ullrich WR, Rockel P (2001) A plasma membrane-bound enzyme of tobacco roots catalyses the formation of nitric oxide from nitrite. Planta 212:835–841

Suzuki N, Koizumi N, Sano H (2001) Screening of cadmium-responsive genes in Arabidopsis thaliana. Plant Cell Environ 24:1177–1188

Swoboda P, Gal S, Hohn B, Puchta H (1994) Intrachromosomal homologous recombination in whole plants. EMBO J. 13:484–489

Tamas L, Dudikova J, Durcekova K, Haluskova L, Huttova J, Mistrik I, Olle M (2008) Alterations of the gene expression, lipid peroxidation, proline and thiol content along the barley root exposed to cadmium. J Plant Physiol 165:1193–1203

Thomine S, Wang R, Ward JM, Crawford NM, Schroeder JI (2000) Cadmium and iron transport by members of a plant metal transporter family in *Arabidopsis* with homology to AtNRAMP genes. Proc Natl Acad Sci USA 97:4991–4996

Thomine S, Lelievre F, Debarbieux E, Schroeder JI, Barbier-Brygoo H (2003) AtNRAMP3, a multispecific vacuolar metal transporter involved in plant responses to iron deficiency. Plant J 34:685–695

Ueno D, Iwashita T, Zhao FJ, Ma JF (2008) Characterization of Cd translocation and identification of the Cd form in xylem sap of the Cd-hyperaccumulator *Arabidopsis halleri*. Plant Cell Physiol 49:540–548

Valko M, Morris H, Cronin MT (2005) Metals toxicity and oxidative stress. Curr Med Chem 12:1161–1208

van de Mortel JE, Schat H, Moerland PD, Loren V, van Themaat E, van der Ent S, Blankestijn H, Ghandilyan A, Tsiatsiani S, Aarts MG (2008) Expression differences for genes involved in lignin, glutathione and sulphate metabolism in response to cadmium in *Arabidopsis thaliana* and the related Zn/Cd-hyperaccumulator *Thlaspi caerulescens*. Plant Cell Environ 31:301–324

van Loon LC, van Strien EA (1999) The families of pathogenesisrelated proteins, their activities, and comparative analysis of PR-1 type proteins. Physiol Mol Plant P 55:85–87

Verbruggen N, Hermans C, Schat H (2009) Mechanisms to cope with arsenic or cadmium excess in plants. Curr Opin Plant Biol 12:364–372

Verkleij J, Schat H (1990) Mechanisms of methal tolerance in higher plants. In: Shaw AJ (ed) Heavy metal tolerance in plants: evolutionary aspects. CRC Press, Boca Raton, pp 179–194

Vitoria AP, Lea PJ, Azevedo RA (2001) Antioxidant enzymes responses to cadmium in radish tissues. Phytochemistry 57:701–710

Wagner GJ (1993) Accumulation of cadmium in crops plants and its consequences to human health. Adv Agron 51:173–212

Wang JW, Wu JY (2005) Nitric oxide is involved in methyl jasmonate-induced defense responses and secondary metabolism activities of Taxus cells. Plant Cell Physiol 46:923–930

Watanabe A, Ito H, Chiba M, Ito A, Shimizu H, Fuji S, Nakamura S, Hattori H, Chino M, Satoh-Nagasawa N, Takahashi H, Sakurai K, Akagi H (2010) Isolation of novel types of *Arabidopsis* mutants with altered reactions to cadmium: cadmium-gradient agar plates are an effective screen for the heavy metal-related mutants. Planta 232:825–836

Weber M, Trampczynska A, Clemens S (2006) Comparative transcriptome analysis of toxic metal responses in *Arabidopsis thaliana* and the Cd($^{2+}$)-hypertolerant facultative metallophyte *Arabidopsis halleri*. Plant Cell Environ 29:950–963

Wilson ID, Neill SJ, Hancock JT (2008) Nitric oxide synthesis and signalling in plants. Plant Cell Environ 31:622–631

Xiang C, Oliver DJ (1998) Glutathione metabolic genes coordinately respond to heavy metals and jasmonic acid in *Arabidopsis*. Plant Cell 10:1539–1550

Xiong J, An L, Lu H, Zhu C (2009) Exogenous nitric oxide enhances cadmium tolerance of rice by increasing pectin and hemicellulose contents in root cell wall. Planta 230:755–765

Yakimova ET, Kapchina-Toteva VM, Laarhoven LJ, Harren FM, Woltering EJ (2006) Involvement of ethylene and lipid signalling in cadmium-induced programmed cell death in tomato suspension cells. Plant Physiol Biochem 44:581–589

Yamaguchi H, Fukuoka H, Arao T, Ohyama A, Nunome T, Miyatake K, Negoro S (2009) Gene expression analysis in cadmium-stressed roots of a low cadmium-accumulating solanaceous plant, *Solanum torvum*. J Exp Bot 61:423–437

Yang MG, Lin XY, Yang XE (1998) Impact of Cd on growth and nutrient accumulation of different plant species. Chin J Appl Ecol 19:89–94

Yeh CM, Hsiao LJ, Huang HJ (2004) Cadmium activates a mitogen-activated protein kinase gene and MBP kinases in rice. Plant Cell Physiol 45:1306–1312

Zhang W, Chen W (2011) Role of salicylic acid in alleviating photochemical damage and autophagic cell death induction of cadmium stress in *Arabidopsis thaliana*. Photochem Photobiol Sci. doi:DOI:10.1039/C0PP00305K

Zhang ZC, Qiu BS (2007) Reactive oxygen species metabolism during the cadmium hyperaccumulation of a new hyperaccumulator *Sedum alfredii* (Crassulaceae). J Environ Sci (China) 19:1311–1317

Zhang J, Hu M, Li JT, Guan JP, Yang B, Shu WS, Liao B (2009) A transcriptional profile of metallophyte *Viola baoshanensis* involved in general and species-specific cadmium-defense mechanisms. J Plant Physiol 166:862–870

Zhao FJ, Jiang RF, Dunham SJ, McGrath SP (2006) Cadmium uptake, translocation and tolerance in the hyperaccumulator *Arabidopsis halleri*. New Phytol 172:646–654

Zhao FJ, Ma JF, Meharg AA, McGrath SP (2008) Arsenic uptake and metabolism in plants. New Phytol 154:29–43

Zhao CR, Ikka T, Sawaki Y, Kobayashi Y, Suzuki Y, Hibino T, Sato S, Sakurai N, Shibata D, Koyama H (2009) Comparative transcriptomic characterization of aluminum, sodium chloride, cadmium and copper rhizotoxicities in *Arabidopsis thaliana*. BMC Plant Biol 9:32

Zhu YL, Pilon-Smits EA, Tarun AS, Weber SU, Jouanin L, Terry N (1999) Cadmium tolerance and accumulation in Indian mustard is enhanced by overexpressing gamma-glutamylcysteine synthetase. Plant Physiol 121:1169–1178

Exploring the Plant Response to Cadmium Exposure by Transcriptomic, Proteomic and Metabolomic Approaches: Potentiality of High-Throughput Methods, Promises of Integrative Biology

Florent Villiers, Véronique Hugouvieux, Nathalie Leonhardt, Alain Vavasseur, Christophe Junot, Yves Vandenbrouck, and Jacques Bourguignon

F. Villiers • V. Hugouvieux
Laboratoire de Physiologie Cellulaire Végétale (LPCV), Commissariat à l'Energie Atomique et aux Energies Alternatives (CEA), iRTSV, Grenoble, France

Université Joseph Fourier (UJF), Grenoble, France

LPCV, CNRS (UMR 5168), Grenoble, France

LPCV, INRA (UMR1200), Grenoble, France
e-mail: flo.villiers@gmail.com; vhugouvieux@cea.fr

N. Leonhardt • A. Vavasseur
CEA, iBEB, Laboratoire des Echanges membranaires et Signalisation, UMR CEA/CNRS 6191, Université Aix Marseille, St Paul lez Durance, France
e-mail: nleonhardt@cea.fr; avavasseur@cea.fr

C. Junot
Laboratoire d'Etude du Métabolisme des Médicaments, CEA, iBiTec-S, CEA-Saclay, France
e-mail: cjunot@cea.fr

Y. Vandenbrouck
Université Joseph Fourier (UJF), Grenoble, France

Laboratoire Biologie à Grande Echelle, CEA, iRTSV, F-38054 Grenoble, France

INSERM, U1038, Grenoble, France
e-mail: yvandenbrouck@cea.fr

J. Bourguignon (✉)
Laboratoire de Physiologie Cellulaire Végétale (LPCV), Commissariat à l'Energie Atomique et aux Energies Alternatives (CEA), iRTSV, Grenoble, France

Université Joseph Fourier (UJF), Grenoble, France

LPCV, CNRS (UMR 5168), Grenoble, France

LPCV, INRA (UMR1200), Grenoble, France

LPCV, iRTSV, CEA, UJF, CNRS (UMR 5168), INRA (UMR1200), Grenoble cedex 9, France
e-mail: jacques.bourguignon@cea.fr

D.K. Gupta and L.M. Sandalio (eds.), *Metal Toxicity in Plants: Perception, Signaling and Remediation*, DOI 10.1007/978-3-642-22081-4_6,

Abstract With the evolution of high-throughput techniques, performing organism-wide analyses of gene expression, protein content and metabolite profile is becoming easier. This new era of biology opens the way for incredibly promising advances but also extremely challenging difficulties in the aim of integrating these datasets into biologically relevant models. To address questions and unravel networks occurring in cadmium stressed plants, researchers have already taken advantage of such tools. We provide here an overview of the discoveries that have been made regarding plant–cadmium interaction using "omics" methods, and show how their use is relevant to compare species and stresses at the whole plant level. At the end, we propose future breakthroughs to be addressed using these techniques, and discuss the up-coming challenges of data mining and model drawing.

Abbreviations

ACO	Aconitase
ALD	Fructose-bisphosphate aldolase
APS	sulfurylase
AtATX1	Homolog of Anti-oxidant 1
AtCCH	Copper chaperone
CAX	Cation exchanger
Cd	Cadmium
CS	Citrate synthase
Cu	Copper
ENO	Enolase (2-phosphoglycerate hydrolase)
GABA	γ-Aminobutyrate
GAPC	Cytosolic glyceraldehyde 3-phosphate dehydrogenase
GLN1	Glutamine synthetase
GLT	Glutamate synthetase
GSH	Glutathione
GSH1	γ-glutamylcysteine synthase
GSH2	Glutathione synthetase
GST	Glutathione S-transferase
HK	Hexose kinase
HMA3	Heavy metal ATP ASE 3
HSPs	Heat shock proteins
ISO	NADP-specific isocitrate dehydrogenase
MDH	Malate dehydrogenase
MetSO	Methionine sulfoxide
NRAMP	Natural resistance-associated macrophage protein
OAS	*O*-acetylserine (thiol)lyase
PC	Phytochelatin
PDH E1	E2 & E3, pyruvate dehydrogenase complex components
PGM	Phosphoglycerate mutase
PK	Pyruvate kinase
Pr	Pathogenesis-related

Prx	Peroxiredoxin
PSI, II	Photosystem I, II
ROS	Reactive oxygen species
RuBisCO	Ribulose bisphosphate carboxylase oxygenase
SBP1	Selenium binding protein 1
SHMT	Serine hydroxymethyltransferase
SOD	Superoxide dismutase
SUC	Succinate dehydrogenase
TCA	Tricarboxylic acid
TPI	Triose phosphate isomerase
Ub	Ubiquitin

1 Introduction

Cadmium (Cd) is a major pollutant of several industrialized or developing countries and its effects on human health are particularly deleterious (Ishihara et al. 2001). Industrial activities and use of sewage sludge account form the principal sources of Cd, and lead to a continuous increase in its concentration in crop fields and water sources. Cd is considered as a non-essential metal, even though it has been reported to act as a cofactor of carbonic anhydrase in *Thalassiosira weissflogii* and other marine diatoms (Lane et al. 2005). Like other non-essential metals, it is assimilated by plant roots via non-specific transport activity (Clemens 2006). Once in the xylem, it spreads over the entire organism where it affects numerous biological functions probably by replacing metal cofactors, by interacting with protein thiols and by inducing the production of reactive oxygen species. In fact, the mechanistic basis of its toxicity is still poorly understood. In order to cope with Cd, plants have evolved various mechanisms leading to its chelation, exclusion and/or stabilization (Clemens 2006; Cobbett and Goldsbrough 2002; Verbruggen et al. 2009). One of the main well-known protection mechanism in plants involves its chelation by phytochelatins, which are cysteine-rich peptides of general formula $(\gamma\text{Glu-Cys})_n$-X [with $n = 2$–11, and $X =$ Gly (PC) or Ser, Ala or Glu (isoPC)], followed by their transport into the vacuoles (Cobbett 2000; Rauser 1990; Zenk 1996). Understanding the molecular activities underlying these processes could allow one to take advantage of them for phytoremediation of contaminated soils or biofortification for instance to avoid Cd in the edible parts of plants (Pilon-Smits 2005; Palmgren 2008).

Transcriptome-, proteome- and metabolome-wide studies have become increasingly used, and a wide variety of techniques allows for quantitative and/or qualitative analyses of samples at the plant, tissue or organelle level (Agrawal et al. 2011; Baginsky et al. 2010; Fukushima et al. 2009). They provide accurate snapshots of their molecular content at a given time and growth conditions, and comparisons of these datasets are able to give an invaluable dimension of dynamic to this profiling. A number of transcriptomic analysis using microarrays have been performed in the last few years to study the impact of Cd on plants (Herbette et al. 2006; Kovalchuk et al. 2005; van de Mortel et al. 2008; Weber et al. 2006; Yamaguchi et al. 2010; Zhao et al. 2009, 2010). Highly informative in terms of gene expression regulation over a time and/or concentration course, they allow the identification of early processes to be switched on and off by the plant during exposure to the metal. Recently, the monitoring of transcript levels of *Solanum torvum* plants subjected to low Cd stress was achieved by high-throughput RNA sequencing technologies (RNA seq) (Yamaguchi et al. 2010). This technique, yet still more expensive than microarray-based analyses, allows one to qualitatively and quantitatively explore transcriptome of species for which no microarray chip is commercially available, or species which their genome has not been annotated or even sequenced.

Proteomic analyses usually involves a pre-fractionation step of the sample, followed by a trypsic digestion of the proteins (Agrawal et al. 2011). Peptides are then analyzed by peptide mass fingerprinting (PMF) or tandem MS (MS/MS), often after decomplexification by chromatography (nanoLC-MS). Protein fractionation prior to MS analyses is, in most cases, based on 1D or, more often, 2D-polyacrylamide gel electrophoresis (PAGE), and this applies to most of the studies analyzing the plant response to Cd at the leaf, root and cell culture levels (Aina et al. 2007; Aloui et al. 2009; Alvarez et al. 2009; Durand et al. 2010; Gianazza et al. 2007; Hajduch et al. 2001; Hradilova et al. 2010; Kieffer et al. 2008, 2009a, b; Lee et al. 2010; Rodriguez-Celma et al. 2010; Roth et al. 2006; Sarry et al. 2006; Semane et al. 2010). These comparative 2D-PAGE studies rely on the densitometry intensities of the protein spots on the gel, measured using dedicated analysis software after classical (colloidal Coomassie blue or silver staining) or fluorescent (DIGE) staining. Protein identification of the different spots is then performed by MS. In addition to these techniques, procedures allowing direct quantification in the mass spectrometer are being developed, and here again several methods have been successfully used. They consist in the simultaneous analysis, during a single MS run, of "heavy" and "light" samples, the "heavy" term referring to as stable isotope (^{15}N or ^{13}C for instance) labelling of the sample prior to protein extraction, during cell culture or plant growth. Other methods, such as iTRAQ, require post-extraction labeling, but allow for multiplexing of up to 8 samples. Whichever method being used, peptides from the different samples conserve identical properties, allowing them to enter the mass spectrometer at the same time when a nanoLC-based separation is performed. Then, simultaneous MS or MS/MS analysis of these peptides permits the discrimination of their origin, and thereby relative quantification. Both quantification methods, e.g. ^{15}N- (Lanquar et al. 2007) and iTRAQ- (Alvarez et al. 2009; Schneider et al. 2009) labeling have been used to evaluate Cd-induced changes in plants, as opposed to label-free methods such as spectral counting or, Accurate Mass and Time (AMT) tag approaches, the latter being particularly promising for quantitative proteomic works.

While transcriptomic and proteomic methodologies have been, to some extent, particularly used to study effects of Cd on plants, metabolomic-based work is much less well represented in the literature. A range of analytical platforms can be used to analyze the maximum number of metabolites in a given sample. These include Gas Chromatography–MS (GC–MS), comprehensive GC × GC–MS, Liquid Chromatography–MS (LC–MS) and variants including Ultra Performance Liquid Chromatography (UPLC), Capillary Electrophoresis–MS (CE–MS), Direct Infusion Mass Spectrometry (DIMS), Fourier Transform InfraRed Spectroscopy (FT–IR) and NMR spectroscopy (Fiehn et al. 2000; Goodacre et al. 2003; Hirai et al. 2004; Sato et al. 2004; Tolstikov et al. 2003; Ward et al. 2003; Wolfender et al.

2003). Both global (without any a priori) analyses and more targeted studies have been undertaken. For example, metabolite analyses performed using NMR revealed modifications of organic and amino acids contents after a Cd exposure (Bailey et al. 2003; Hediji et al. 2010) while a global approach performed by LC-MS/MS pointed out that the main variation at the metabolite level came from the presence of several types of chelating peptides (phytochelatins, PC) in Cd-stressed *Arabidopsis thaliana* cells (Sarry et al. 2006). Targeted metabolite studies revealed new insights into the regulation of the biosynthesis of glutathione (GSH) and PC biosynthesis in *A. thaliana* cells (Ducruix et al. 2006) whereas MS analyses of the phloem sap of *Brassica napus* suggested a role of the thiol-peptides (GSH and PCs) in the long-distance transport of Cd (Mendoza-Cozatl et al. 2008). With the aim of understanding targets of Cd in plants, as well as general and specific defence mechanisms it induces as a response to its toxicity, researchers have taken advantage of these "omics" techniques to draw the complex signaling and metabolic network resulting from Cd contamination. The picture, however, is still incomplete, and some key aspects of the plant defence mechanisms remain unknown. We present here a summary of the different results, from "ome"-wide methodology, that participate in the establishment of the current state of the art of the plant–Cd interaction. Other chapters intend to suggest new additional directions to be considered for further improvement of our knowledge. We particularly focus on the interest and promises of combining and cross-analyzing these data in a system biology-oriented framework to decipher actions and reactions of the plant–Cd conflict.

2 Global Response of Plants to Cadmium Exposure

2.1 Overview of a Cadmium Exposure in Plants as Evaluated by Transcriptomic, Proteomic and Metabolomic Approaches

2.1.1 Primary Metabolism

Even though its manifestation and amplitude importantly differs between plant species and stress severity, the harmful effect of Cd on the photosynthetic machinery is constantly reported as a major cause of its toxicity on plants (DalCorso et al. 2008; Sanità di Toppi and Gabbrielli 1999). Observed at both macroscopic (morphological) (Semane et al. 2010) and molecular (protein) levels, Cd exposure induces a strong disorganization of the light collecting antennas (Fagioni et al. 2009), electron transfer chain (Durand et al. 2010; Kieffer et al. 2009b), Calvin cycle (Durand et al. 2010) and membranes (Semane et al. 2010), as established by proteomic approaches. Kieffer and co-workers (2008) reported for instance a down regulation of several enzymes (chloroplast ferredoxin-NADP$^+$ oxidoreductase, Cyt B6-f complex iron-sulfur subunit, oxygen-evolving enhancer protein 1 and 2) on poplar, and Hajduch et al. (2001) observed changes in RuBisCO levels as well as other enzymes of the Calvin cycle on rice leaves, later confirmed on poplar (Durand et al. 2010) or spinach leaves (Fagioni and Zolla 2009). Interestingly these changes at the protein levels are correlated with a global down regulation of the expression of genes involved in photosynthesis, meaning that not only is this reduction of protein accumulation the result of an enhanced degradation, but also the consequence of an active repression of their synthesis. Microarray analysis performed by Herbette et al. (2006) highlighted a transcriptional repression of genes coding for subunits of the PSI (psI-G, psI-K, psI-N, psaE, psaD, etc.), PSII (psbQ, psbP, psbW), enzymes of the chlorophyll biosynthetic pathway (Mg chelatases, 4–aminobutyrate transaminases, hydroxymethylbilane synthase, etc.), and enzymes of the Calvin cycle (Transketolase, Rubisco, Ribose-5-phosphate isomerase, fructose-bisphosphate aldolase, etc.) (schematized in Fig. 1). Overall, understanding the effect of Cd on photosynthesis is of prime importance to decipher molecular mechanisms leading to Cd resistance as the ability of plants to control the damage that is done to this Achilles' heel has been proposed to be a major contributor to their sensitivity/tolerance (Bah et al. 2010). Indeed, destabilization of the photosynthetic apparatus by Cd is thought to lead to an important number of

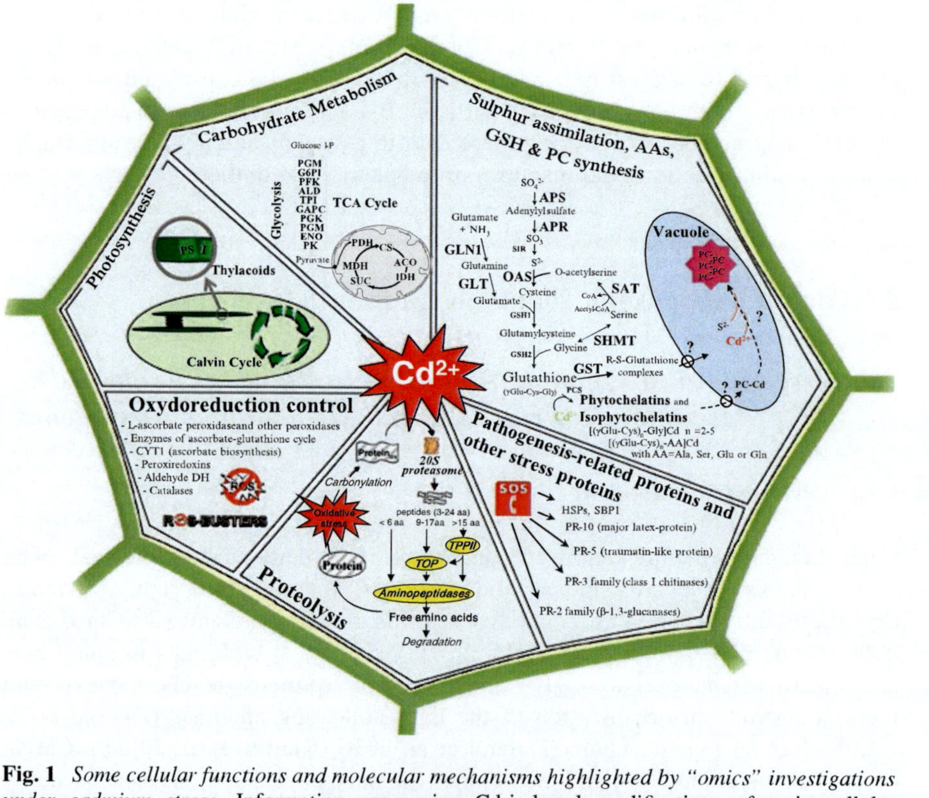

Fig. 1 *Some cellular functions and molecular mechanisms highlighted by "omics" investigations under cadmium stress*. Information concerning Cd-induced modifications of main cellular functions or metabolic pathways presented here (photosynthesis; carbohydrate metabolism, sulphur assimilation, amino acids (*AAs*), Glutathione (*GSH*) and phytochelatins (*PC*) syntheses; pathogenesis-related proteins and other stress proteins; proteolysis and oxidoreduction control) are further detailed in the text

subsequent modifications of the plant physiology, including carbohydrate metabolism re-organization or oxidative stress.

Plant response to Cd is, in contrast, less clear at the carbohydrate metabolism level, yet important modifications of its constituents are invariably mentioned. On the one hand, the shutdown of the photosynthetic machinery calls for an up-regulation of other sources of reducing power and ATP, such as glycolysis or TCA cycle; but on the other hand, supplying these pathways with necessary carbohydrates is becoming precarious as the main source of carbon has dried up. In poplar leaves, Kieffer et al. (2008) showed that an α–amylase was up-regulated, as well as a fructose kinase 2, involved in starch degradation and the phosphorylation of fructose prior to its entry in the glycolysis pathway. Similarly, genes encoding α– and β–amylases were shown to be up-regulated at the transcriptional level (Herbette et al. 2006). In parallel, numerous enzymes of this pathway (hexose kinase, fructose-bisphosphate aldolase, triose phosphate isomerase, cytosolic glyceraldehyde 3-phosphate dehydrogenase, phosphoglycerate mutase, 2-phosphoglycerate hydrolase, pyruvate kinase) were shown to be up-regulated in different proteomic works performed on *Arabidopsis* (Roth et al. 2006; Sarry et al. 2006), rice (Aina et al. 2007; Lee et al. 2010), poplar (Durand et al. 2010; Kieffer et al. 2009a, b), tomato (Rodriguez-Celma et al. 2010) or flax (Hradilova et al. 2010) (Fig. 1). Accumulation of these enzymes was also correlated with an up-regulation of several enzymes of the TCA cycle (pyruvate dehydrogenase complex enzymes 1, 2 and 3, citrate synthase, aconitase, NADP-specific isocitrate dehydrogenase, succinate dehydrogenase or malate dehydrogenase) (Kieffer et al. 2009b; Sarry et al. 2006; Semane et al. 2010; Visioli et al. 2010). However, even though some enzymes of this pathway (glucose-6-phosphate dehydrogenase, transketolase) seem to be up-regulated at the protein level (Kieffer et al. 2009b; Sarry et al. 2006), transcriptomic analysis revealed a repression of the expression for most of the enzymes, including 2-glucose-6-phosphate dehydrogenases and the transketolase, but also the 3-hydroxyisobutyrate dehydrogenase and several phosphogluconate dehydrogenases and gluconolactonases (Herbette et al. 2006). These contradictory results may reflect differences in time of exposure and concentration of Cd employed, as these pathways are, though dependent upon carbon source availability, required for preserving suitable levels of reducing molecules and carbon skeletons that are necessary for the synthesis of amino acids and molecules involved in Cd^{2+} chelation such as GSH and PC.

It is noticeable that enzymes of the sulfur assimilation pathway, all the way down to cysteine synthesis and GSH production, are actively up-regulated in order to sustain the production of PC (Figs. 1 and 2). Herbette et al. (2006) reported a transcriptional activation of 6 sulfate transporters, out of 14, in response to Cd, including the high-affinity Sultr1:1 at as early as 2 h of a 50-μM Cd exposure (Fig. 2). In addition to an increased sulfur uptake from the medium, intracellular stocks are also remobilized and Cd exposure leads to an overexpression of the vacuolar Sultr4:1 transporter, as well as others involved in long-distance transport (Sultr2:1, Sultr3:5). Genes coding for enzymes involved in the subsequent reduction of sulfur in sulfide (ATP sulfurylase, APS reductase) are also activated early. Interestingly, an ATP sulfurylase was also found up-regulated in the Cd^{2+}-hypertolerant facultative

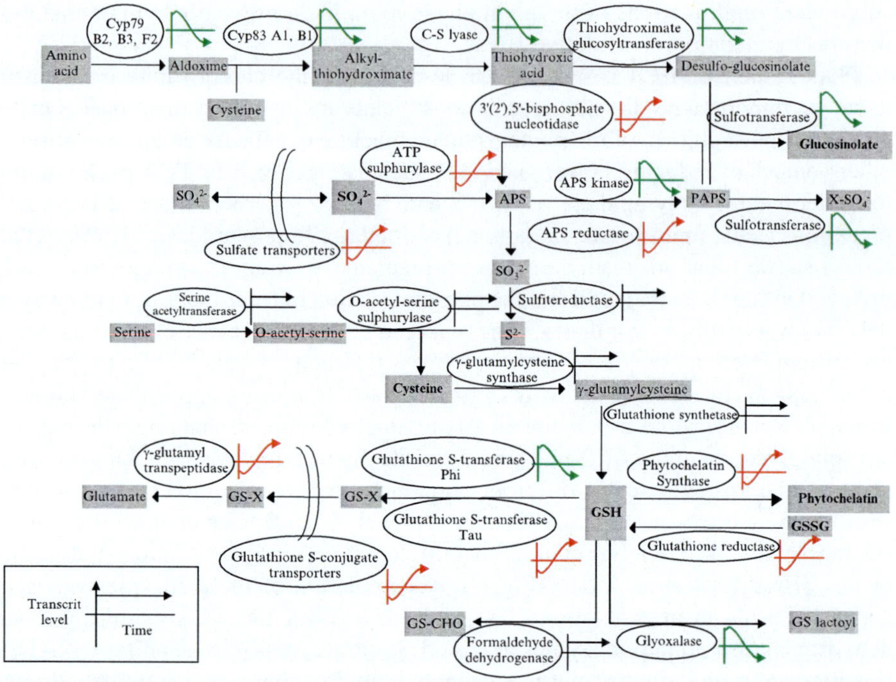

Fig. 2 *Simplified and schematic representation of transcriptional changes of genes involved in sulphur and glutathione metabolisms in roots*. The metabolites are represented in *gray boxes*; *arrows* represent enzymatic reactions; when determined, changes in gene expression are indicated schematically to the right of each gene

metallophyte *A. halleri* at 125 μM Cd^{2+} after 2 h of exposure and was present in the five genes that constitute the *A. halleri* core response (Weber et al. 2006). In *Thlaspi caerulescens*, another Zn/Cd hyperaccumulator, an ATP sulfurylase was also up-regulated under Zn-deficiency and high Cd exposure (van de Mortel et al. 2008). This gene was also strongly up-regulated in *Solanum torvum* roots stressed with a mild Cd^{2+} treatment (0.1 μM) for 3 days (Yamaguchi et al. 2010).

During Cd stress, the up-regulation of enzymes involved in the sulfur assimilation pathway, which ultimately leads to cysteine and enzymes involved in its amino acid precursor biosynthesis, was also confirmed by proteomic works with the accumulation of ATP sulfurylase, serine hydroxymethyltransferase (both mitochondrial and cytosolic forms), components of the glycine decarboxylase complex and *O*-acetylserine (thiol)lyase (Fagioni and Zolla 2009; Kieffer et al. 2009a; Roth et al. 2006; Sarry et al. 2006). In addition to these enzymes, others belonging to the amino acid metabolism and in particular to the synthesis of the other aminoacid precursors of glutathione, were also shown to be upregulated. This includes a glutamine synthetase, a glutamate synthase, an alanine aminotransferase, an aspartate aminotransferase, or the alanine:glyoxylate aminotransferase 2, which produces glycine from alanine (Aina et al. 2007; Kieffer et al. 2008, 2009a, b; Roth et al. 2006; Sarry et al. 2006; Semane et al. 2010).

Interestingly, Herbette et al. (2006) observed that Cd treatment induced an early and sustained down-regulation of many genes involved in the biosynthesis pathway of glucosinolates (Fig. 2), sulfur-rich compounds which play an important role in defense against herbivores and pathogens (Grubb and Abel 2006; Halkier and Gershenzon 2006). The down-regulation by Cd of most of the genes involved in this biosynthesis pathway observed by Herbette et al. (2006) suggests that, to support the sulfate demand for GSH and PC biosynthesis, the synthesis of glucosinolate is limited. In agreement with this hypothesis, the down-regulation of two genes encoding APS kinases and two genes encoding sulfotransferases were also observed, whereas three genes encoding 3'(2'),5'–biphosphate nucleotidase were up-regulated (Fig. 2).

2.1.2 Defence Mechanisms

Particularly highlighted in transcriptomic and proteomic studies, the deleterious effects of Cd on the plant redox status account for an important part of its toxicity. As opposed to iron, which mediates ROS production via the Haber-Weiss/Fenton reactions, Cd induces a redox imbalance by disorganizing the photosynthetic apparatus (Herbette et al. 2006) and by interfering with antioxidant enzymes and molecules (Clemens 2006; Sanità di Toppi and Gabbrielli 1999). A repression of several ROS-coping proteins, including, for instance, enzymes of the ascorbate biosynthesis pathway, can be observed (Kieffer et al. 2008), as well as the degradation of a Cu/Zn superoxide dismutase (Alvarez et al. 2009; Kieffer et al. 2008, 2009a). In contrast, a Fe-superoxide dismutase is up-regulated (Herbette et al. 2006; Kieffer et al. 2009a; Semane et al. 2010), together with different catalases and peroxidases (Alvarez et al. 2009; Herbette et al. 2006; Kieffer et al. 2008; Lee et al. 2010; Sarry et al. 2006), monodehydroascorbate reductase (Sarry et al. 2006; Alvarez et al. 2009; Herbette et al. 2006) and glutathione reductase (Herbette et al. 2006; Lee et al. 2010), transcriptional and protein levels being remarkably well correlated. In relation to this, the overexpression of numerous enzymes involved in the reduction of various oxidized molecules is particularly important. Peroxiredoxins (Ahsan et al. 2007; Durand et al. 2010; Sarry et al. 2006), phospholipid hydroperoxide glutathione peroxidase (Sarry et al. 2006), aldehyde dehydrogenases (Kieffer et al. 2009a; Sarry et al. 2006), formate dehydrogenase (Durand et al. 2010; Hradilova et al. 2010; Sarry et al. 2006), glyoxalase I (Ahsan et al. 2007), 2–nitropropane dioxygenase (Alvarez et al. 2009; Roth et al. 2006), NADPH oxidoreductase (Sarry et al. 2006), several quinone or other oxidoreductases (Kieffer et al. 2008, 2009a; Sarry et al. 2006; Semane et al. 2010; Visioli et al. 2010) were shown to be up-regulated under Cd treatment. In *Brassica juncea*, a methionine sulfoxide reductase, probably involved in the reduction of oxidized sulfur-containing residues of proteins (Davies 2005), was also reported to be induced by Cd (Alvarez et al. 2009). Interestingly, metabolite analyses performed by GC-MS or ^{1}H-NMR (Hediji et al. 2010; Sun et al. 2010) revealed a significant increase of antioxidant molecules including α-tocopherols, confirming the important role of antioxidant defences in the mechanisms of plant resistance to Cd (Collin et al. 2008).

This deleterious effect of Cd on proteins is further exemplified by the overexpression of a number of chaperones and HSPs, such as HSP70, DnaK subfamily proteins and HSP60 at the protein level (Ahsan et al. 2007; Hradilova et al. 2010; Kieffer et al. 2009b; Rodriguez-Celma et al. 2010; Sarry et al. 2006), and a disulfide isomerase (Kieffer et al. 2009b; Sarry et al. 2006), involved in protein correct folding. Up-regulation of HSPs, in particular small HSPs, was also highlighted at the transcription level (Herbette et al. 2006; Weber et al. 2006) (Fig. 1). However, and despite the overproduction of this important set of repairing enzymes, Cd-induced damages on proteins might be irreversible, as it is often the case when carbonylation occurs, for instance upon oxidative stress (Moller et al. 2007). In such a case, degradation of affected proteins is required, which importance was highlighted by in vitro assays (Polge et al. 2009). Polge et al. (2009) also showed that, in response to Cd, the 20S proteasome proteolytic pathway (i.e., the 20S proteasome and the peptidases acting sequentially downstream from the proteasome: tripeptidyl-peptidase II, thimet oligopeptidase and leucine aminopeptidase) is up-regulated at both RNA and protein activity levels in *Arabidopsis* leaves and may play a major role in degrading oxidized proteins generated by the Cd stress (Fig. 1). Experiments performed on yeast also demonstrated that mutants of the proteasome pathway are more sensitive to Cd than wt (Vido et al. 2001). Subunits of the 20S and 26S proteasomes, ubiquitin (Ub) family proteins, Ub-activating enzyme E1, Ub-conjugating enzyme E2, Ub-protein ligase E3, subtilase (serine-type endopeptidase), aspartyl protease family protein and cytosol aminopeptidase were shown to be up-regulated under Cd stress (Ahsan et al. 2007; Aina et al. 2007; Alvarez et al. 2009; Kieffer et al. 2008; Lee et al. 2010; Rodriguez-Celma et al. 2010; Roth et al. 2006; Sarry et al. 2006; Semane et al. 2010). In addition to proteins, other molecules might be affected by Cd and therefore require sequestration and degradation. The accumulation of several glutathione *S*-transferases (GST), mainly belonging to the τ- and φ-families, involved in the detoxification of toxic compounds by conjugation with GSH (Edwards et al. 2000; Marrs 1996), is often mentioned in the literature. Proteomic works performed in *Arabidopsis* revealed that AtGSTF2, 6, 7, 8, 9, 10 and AtGSTU8, 19 and 28 (Roth et al. 2006; Semane et al. 2010) were overexpressed by Cd. OsGSTU3, 4 and 12 were shown to be Cd-responsive in rice (Lee et al. 2010), and induction of various GSTs was also reported from studies conducted on soybean cells (Sobkowiak and Deckert 2006) and poplar (Kieffer et al. 2008, 2009b). Again, a good correlation exists between proteomic and transcriptomic data as several transcriptomic analyses also report up-regulated GST (Herbette et al. 2006; Kovalchuk et al. 2005; van de Mortel et al. 2008; Zhao et al. 2009) (Fig. 1).

Such detoxification processes require an important production of glutathione. While the sulfur assimilation/cysteine synthesis pathway is globally up-regulated upon Cd stress, enzymes involved in GSH synthesis (GSH1 and GSH2), in contrast, remain stable at both transcriptional and protein levels (Herbette et al. 2006; May et al. 1998; Weber et al. 2006). This strongly suggests that amino acid availability, rather than enzyme activity, is limiting in the production of GSH, used for both conjugation of toxic compounds and production of PC. While the different isoforms of PC (defined by their C-terminal amino acid) were thought to be species specific (Zenk 1996), metabolic profiling performed on *Arabidopsis* cells actually revealed that all the known isoforms were produced in this same organism, and even specified different times and levels of production for each of them (Ducruix et al. 2006, 2008; Sarry et al. 2006). Further metabolomic approaches highlighted that, as suggested by transcriptomic and proteomic works, amino acid availability, rather than enzyme activity, was critical for GSH and PC synthesis. Indeed, severe Cd stress led to the accumulation of the GSH immediate precursor: the dipeptide γ-GluCys, immediately consumed when glycine was added to the medium (Ducruix et al. 2006). This research elegantly illustrates the meaningfulness of combining different "omics" approaches (e.g., proteomics and metabolomics), which in this case pictures strengths and weaknesses of the PC-based defence mechanism in plants.

Interestingly, transcriptomic studies have shown that genes involved in cell wall metabolism were modulated in response to Cd in roots and leaves (Herbette et al. 2006; van de Mortel et al. 2008). Genes involved in lignin biosynthesis such as cinnamoyl-coA reductases, cinamyl-alcohol dehydrogenases, 4–Coumarate: CoA ligase 2 were up-regulated (Herbette et al. 2006; van de Mortel et al. 2008) whereas several genes coding for expensins and pectinesterases involved in cell expansion were repressed (Herbette et al. 2006), suggesting that components to strengthen the cell wall were modified probably to protect the plants from Cd stress. The modification of the cell wall at the root level could be a specific detoxification mechanism that limits the entry of toxic metals in *Arabidopsis*, similar to what has been observed in *Phragmites australis* (Ederli et al. 2004). Proteomic approaches combined with genetic and biochemical analysis performed on candidate proteins that accumulates in *Arabidopsis* cells treated with Cd (Sarry et al. 2006) have highlighted potential new detoxification mechanisms that plants use to face Cd toxicity (Dutilleul et al. 2008). This was the case of the *Arabidopsis* homologue to the human selenium binding protein, SBP1. SBP1 overexpression was shown to protect *Arabidopsis* from the toxic effects of Cd and binding studies indicated direct interactions between Cd and the protein (Dutilleul et al. 2008). Expression analysis performed on *Arabidopsis SBP1::LUCIFERASE* lines and by immunodetection assays demonstrated that *SBP1* expression was not specifically induced in response to Cd stress but enhanced in response to stresses provoking a cellular sulfur demand (Hugouvieux et al. 2009). In line with these results, sulfur starvation increased SBP1 expression and SBP1 overexpression enhanced tolerance to stresses that require GSH for detoxification (Hugouvieux et al. 2009).

2.2 Transcriptomic Analysis Allow for Large Scale Comparisons Between Species and/or Treatment

Transcriptomics has been used to analyze the similarities and differences in the metal response among different plant species. In particular, comparative transcriptomic analysis on *Arabidopsis thaliana* and in the Zn^{2+} and Cd^{2+}-hypertolerant facultative metallophyte *Arabidopsis halleri* were performed in order to identify response elements shared by these two species and highlight specific aspects of *A. halleri* that might lead to a better understanding of the metal-tolerance mechanisms (for more details see chapters of Marmiroli and colleagues). In pioneering works, Becher et al. (2004) and Weber et al. (2004) have shown that metal-homeostasis genes are constitutively expressed at much higher levels in *A. halleri* than in *A. thaliana* and independently from micronutrient status. It therefore appears that the effect of Cd on the *A. halleri* root transcriptome is much smaller than on that of *A. thaliana* (Weber et al. 2004). Indeed, after 2 h of treatment, only five genes were more than twofold up-regulated in *A. halleri*, versus 112 genes in *A. thaliana*. Four genes were up-regulated in the two species, constituting the "Cd^{2+} core response". A reduced Cd^{2+} uptake capacities by roots and its higher faculty to sequestrate Cd^{2+} could contribute, at least in part, to Cd^{2+} tolerance of *A. halleri* (Weber et al. 2006). More recently, Courbot et al. (2007) have shown that a QTL involved with Zn and Cd tolerance co-localizes in *A. halleri* with the plasma membrane P_{1B}-ATPase, HMA4, and it was demonstrated that the metal tolerance was due to both a triplication and altered cis regulation of the HMA4 gene (Hanikenne et al. 2008) (see also chapter on HMAs by Leonhardt et al.). Weber et al. (2006) have also shown that a short-term Cd exposure could provoke an apparent Zn^{2+} deficiency as judged by a high induction of the AtZIP9 uptake system and known as a marker of Zn deficiency. Another toxic effect of Cd could be an increased protein denaturation as illustrated by a high induction of several HSPs, and also mentioned by several proteomic data and other transcriptomic works. A transcriptomic study between *Arabidopsis thaliana* and the Zn/Cd-hyperaccumulator *Thlaspi caerulescens* exposed to different concentrations of Cd and Zn was performed by Aarts and colleagues (van de Mortel et al. 2008). Among the results they obtained, and as mentioned for *A. halleri*, a number of genes induced by Cd exposure were constitutively highly expressed in *T. caerulescens*. They also showed that Cd appears to induce Fe deficiency in *Arabidopsis*; an effect on lignin biosynthesis in both species as mentioned above and an effect on sulfate assimilation only in *T. caerulescens*. The discrepancy observed with other transcriptomic works could be explain by the difference in exposure times, since Herbette et al. (2006) mentioned that gene expression in response to Cd was more time-regulated than dose-regulated in *Arabidopsis*. Comparative transcriptomic analyses between plants exposed to different stresses were also established by other groups. Comparative microarray studies of seedlings exposed to Cd, Al, Cu and NaCl (applied at a similar degree of severity as estimated by root growth inhibition) were performed in order to identify the specific and generic responses they trigger (Zhao et al. 2009, 2010). Among the commonly up-regulated genes, they identified genes encoding ROS-scavenging enzymes such as GST and peroxidases, several Ca-binding proteins including calmodulin-like proteins, several enzymes involved in the tryptophane synthesis, the alanine amino-transferase 1 (ALAAT1), the trehalose–6–phosphate phosphatase and the arginine decarboxylase, these latter two enzymes being involved in the biosynthesis of trehalose and polyamine synthesis, respectively, which are two stress-responsive metabolites. Interestingly, when the trehalose synthesis pathway was analyzed using the Kappa-view tool (Tokimatsu et al. 2005), an activation of the trehalose pathway was predicted, correlating with an increase of root trehalose content that was experimentally determined in all treatments (Zhao et al. 2010). In the specific group of genes that were up-regulated by Cd, Zhao et al. (2009, 2010) identified a GDP-mannose pyrophosphorylase (CYT1) involved in the ascorbate biosynthesis and critical for ROS-related stress resistance, heat shock proteins and pathogenesis-related (PR) proteins. Interestingly, several proteomic studies also revealed that some of the mechanisms involved in the plant response to a Cd stress are common, with inducible defence mechanisms activated upon pathogen attack. These include in particular the production of different PR proteins normally associated with the oxidative burst during biotic stress responses. This common feature may be due to the fact that signaling pathways are probably shared, as application of some hormones (e.g. ethylene, jasmonate and salycilic acid) are known to mimic the effect of a pathogen attack (Sels et al. 2008). Among the PR proteins induced by Cd are a major latex-protein, a PR-10 family member (Roth et al. 2006), also induced in response to salt stress and bacteria (Jiang et al. 2007; Jones et al. 2006), a traumatin-like protein, a PR-5 family member (Kieffer et al. 2008; Semane et al. 2010), class I chitinases, members of the PR-3 family and several β-1,3-glucanases belonging to the PR-2 family (Kieffer et al. 2008) (Fig. 1).

3 Future Directions

3.1 Subcellular-Level Analysis of the Cd Response

The major advantage of proteome-wide analyses of whole plants or tissues, which is a large coverage of most of the major proteins for a "no a priori" study, also constitutes a considerable drawback as peptides from minor proteins often get hidden during MS runs by other, more represented, peptides. This is particularly exemplified in the case of the vacuole, which overall protein content is suggested to represent less than 2%, in mass, of the total cellular proteome. While extremely important in the processes leading to Cd sequestration and tolerance, its protein constitution was, until recently, rather unknown and the vacuolar actors of the Cd tolerance poorly identified. The lack of exhaustiveness resulting from an under representation of a particular organelle proteome can be more or less easily circumvented by deciding to focus on this single subcellular compartment, and therefore doing a sub-fractionation of the plant material prior to protein extraction (Agrawal et al. 2011). First proteomic studies of the vacuole of *Arabidopsis* appeared between 2004 and 2007, and they led to the establishment of a list of about 1100 non-redundant proteins, including more than 120 transporters (Carter et al. 2004; Jaquinod et al. 2007a, b; Shimaoka et al. 2004; Szponarski et al. 2004).

Others have been completed, performed on barley (Endler et al. 2006) or cauliflower (Schmidt et al. 2007), which opened the way to differential proteomic analyses of this organelle under Cd stress. In this matter, Schneider et al. (2009) have used iTRAQ-labeled tonoplast samples isolated from barley subjected to 0, 20 or 200 μM Cd. Out of 56 transporters identified, 6 were found to be differentially expressed. Surprisingly, only one – an ABC-type MRP-like transporter homologous to AtMRP3 – was upregulated at 20 and 200 μM Cd, while others were only affected at 20 μM. It is interesting to notice that AtMRP3 was previously shown to be upregulated by Cd (Bovet et al. 2003), and to partly rescue the Cd-sensitive phenotype of DTY168 yeasts (Tommasini et al. 1998). A CAX1a-like transporters was found to be over-expressed at 20 μM (× 1.4), confirming that this kind of transporter might play an important role in Cd^{2+} detoxification, and a member of the natural resistance-associated macrophage protein (NRAMP) family (× 2.3), which homologous proteins AtNRAMP3 and 4 are known actors of the plant metal homeostasis (Lanquar et al. 2005, 2010). In addition, an isoform of the vacuolar H^+-pyrophosphatase was overexpressed (× 2) as well as a γ-TIP-like aquaporin (× 1.9). Such organelle-targeted differential studies were also conducted on the thylakoid membrane (Fagioni et al. 2009), where a disruption of the PSI was observed, concomitant with the diminution of chlorophyll a and b contents. In contrast, the PSII was less affected and no change was observed on the cytochrome *b6/f* and ATP-synthase complex organization. Changes at the plasma membrane level were also studied by ^{15}N labeling of cell cultures (Lanquar et al. 2007), and evidenced the overexpression by 2.5-fold of the ABC transporter AtPDR8, shown to be an extrusion pump conferring Cd resistance (Kim et al. 2007). An ammonium transporter (AtAMT1.1) was also shown to be fivefold upregulated, and seems to confirm the importance of side mechanisms, such as nutrient uptake to sustain amino acid production, as probable key players in the tolerance mechanisms. While very promising, these results did not lead to the clear identification of major, vacuolar actors of Cd tolerance, such as the ABCC1 and 2 ABC-type transporters recently characterized for their PC−arsenic transport activity in the vacuole (Song et al. 2010). The evaluation of their implication in the transport of PC−Cd complexes towards the lumen of the vacuole needs to be assessed, and may, after decades of research, finally identify the transporter responsible for this translocation. However it is particularly noticeable that, in the quest of this Holy Grail that is (or used to be?) the plant vacuolar PC−Cd transporter, authors have emphasized on the dynamic of membrane proteins. The yeast homologous *Sp*HMT1 or *Sc*YCF1, on the other hand, are not described to be up-regulated upon Cd stress, their activity being probably not limiting in the processes of vacuolar sequestration. Catalytic activities subsequent to Cd internalization in the vacuole and allowing stable and safe storage of this toxic are in contrast rather unknown, as well as a plausible recycling of the PCs, as suggested by some authors. Preliminary results of experiments aiming to identify soluble proteins differentially expressed under Cd stress revealed an over-expression of some of them, weakly expressed in control conditions, which are currently in the process of being characterized (Villiers F and Bourguignon J, unpublished data). Altogether, these results confirm an organelle-targeted proteomic approach as very valuable, but also point out the need for researchers to expend their digging area, to soluble proteins for instance, or to activity regulation other than over-expression, such as post-translational modifications (PTM).

3.2 *Post-translational Modifications: Phosphoproteomic Studies*

PTM studies are an important challenge for the future and proteomic analysis of most PTM, except for protein phosphorylation, is still in the initial phases in plants (Ytterberg and Jensen 2010). Phosphoproteomic analysis of *Arabidopsis* upon various types of stimuli has tremendously increased in the literature over the last few years (Benschop et al. 2007; Kersten et al. 2009; Li et al. 2009; Niittylae et al. 2007), and has allowed the identification of numerous proteins as important actors of the processes being considered (Schulze 2010). These techniques have also been successfully applied to isolated organelles such as vacuoles (Endler et al. 2009; Whiteman et al. 2008a, b) or plasma membrane (Hem et al. 2007; Nuhse et al. 2007; Whiteman et al. 2008a), highlighting hundreds of phosphorylation sites. In parallel, other works have pointed out the phosphorylation status of different proteins such as transporters (Kobayashi et al. 2007) as essential in the regulation of aluminum tolerance in *Arabidopsis*. This should soon lead people interested in understanding the regulation of Cd tolerance to perform phosphorylation profiling of whole protein extract, as well as more tissue- or organelle-specific samples, from Cd-challenged plants.

3.3 *Systems Biology and Integrated Analysis*

As mentioned above, post-genomic biology has allowed advances in our understanding of plant biological systems by mining genomic data, making use of high-throughput methods (transcriptomics, proteomics, metabolomics, screening, etc.), and resources for mutants functional studies. This new era of integrative biology enables scientists to determine how the interconnected networks of genes, proteins and metabolites (substrates, other ligands, etc.) work together in complex cellular processes which can now be followed in the cell owing to revolutionary imaging technologies (Mano et al. 2009). However, the functioning of a plant as a system is not the result of a simple network, but rather a combination of multiple, intertwined, dynamic, and linear or nonlinear interactions between its components (DNA, RNA, proteins, metabolites, organelles, cell types, organs, etc.). It is now well accepted that the comprehensive knowledge of how a plant system functions in fluctuating environmental conditions cannot be achieved by the characterization of its elements one by one, or a single class of elements in isolation of the others (Ruffel et al. 2010). One of the most challenging issues for biology in this twenty-first century is deciphering the diversity and complexity of biological processes in plant science as well as in other fields of fundamental biology. Addressing this priority is a multistep procedure as biological processes are first studied at a molecular scale, while seeking to elucidate the fine organization of the proteins involved, in order to monitor their dynamic interactions with other partners and to understand their reactivity. Then, the molecular information is integrated on a larger scale (organelle, cell and organism) and related to biological functions. A nice example of this approach is illustrated by the work of Curien and colleagues (2009). Combining expertise on molecular description of proteins, their interactions and their dynamics together with those on integrative biology, including well-thought modeling, should ultimately lead to a paradigm shift in how we understand complex biological processes, as a step towards systems biology and its technological counterpart, synthetic biology (Joyard and McCormick 2010).

3.4 Data Mining and Integration: The Bioinformatics Challenge

Because of their size, diversity and complexity, all the data generated by such multi-omics projects (DNA chips, qPCR, MS data as well as imaging and biochemical data) should be further integrated in order to search for the key information they contain. Several bottlenecks are already identified. Properly integrating data into dedicated knowledge bases (KBs) as well as designing user-oriented tools to efficiently query and visualize their content are two necessary steps towards higher level analysis, like network reconstruction or dynamic modeling of physiological states. In such a framework, standards are also necessary to ensure data quality and integrity in terms of capture, curation, verification and storage. For instance, depositing data in public repositories is necessary but not sufficient and the setting up of guidelines for reporting (namely "minimum information") and standards is an important step to ensure that submitted omics data are sufficient for non-ambiguous interpretation and querying by other scientists (Chervitz et al. 2011). Within this context, considerable efforts have been carried out to make these datasets publicly available through databases and web interfaces (for a recent list of plant database resources, see Mano et al. 2009). As such, these webtools often represent a good starting point for planning experiments or generating hypotheses in silico or exploring your own dataset by comparative approach (for review see Brady and Provart (2009)). A supplementary challenge lies in techniques of data mining for high-throughput data which require a high level of expertise and represent a constantly evolving methodological research field for functional genomics and biological networks studies. Among these resources, transcriptome datasets represent the most comprehensive of all of the large-scale data types and have led to the development of many tools for querying these both in a directed manner and correlatively. Hence, several co-expression gene databases and webtools dedicated to *Arabidopsis* have been developed for functional inference purposes, mainly by calculating standard Pearson correlation coefficients between gene pairs (for review, see Usadel et al. 2009). In addition, on the basis of a more sophisticated methodological framework (Bansal et al. 2007), a compendium of public microarrays has been shown to be valuable either to explore the whole transcriptional network (Ma et al. 2007) or to focus on a smaller subpart such as cell wall biosynthesis (Persson et al. 2005), glucosinolate metabolic pathway (Hirai et al. 2007) or amino acid metabolism in response to abiotic stress (Less and Galili 2008). To go further, some bioinformatic programs and omics datasets can be smartly combined into a strategy to elucidate the function of a gene of interest, including modes of regulation and synthesis, and potential interacting partners, and last but not least processes that are regulated by its gene products. With the growing number of programs, webtools and various datasets available, examples of combinatory applications of bioinformatic tools that lead to the generation of testable hypotheses (sometimes with subsequent confirmation) have been recently proposed with a focus on a cross-talk between nitrate response and hormone signals for the control of the plant development (Nero et al. 2009). Regarding the field of plant response to Cd, the number of published integrated analysis remains very limited. Owing to the low amount of "omics" data produced under these experimental perturbations, a set of genes known to be deregulated in Cd stress condition was used to select co-expression graph through a large compendium of microarrays datasets encompassing various experimental conditions including biotic and abiotic stresses. The resulting transcriptional modules are assessed by GO term enrichment and promoter analysis to identify shared combinatorial control (Boyer et al. 2009). Our preliminary results highlight a potential dialog between the sulfur/nitrate assimilation pathways, the biosynthesis of glutathione amino-acid precursors and the glucosinolate biosynthesis pathway. Further studies are under investigation for confirmation and targeted experiments (Boyer et al., in preparation). This may lead to new findings of gene regulation under stress condition in plant.

4 Conclusion

Finally, if data generation is no longer the limiting factor in advancing biological research, their integration, analysis, and interpretation have become key bottlenecks and challenges that biologists conducting genomic research face daily. While advanced data analysis tools for exploiting genomic data are rapidly emerging, the narrow specialization of most current software tools forces biologists to employ many tools to analyze the data in a single biological study. Despite the improved integration in software tools, the sheer number and diversity of analytical and visualization tools available might be bewildering to biologists who wish to apply them correctly. In this regard, the setup of a sustained and open dialog between computer scientists, statisticians and biologists actually represents one of the key steps for successful collaboration in plant systems biology (Shasha 2003). By integrating selected genomic data, visualization and analysis tools into a single Web-accessible software platform, the Virtual Plant project is promising for bridging the gap between biologists and bioinformaticians by providing an intuitive way to integrate and mine diverse data sources (Katari et al. 2010).

Acknowledgments We would like to thank Dr Jérôme Garin and the members of his laboratory for fruitful collaboration on proteomics and Dr Eric Quémeneur and the CEA programme of toxicology for their support.

References

Agrawal GK, Bourguignon J, Rolland N, Ephritikhine G, Ferro M, Jaquinod M, Alexiou KG, Chardot T, Chakraborty N, Jolivet P, Doonan JH, Rakwal R (2011) Plant organelle proteomics: collaborating for optimal cell function. Mass Spectrum Rev. doi:2010.1002/mas.20301

Ahsan N, Lee SH, Lee DG, Lee H, Lee SW, Bahk JD, Lee BH (2007) Physiological and protein profiles alternation of germinating rice seedlings exposed to acute cadmium toxicity. C R Biol 330:735–746

Aina R, Labra M, Fumagalli P, Vannini C, Marsoni M, Cucchi U, Bracale M, Sgorbati S, Citterio S (2007) Thiol-peptide level and proteomic changes in response to cadmium toxicity in *Oryza sativa* L. roots. Environ Exp Bot 59:381–392

Aloui A, Recorbet G, Gollotte A, Robert F, Valot B, Gianinazzi-Pearson V, Aschi-Smiti S, Dumas-Gaudot E (2009) On the mechanisms of cadmium stress alleviation in *Medicago truncatula* by arbuscular mycorrhizal symbiosis: a root proteomic study. Proteomics 9:420–433

Alvarez S, Berla BM, Sheffield J, Cahoon RE, Jez JM, Hicks LM (2009) Comprehensive analysis of the *Brassica juncea* root proteome in response to cadmium exposure by complementary proteomic approaches. Proteomics 9:2419–2431

Baginsky S, Hennig L, Zimmermann P, Gruissem W (2010) Gene expression analysis, proteomics, and network discovery. Plant Physiol 152:402–410

Bah AM, Sun HY, Chen F, Zhou J, Dai HX, Zhang GP, Wu FB (2010) Comparative proteomic analysis of *Typha angustifolia* leaf under chromium, cadmium and lead stress. J Hazard Mater 184:191–203

Bailey NJC, Oven M, Holmes E, Nicholson JK, Zenk MH (2003) Metabolomic analysis of the consequences of cadmium exposure in *Silene cucubalus* cell cultures via H-1 NMR spectroscopy and chemometrics. Phytochemistry 62:851–858

Bansal M, Belcastro V, Ambesi-Impiombato A, di Bernardo D (2007) How to infer gene networks from expression profiles. Mol Syst Biol 3: Article number 78. doi:10.1038/msb4100120

Becher M, Talke IN, Krall L, Kramer U (2004) Cross-species microarray transcript profiling reveals high constitutive expression of metal homeostasis genes in shoots of the zinc hyperaccumulator *Arabidopsis halleri*. Plant J 37:251–268

Benschop JJ, Mohammed S, O'Flaherty M, Heck AJR, Slijper M, Menke FLH (2007) Quantitative phosphoproteomics of early elicitor signaling in *Arabidopsis*. Mol Cell Proteomics 6:1198–1214

Bovet L, Eggmann T, Meylan-Bettex M, Polier J, Kammer P, Marin E, Feller U, Martinoia E (2003) Transcript levels of AtMRPs after cadmium treatment: induction of AtMRP3. Plant Cell Environ 26:371–381

Boyer F, Combes F, Bourguignon J, Vandenbrouck Y (2009) Extracting transcriptional regulatory modules from a large microarray compendium using a graph-based approach: application to the response of *Arabidopsis thaliana* under stress conditions. TIC Signal Process Proc Ser 48:23–27

Brady SM, Provart NJ (2009) Web-queryable large-scale data sets for hypothesis generation in plant biology. Plant Cell 21:1034–1051

Carter C, Pan SQ, Jan ZH, Avila EL, Girke T, Raikhel NV (2004) The vegetative vacuole proteorne of *Arabidopsis thaliana* reveals predicted and unexpected proteins. Plant Cell 16:3285–3303

Chervitz SA, Deutsch EW, Field D, Parkinson H, Quackenbush J, Rocca-Serra P, Sansone SA, Stoeckert CJ Jr, Taylor CF, Taylor R, Ball CA (2011) Data standards for Omics data: the basis of data sharing and reuse. Methods Mol Biol 719:31–69

Clemens S (2006) Toxic metal accumulation, responses to exposure and mechanisms of tolerance in plants. Biochimie 88:1707–1719

Cobbett CS (2000) Phytochelatin biosynthesis and function in heavy-metal detoxification. Curr Opin Plant Biol 3:211–216

Cobbett C, Goldsbrough P (2002) Phytochelatins and metallothioneins: roles in heavy metal detoxification and homeostasis. Annu Rev Plant Biol 53:159–182

Collin VC, Eymery F, Genty B, Rey P, Havaux M (2008) Vitamin E is essential for the tolerance of *Arabidopsis thaliana* to metal-induced oxidative stress. Plant Cell Environ 31:244–257

Courbot M, Willems G, Motte P, Arvidsson S, Roosens N, Saumitou-Laprade P, Verbruggen N (2007) A major quantitative trait locus for cadmium tolerance in *Arabidopsis halleri* colocalizes with HMA4, a gene encoding a heavy metal ATPase. Plant Physiol 144:1052–1065

Curien G, Bastlen O, Robert-Genthon M, Cornish-Bowden A, Cardenas ML, Dumas R (2009) Understanding the regulation of aspartate metabolism using a model based on measured kinetic parameters. Mol Syst Biol 5: Article number 271. doi:210.1038/msb.2009.1029

DalCorso G, Farinati S, Maistri S, Furini A (2008) How plants cope with cadmium: staking all on metabolism and gene expression. J Integr Plant Biol 50:1268–1280

Davies MJ (2005) The oxidative environment and protein damage. Biochim Biophys Acta 1703:93–109

Ducruix C, Junot C, Fievet JB, Villiers F, Ezan E, Bourguignon J (2006) New insights into the regulation of phytochelatin biosynthesis in *A. thaliana* cells from metabolite profiling analyses. Biochimie 88:1733–1742

Ducruix C, Vailhen D, Werner E, Fievet JB, Bourguignon J, Tabet JC, Ezan E, Junot C (2008) Metabolomic investigation of the response of the model plant *Arabidopsis thaliana* to cadmium exposure: evaluation of data pretreatment methods for further statistical analyses. Chemo Intell Lab Syst 91:67–77

Durand TC, Sergeant K, Planchon S, Carpin S, Label P, Morabito D, Hausman JF, Renaut J (2010) Acute metal stress in *Populus tremula* x *P. alba* (717-1B4 genotype): leaf and cambial proteome changes induced by cadmium(2+). Proteomics 10:349–368

Dutilleul C, Jourdain A, Bourguignon J, Hugouvieux V (2008) The *Arabidopsis* putative selenium-binding protein family: expression study and characterization of SBP1 as a potential new player in cadmium detoxification processes. Plant Physiol 147:239–251

Ederli L, Reale L, Ferranti F, Pasqualini S (2004) Responses induced by high concentration of cadmium in *Phragmites australis* roots. Physiol Plant 121:66–74

Edwards R, Dixon DP, Walbot V (2000) Plant glutathione S-transferases: enzymes with multiple functions in sickness and in health. Trends Plant Sci 5:193–198

Endler A, Meyer S, Schelbert S, Schneider T, Weschke W, Peters SW, Keller F, Baginsky S, Martinoia E, Schmidt UG (2006) Identification of a vacuolar sucrose transporter in barley and *Arabidopsis* mesophyll cells by a tonoplast proteomic approach. Plant Physiol 141:196–207

Endler A, Reiland S, Gerrits B, Schmidt UG, Baginsky S, Martinoia E (2009) In vivo phosphorylation sites of barley tonoplast proteins identified by a phosphoproteomic approach. Proteomics 9:310–321

Fagioni M, Zolla L (2009) Does the different proteomic profile found in apical and basal leaves of Spinach reveal a strategy of this plant toward cadmium pollution response? J Proteome Res 8:2519–2529

Fagioni M, Damici GM, Timperio AM, Zolla L (2009) Proteomic analysis of multiprotein complexes in the thylakoid membrane upon cadmium treatment. J Proteome Res 8:310–326

Fiehn O, Kopka J, Dormann P, Altmann T, Trethewey RN, Willmitzer L (2000) Metabolite profiling for plant functional genomics. Nat Biotechnol 18:1157–1161

Fukushima A, Kusano M, Redestig H, Arita M, Saito K (2009) Integrated omics approaches in plant systems biology. Curr Opin Chem Biol 13:532–538

Gianazza E, Wait R, Sozzi A, Regondi S, Saco D, Labrac M, Agradi E (2007) Growth and protein profile changes in *Lepidium sativum* L. plantlets exposed to cadmium. Environ Exp Bot 59:179–187

Goodacre R, York EV, Heald JK, Scott IM (2003) Chemometric discrimination of unfractionated plant extracts analyzed by electrospray mass spectrometry. Phytochemistry 62:859–863

Grubb CD, Abel S (2006) Glucosinolate metabolism and its control. Trends Plant Sci 11:89–100

Hajduch M, Rakwal R, Agrawal GK, Yonekura M, Pretova A (2001) High resolution two dimensional electrophoresis separation of proteins from metal-stressed rice (*Oryza sativa* L.) leaves: drastic reductions/fragmentation of ribulose-1,5-bisphosphate carboxylase/oxygenase and induction of stress-related proteins. Electrophoresis 22:2824–2831

Halkier BA, Gershenzon J (2006) Biology and biochemistry of glucosinolates. Annu Rev Plant Biol 57:303–333

Hanikenne M, Talke IN, Haydon MJ, Lanz C, Nolte A, Motte P, Kroymann J, Weigel D, Kramer U (2008) Evolution of metal hyperaccumulation required cis-regulatory changes and triplication of HMA4. *Nature* **453**: 391–395 (15 May 2008) doi: 10.1038/nature06877; Received 20 December 2007; Accepted 28 February 2008; Published online 20 April 2008

Hediji H, Djebali W, Cabasson C, Maucourt M, Baldet P, Bertrand A, Zoghlami LB, Deborde C, Moing A, Brouquisse R, Chaibi W, Gallusci P (2010) Effects of long-term cadmium exposure on growth and metabolomic profile of tomato plants. Ecotox Environ Safe 73:1965–1974

Hem S, Rofidal V, Sommerer N, Rossignol M (2007) Novel subsets of the *Arabidopsis* plasmalemma phosphoproteome identify phosphorylation sites in secondary active transporters. Biochem Biophys Res Commun 363:375–380

Herbette S, Taconnat L, Hugouvieux V, Piette L, Magniette MLM, Cuine S, Auroy P, Richaud P, Forestier C, Bourguignon J, Renou JP, Vavasseur A, Leonhardt N (2006) Genome-wide transcriptome profiling of the early cadmium response of *Arabidopsis* roots and shoots. Biochimie 88:1751–1765

Hirai MY, Yano M, Goodenowe DB, Kanaya S, Kimura T, Awazuhara M, Arita M, Fujiwara T, Saito K (2004) Integration of transcriptomics and metabolomics for understanding of global responses to nutritional stresses in *Arabidopsis thaliana*. Proc Natl Acad Sci U S A 101:10205–10210

Hirai MY, Sugiyama K, Sawada Y, Tohge T, Obayashi T, Suzuki A, Araki R, Sakurai N, Suzuki H, Aoki K, Goda H, Nishizawa OI, Shibata D, Saito K (2007) Omics-based identification of *Arabidopsis* Myb transcription factors regulating aliphatic glucosinolate biosynthesis. Proc Nat Acad Sci U S A 104:6478–6483

Hradilova J, Rehulka P, Rehulkova H, Vrbova M, Griga M, Brzobohaty B (2010) Comparative analysis of proteomic changes in contrasting flax cultivars upon cadmium exposure. Electrophoresis 31:421–431

Hugouvieux V, Dutilleul C, Jourdain A, Reynaud F, Lopez V, Bourguignon J (2009) *Arabidopsis* putative selenium-binding protein1 expression is tightly linked to cellular sulfur demand and can reduce sensitivity to stresses requiring glutathione for tolerance. Plant Physiol 151:768–781

Ishihara T, Kobayashi E, Okubo Y, Suwazono Y, Kido T, Nishijyo M, Nakagawa H, Nogawa K (2001) Association between cadmium concentration in rice and mortality in the Jinzu River basin, Japan. Toxicology 163:23–28

Jaquinod M, Villiers F, Kieffer-Jaquinod S, Hugouvieux V, Bruley C, Garin J, Bourguignon J (2007a) A proteomics approach highlights a myriad of transporters in the *Arabidopsis thaliana* vacuolar membrane. Plant Signal Behav 2:413–415

Jaquinod M, Villiers F, Kieffer-Jaquinod S, Hugouvieux V, Bruley C, Garin J, Bourguignon J (2007b) A proteomics dissection of *Arabidopsis thaliana* vacuoles isolated from cell culture. Mol Cell Proteomics 6:394–412

Jiang Y, Yang B, Harris NS, Deyholos MK (2007) Comparative proteomic analysis of NaCl stress-responsive proteins in *Arabidopsis* roots. J Exp Bot 58:3591–3607

Jones AM, Thomas V, Bennett MH, Mansfield J, Grant M (2006) Modifications to the *Arabidopsis* defense proteome occur prior to significant transcriptional change in response to inoculation with *Pseudomonas syringae*. Plant Physiol 142:1603–1620

Joyard J, McCormick S (2010) Plant systems biology. Plant Physiol 152:401–401

Katari MS, Nowicki SD, Aceituno FF, Nero D, Kelfer J, Thompson LP, Cabello JM, Davidson RS, Goldberg AP, Shasha DE, Coruzzi GM, Gutierrez RA (2010) Virtual plant: a software platform to support systems biology research. Plant Physiol 152:500–515

Kersten B, Agrawal GK, Durek P, Neigenfind J, Schulze W, Walther D, Rakwal R (2009) Plant phosphoproteomics: an update. Proteomics 9:964–988

Kieffer P, Dommes J, Hoffmann L, Hausman JF, Renaut J (2008) Quantitative changes in protein expression of cadmium-exposed poplar plants. Proteomics 8:2514–2530

Kieffer P, Planchon S, Oufir M, Ziebel J, Dommes J, Hoffmann L, Hausman JF, Renaut J (2009a) Combining proteomics and metabolite analyses to unravel cadmium stress-response in poplar leaves. J Proteome Res 8:400–417

Kieffer P, Schroder P, Dommes J, Hoffmann L, Renaut J, Hausman JF (2009b) Proteomic and enzymatic response of poplar to cadmium stress. J Proteomics 72:379–396

Kim DY, Bovet L, Maeshima M, Martinoia E, Lee Y (2007) The ABC transporter AtPDR8 is a cadmium extrusion pump conferring heavy metal resistance. Plant J 50:207–218

Kobayashi Y, Hoekenga OA, Itoh H, Nakashima M, Saito S, Shaff JE, Maron LG, Pineros MA, Kochian LV, Koyama H (2007) Characterization of AtALMT1 expression in aluminum-inducible malate release and its role for rhizotoxic stress tolerance in *Arabidopsis*. Plant Physiol 145:843–852

Kovalchuk I, Titov V, Hohn B, Kovalchuka O (2005) Transcriptome profiling reveals similarities and differences in plant responses to cadmium and lead. Mutat Res-Fundam Mol Mech Mutagen 570:149–161

Lane TW, Saito MA, George GN, Pickering IJ, Prince RC, Morel FMM (2005) A cadmium enzyme from a marine diatom. Nature 435:42–42

Lanquar V, Lelievre F, Bolte S, Hames C, Alcon C, Neumann D, Vansuyt G, Curie C, Schroder A, Kramer U, Barbier-Brygoo H, Thomine S (2005) Mobilization of vacuolar iron by AtNRAMP3 and AtNRAMP4 is essential for seed germination on low iron. EMBO J 24:4041–4051

Lanquar V, Kuhn L, Lelievre F, Khafif M, Espagne C, Bruley C, Barbier-Brygoo H, Garin J, Thomine S (2007) N-15-Metabolic labeling for comparative plasma membrane proteomics in *Arabidopsis* cells. Proteomics 7:750–754

Lanquar V, Ramos MS, Lelievre F, Barbier-Brygoo H, Krieger-Liszkay A, Kramer U, Thomine S (2010) Export of vacuolar manganese by AtNRAMP3 and AtNRAMP4 is required for optimal photosynthesis and growth under manganese deficiency. Plant Physiol 152:1986–1999

Lee K, Bae DW, Kim SH, Han HJ, Liu X, Park HC, Lim CO, Lee SY, Chung WS (2010) Comparative proteomic analysis of the short-term responses of rice roots and leaves to cadmium. J Plant Physiol 167:161–168

Less H, Galili G (2008) Principal transcriptional programs regulating plant amino acid metabolism in response to abiotic stresses. Plant Physiol 147:316–330

Li H, Wong WS, Zhu L, Guo HW, Ecker J, Li N (2009) Phosphoproteomic analysis of ethylene-regulated protein phosphorylation in etiolated seedlings of *Arabidopsis* mutant ein2 using two-dimensional separations coupled with a hybrid quadrupole time-of-flight mass spectrometer. Proteomics 9:1646–1661

Ma SS, Gong QQ, Bohnert HJ (2007) An Arabidopsis gene network based on the graphical Gaussian model. Genome Res 17:1614–1625

Mano S, Miwa T, Nishikawa S, Mimura T, Nishimura M (2009) Seeing is believing: on the use of image databases for visually exploring plant organelle dynamics. Plant Cell Physiol 50:2000–2014

Marrs KA (1996) The functions and regulation of glutathione S-transferases in plants. Annu Rev Plant Physiol Plant Mol Biol 47:127–158

May MJ, Vernoux T, Sanchez-Fernandez R, Van Montagu M, Inze D (1998) Evidence for posttranscriptional activation of gamma-glutamylcysteine synthetase during plant stress responses. Proc Natl Acad Sci U S A 95:12049–12054

Mendoza-Cozatl DG, Butko E, Springer F, Torpey JW, Komives EA, Kehr J, Schroeder JI (2008) Identification of high levels of phytochelatins, glutathione and cadmium in the phloem sap of Brassica napus: a role for thiol-peptides in the long-distance transport of cadmium and the effect of cadmium on iron translocation. Plant J 54:249–259

Moller IM, Jensen PE, Hansson A (2007) Oxidative modifications to cellular components in plants. Annu Rev Plant Biol 58:459–481

Nero D, Krouk G, Tranchina D, Coruzzi GM (2009) A system biology approach highlights a hormonal enhancer effect on regulation of genes in a nitrate responsive "biomodule". BMC Syst Biol 3:59. doi:10.1186/1752-0509-1183-1159

Niittylae T, Fuglsang AT, Palmgren MG, Frommer WB, Schulze WX (2007) Temporal analysis of sucrose-induced phosphorylation changes in plasma membrane proteins of *Arabidopsis*. Mol Cell Proteomics 6:1711–1726

Nuhse TS, Bottrill AR, Jones AME, Peck SC (2007) Quantitative phosphoproteomic analysis of plasma membrane proteins reveals regulatory mechanisms of plant innate immune responses. Plant J 51:931–940

Palmgren MG, Clemens S, Williams LE, Kraemer U, Borg S, Schjorring JK, Sanders D (2008) Zinc biofortification of cereals: problems and solutions. Trends Plant Sci 13:464–473

Persson S, Wei HR, Milne J, Page GP, Somerville CR (2005) Identification of genes required for cellulose synthesis by regression analysis of public microarray data sets. Proc Natl Acad Sci U S A 102:8633–8638

Pilon-Smits (2005) Phytoremediation. Ann Rev Plant Biol 56:15–29

Polge C, Jaquinod M, Holzer F, Bourguignon J, Walling L, Brouquisse R (2009) Evidence for the existence in *Arabidopsis thaliana* of the proteasome proteolytic pathway: activation in response to cadmium. J Biol Chem 284:35412–35424

Rauser WE (1990) Phytochelatins. Annu Rev Biochem 59:61–86

Rodriguez-Celma J, Rellan-Alvarez R, Abadia A, Abadia J, Lopez-Millan AF (2010) Changes induced by two levels of cadmium toxicity in the 2-DE protein profile of tomato roots. J Proteomics 73:1694–1706

Roth U, von Roepenack-Lahaye E, Clemens S (2006) Proteome changes in *Arabidopsis thaliana* roots upon exposure to Cd^{2+}. J Exp Bot 57:4003–4013

Ruffel S, Krouk G, Coruzzi GM (2010) A systems view of responses to nutritional cues in *Arabidopsis*: toward a paradigm shift for predictive network modeling. Plant Physiol 152:445–452

Sanità di Toppi L, Gabbrielli R (1999) Response to cadmium in higher plants. Environ Exp Bot 41:105–130

Sarry JE, Kuhn L, Ducruix C, Lafaye A, Junot C, Hugouvieux V, Jourdain A, Bastien O, Fievet JB, Vailhen D, Amekraz B, Moulin C, Ezan E, Garin J, Bourguignon J (2006) The early responses of *Arabidopsis thaliana* cells to cadmium exposure explored by protein and metabolite profiling analyses. Proteomics 6:2180–2198

Sato S, Soga T, Nishioka T, Tomita M (2004) Simultaneous determination of the main metabolites in rice leaves using capillary electrophoresis mass spectrometry and capillary electrophoresis diode array detection. Plant J 40:151–163

Schmidt UG, Endler A, Schelbert S, Brunner A, Schnell M, Neuhaus HE, Marty-Mazars D, Marty F, Baginsky S, Martinoia E (2007) Novel tonoplast transporters identified using a proteomic approach with vacuoles isolated from cauliflower buds. Plant Physiol 145:216–229

Schneider T, Schellenberg M, Meyer S, Keller F, Gehrig P, Riedel K, Lee Y, Eberl L, Martinoia E (2009) Quantitative detection of changes in the leaf-mesophyll tonoplast proteome in dependency of a cadmium exposure of barley (*Hordeum vulgare* L.) plants. Proteomics 9:2668–2677

Schulze WX (2010) Proteomics approaches to understand protein phosphorylation in pathway modulation. Curr Opin Plant Biol 13:280–287

Sels J, Mathys J, De Coninck BM, Cammue BP, De Bolle MF (2008) Plant pathogenesis-related (PR) proteins: a focus on PR peptides. Plant Physiol Biochem 46:941–950

Semane B, Dupae J, Cuypers A, Noben JP, Tuomainen M, Tervahauta A, Karenlampi S, Van Belleghem F, Smeets K, Vangronsveld J (2010) Leaf proteome responses of *Arabidopsis thaliana* exposed to mild cadmium stress. J Plant Physiol 167:247–254

Shasha DE (2003) Plant systems biology: lessons from a fruitful collaboration. Plant Physiol 132:415–416

Shimaoka T, Ohnishi M, Sazuka T, Mitsuhashi N, Hara-Nishimura I, Shimazaki KI, Maeshima M, Yokota A, Tomizawa KI, Mimura T (2004) Isolation of intact vacuoles and proteomic analysis of tonoplast from suspension-cultured cells of *Arabidopsis thaliana*. Plant Cell Physiol 45:672–683

Sobkowiak R, Deckert J (2006) Proteins induced by cadmium in soybean cells. J Plant Physiol 163:1203–1206

Song WY, Park J, Mendoza-Cozatl DG, Suter-Grotemeyer M, Shim D, Hortensteiner S, Geisler M, Weder B, Rea PA, Rentsch D, Schroeder JI, Lee Y, Martinoia E (2010) Arsenic tolerance in *Arabidopsis* is mediated by two ABCC-type phytochelatin transporters. Proc Natl Acad Sci U S A 107:21187–21192

Sun XM, Zhang JX, Zhang HJ, Ni YW, Zhang Q, Chen JP, Guan YF (2010) The responses of *Arabidopsis thaliana* to cadmium exposure explored via metabolite profiling. Chemosphere 78:840–845

Szponarski W, Sommerer N, Boyer JC, Rossignol M, Gibrat R (2004) Large-scale characterization of integral proteins from *Arabidopsis* vacuolar membrane by two-dimensional liquid chromatography. Proteomics 4:397–406

Tokimatsu T, Sakurai N, Suzuki H, Ohta H, Nishitani K, Koyama T, Umezawa T, Misawa N, Saito K, Shibata D (2005) KaPPA-view: a web-based analysis tool for integration of transcript and metabolite data on plant metabolic pathway maps. Plant Physiol 138:1289–1300

Tolstikov VV, Lommen A, Nakanishi K, Tanaka N, Fiehn O (2003) Monolithic silica-based capillary reversed-phase liquid chromatography/electrospray mass spectrometry for plant metabolomics. Anal Chem 75:6737–6740

Tommasini R, Vogt E, Fromenteau M, Hortensteiner S, Matile P, Amrhein N, Martinoia E (1998) An ABC-transporter of *Arabidopsis thaliana* has both glutathione-conjugate and chlorophyll catabolite transport activity. Plant J 13:773–780

Usadel B, Obayashi T, Mutwil M, Giorgi FM, Bassel GW, Tanimoto M, Chow A, Steinhauser D, Persson S, Provart NJ (2009) Co-expression tools for plant biology: opportunities for hypothesis generation and caveats. Plant Cell Environ 32:1633–1651

van de Mortel JE, Schat H, Moerland PD, Van Themaat EVL, Van der Ent S, Blankestijn H, Ghandilyan A, Tsiatsiani S, Aarts MGM (2008) Expression differences for genes involved in lignin, glutathione and sulphate metabolism in response to cadmium in *Arabidopsis thaliana* and the related Zn/Cd-hyperaccumulator *Thlaspi caerulescens*. Plant Cell Environ 31:301–324

Verbruggen N, Hermans C, Schat H (2009) Mechanisms to cope with arsenic or cadmium excess in plants. Curr Opin Plant Biol 12:364–372

Vido K, Spector D, Lagniel G, Lopez S, Toledano MB, Labarre J (2001) A proteome analysis of the cadmium response in *Saccharomyces cerevisiae*. J Biol Chem 276:8469–8474

Visioli G, Marmiroli M, Marmiroli N (2010) Two-dimensional liquid chromatography technique coupled with mass spectrometry analysis to compare the proteomic response to cadmium stress in plants. J Biomed Biotechnol. doi:567510.561155/562010/567510

Ward JL, Harris C, Lewis J, Beale MH (2003) Assessment of H-1 NMR spectroscopy and multivariate analysis as a technique for metabolite fingerprinting of *Arabidopsis thaliana*. Phytochemistry 62:949–957

Weber M, Harada E, Vess C, von Roepenack-Lahaye E, Clemens S (2004) Comparative microarray analysis of *Arabidopsis thaliana* and *Arabidopsis halleri* roots identifies nicotianamine synthase, a ZIP transporter and other genes as potential metal hyperaccumulation factors. Plant J 37:269–281

Weber M, Trampczynska A, Clemens S (2006) Comparative transcriptome analysis of toxic metal responses in *Arabidopsis thaliana* and the Cd^{2+}-hypertolerant facultative metallophyte *Arabidopsis halleri*. Plant Cell Environ 29:950–963

Whiteman SA, Nuhse TS, Ashford DA, Sanders D, Maathuis FJM (2008a) A proteomic and phosphoproteomic analysis of Oryza sativa plasma membrane and vacuolar membrane. *Plant J* 56:146–156

Whiteman SA, Serazetdinova L, Jones AME, Sanders D, Rathjen J, Peck SC, Maathuis FJM (2008b) Identification of novel proteins and phosphorylation sites in a tonoplast enriched membrane fraction of Arabidopsis thaliana. *Proteomics* 8:3536–3547

Wolfender JL, Ndjoko K, Hostettmann K (2003) Liquid chromatography with ultraviolet absorbance-mass spectrometric detection and with nuclear magnetic resonance spectroscopy: a powerful combination for the on-line structural investigation of plant metabolites. J Chromatogr A 1000:437–455

Yamaguchi H, Fukuoka H, Arao T, Ohyama A, Nunome T, Miyatake K, Negoro S (2010) Gene expression analysis in cadmium-stressed roots of a low cadmium-accumulating solanaceous plant, *Solanum torvum*. J Exp Bot 61:423–437

Ytterberg AJ, Jensen ON (2010) Modification-specific proteomics in plant biology. J Proteomics 73:2249–2266

Zenk MH (1996) Heavy metal detoxification in higher plants - a review. Gene 179:21–30

Zhao CR, Ikka T, Sawaki Y, Kobayashi Y, Suzuki Y, Hibino T, Sato S, Sakurai N, Shibata D, Koyama H (2009) Comparative transcriptomic characterization of aluminum, sodium chloride, cadmium and copper rhizotoxicities in *Arabidopsis thaliana*. BMC Plant Biol 9:32. doi:10.1186/1471-2229-1189-1132

Zhao CR, Sawaki Y, Sakurai N, Shibata D, Koyama H (2010) Transcriptomic profiling of major carbon and amino acid metabolism in the roots of *Arabidopsis thaliana* treated with various rhizotoxic ions. Soil Sci Plant Nutr 56:150–162

Proteomics as a Toolbox to Study the Metabolic Adjustment of Trees During Exposure to Metal Trace Elements

Kjell Sergeant, Jenny Renaut, and Jean-François Hausman

Abstract Although trees form an important part of the global environment, the use of proteomics in the study of environmental adaptation of trees, and specifically in the study of heavy metal exposure, is currently limited. This lack of proteomics studies is mainly due to the characteristics of trees making them difficult species to work with but can also partially be attributed to the limited genetic information that is available for most tree species. In this chapter, the tools and their potential, but also their limits, that are currently used in the proteome studies of trees will be discussed.

1 Introduction

Although often regarded as a recent, industry-related problem, environmental pollution with heavy metals and its link with human activity predate the industrial age (Cooke et al. 2009). Nonetheless, the increasing pressure contributed by the growing world population makes environmental pollution, which results in the deterioration of available natural resources and/or threatening the natural habitat of organisms, a major concern for current and future generations. Although everyone has their own clear personal idea of what constitutes pollution, it is particularly difficult to define. While a streetlight for most of us is an aid in helping to get home safely, the amateur astronomer will consider it as light pollution. Therefore, thresholds, i.e., values above which the concentration of a specific component becomes a problem in specific conditions, have been established by national or supranational legislation. Some metal trace elements, such as Zn, Fe, Cu, Mn, Mb and Co, are essential for biological systems and must be present in a certain

K. Sergeant • J. Renaut (✉) • J.-F. Hausman
Department Environment and Agro-biotechnology, Centre de Recherche Public – Gabriel Lippmann, Belvaux, Luxembourg
e-mail: sergeant@lippmann.lu; renaut@lippmann.lu; hausman@lippmann.lu

D.K. Gupta and L.M. Sandalio (eds.), *Metal Toxicity in Plants: Perception, Signaling and Remediation*, DOI 10.1007/978-3-642-22081-4_7,

concentration range to ensure the proper functioning of cellular metabolism in living organisms (Garbisu and Alkorta 2003). Despite the fact that these metals are essential parts of the molecules that ensure the proper functioning of cells, at higher concentrations they can act in a deleterious manner. Other heavy metals, most notably Pb, Cd and Hg, have no or very little known positive effects, although some particular organisms living in a very specific environment can use these metals as cofactor (Lane and Morel 2000). The natural geographical distribution of metals in the environment is limited; however, human, industrial activity has made metal contamination an acute problem on a much larger scale (Kidd et al. 2009). The current emphasis on heavy metal pollution stems from their persistence in the environment, but also from the potential bioaccumulation and biomagnification of these molecules in the food chain, exacerbating their impact on higher trophic levels (Chen et al. 2000; Devkota and Schmidt 2000), including humans (Järup 2003). The effects resulting from the excessive accumulation of metals in plants are diverse (Cheng 2002). First of all, the presence of non-essential metals or of an excessive concentration of essential metal trace elements can disturb the chemical homeostasis of the plant cells due to effects on the uptake and transport of other essential elements (Brune and Dietz 1995; Perfus-Barbeoch et al. 2002). The chemical and/or physical resemblance between pollutant metal trace elements and essential cellular components is another way by which heavy metals can interfere with cellular function; for instance, the displacement of Zn, an essential cofactor, in the active site of enzymes by the toxic Cd (Vallee and Auld 1990; Clemens 2006). The generally observed phenotypes of plants exposed to deleterious concentrations of metal trace elements are chlorosis, local necroses, stunting, wilting, growth reduction and, in extreme cases, death (Baryla et al. 2001). The macro- and microscopic effects of heavy metal pollution, and other biotic and abiotic stresses, on trees were recently reviewed (Gunthardt-Goerg and Vollenweider 2007). However, as noted in this review and references therein studies combining metal detection and structural characterization are scarce, particularly in trees.

The different tactics used by plants to ensure the uptake of metals from the soil solution, and their transport to the above-ground parts as well as the cellular responses of the plant, are important in determining the effects of metal pollution (Dalcorso et al. 2010). To do this, it is important to obtain information not only on the specific proteins and molecular adaptations that plants possess to sequester metals but also on the molecular mechanisms of transport to the apoplast of the roots (Verbruggen et al. 2009), these mechanisms and the proteins involved are the focal point of other chapters in this book. Equally important is the description of the allocation of these components to the different plant parts. This is not only determined by the genetic background of the tree but, as illustrated by a number of studies done throughout the years on sycamore trees growing on polluted substrates (Lepp and Eardley 1978; Turner and Dickinson 1993), largely depends on the nature of the trace element studied (Laureysens et al. 2004). A general observation is that accumulation of metals is mainly situated in actively growing tissues such as young leaves and shoots (Pulford and Watson 2003). Different recent studies confirm this for poplar (Kieffer et al. 2008; Durand et al. 2010a) or willow (Robinson et al. 2002; Vandecasteele et al. 2005). Likewise, studies on adult trees growing at polluted sites, contrary to most studies that are done after relative short exposure of plants in hydropony or pot experiments, have confirmed this propensity (Unterbrunner et al. 2007; Migeon et al. 2009). The latter study, the analysis of samples from 25 different species, furthermore indicates that members of the *Salicaceae* family have the highest accumulation potential. What is the most obvious from these studies is that leaves and thin roots accumulate the highest concentrations, with the concentrations in wood being relatively low for most species. However, metal accumulation is a very complex trait; for instance, a marked increase in metal accumulation in leaves of willow has been noted just before senescence (Riddell-Black 1994).

Apart from the search for basic knowledge on the way trees deal with heavy metal pollution, there are two main reasons why the scientific community recently started focusing on the heavy metal tolerance/accumulation potential of woody plants. These are the potential to produce energy in the form of wood (biomass) on soils unsuited for agriculture and the potential use of woody plants for the remediation of metal-polluted soils, or a combination of both (Witters et al. 2009; Capuana 2011). Phytoremediation, the clean-up of contaminated soils using the potential of plants to accumulate and/or metabolize pollutants, is now recognized as an alternative to chemical and physical methods for soil remediation (Cunningham and Ow 1996; Lombi et al. 2001; Dickinson et al. 2008). Different approaches to phytoremediation have been developed with the two extremes (Kidd et al. 2009) known as phytoextraction and phytostabilization: the removal of pollutants through transfer to aerial harvestable plant parts and the immobilization of the pollutants in the soil, respectively (Vangronsveld et al. 1995; Geebelen et al. 2006; Mench et al. 2006; Ruttens et al. 2010). Given the different aims of these strategies, different plants have to be used. For phytoextraction, two groups of plants can in principle be used, on the one hand the so-called hyperaccumulators such as *Thlaspi* sp. and *Arabidopsis halleri* (Cosio et al. 2004) and on the other hand plants that combine a moderate accumulation potential with a high production of biomass that can be used to supplement the energy budgets of the exploitation (Dickinson and Pulford 2005; Lewandowski et al. 2006; French et al. 2006). Ligneous species such as poplar, alder and willow fall into the latter category and (very) short rotation coppices of willow are used in industrial plants for the treatment of sewage sludge (Mirck et al. 2005). Despite the potential of phytoextraction, the realism of the effort remains to be proven. Several studies have indicated that the risk associated with the reintroduction of pollutants in circulation, including throphic cycles, might outweigh the benefits (Scott Angle and Linacre 2005), certainly when the long time needed for full reclamation of even mildly polluted soils is considered (Mertens et al. 2004; Ernst 2005; Van Nevel et al. 2007). In this view, phytostabilization or phytostabilization aided by chemical means (strict control of pH, liming agents, etc.,) might be the preferable approach to contain heavy metal pollution within predefined boundaries (Alkorta et al. 2010). Forests are of vital ecological, recreational, social, and economical importance. Although the most striking economical impact of forests is as a source of wood for construction or combustion, forests also provide industrial base material (fibers, cork, etc.), food and other, often locally important, products (FAO 2009). However, the biggest impact of forests lies not in economical but in ecological value, sustaining the majority of the world's biological diversity and sequestering large quantities of carbon (Dixon et al. 1994). It has been realized that research can help in maintaining our forests and at the same time contribute to an increase in forest production. Nonetheless the use of omics tools in trees is still limited when compared to more accessible and easy-to-handle herbaceous species (Ahsan et al. 2009). The reasons for this lack of molecular studies on trees are mainly related to the problems associated with dealing with large, slow-growing plants. The large genomic size of most trees and their long life cycle put another practical hurdle in the path of molecular studies in trees. The number of publications to make a comprehensive overview of the proteome studies done on forest trees is therefore rather limited, as has recently been shown (Abril et al. 2011).

2 Proteomics of Woody Species

2.1 *Proteomics: General Considerations*

In recent years, proteomics or the characterization of all proteins present in a cell/tissue/organism has become the cornerstone of research destined at understanding the adaptation of organisms to their environment (Chen and Harmon 2006; Cox and Mann 2007). The general workflow in proteomics experiments is largely independent of the organism studied or the biological question addressed, and includes all or most of the following steps: experimental design, sampling, tissue/cell or organelle preparation, protein extraction/fractionation/purification, separation, (relative) quantification with statistical validation, MS analysis, protein identification and biological interpretation. A simplified overview of this is given in Fig. 1. In the last 10 years, excellent reviews and monographs on the fundamentals, concepts, applications, power and limitations of proteomics have appeared (Cox and Mann 2007; Cravatt et al. 2007; Han et al. 2008; Mann 2009), some of them dealing with plants (Canovas et al. 2004; Rossignol et al. 2006; Jorrin-Novo et al. 2009), and forest tree species (Pawlowski 2010). In this general scheme there are three steps that are critical and which pose specific problems in the context of the study of woody species. Perhaps the most critical step is the extraction of proteins from their biological matrix. Given the dominance of 2D-gel-based studies in plant proteomics this step will be discussed more in detail. Finally, the real bottleneck in the study of the proteome of woody species is the lack of comprehensive sequence databases; therefore, a large part of the text will be allocated to a discussion on the tools used for the identification of proteins from poorly characterized species.

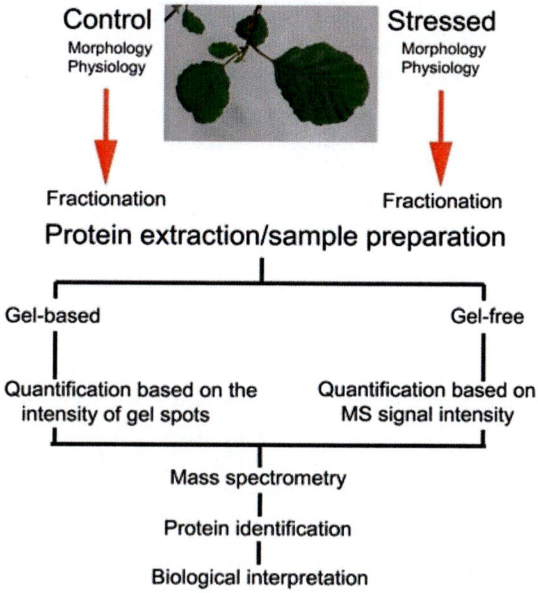

Fig. 1 General overview of a proteomics experiment; samples are harvested from plants exposed to two different conditions, after fractionation (if needed), the proteins from the samples are extracted and used either in a gel-based or gel-free approach. After relative quantification of the abundance of the proteins in the samples, identification is obtained by mass spectrometric analysis

2.1.1 Protein Extraction and Sample Preparation

The most critical step of each proteome experiment is the extraction of proteins from their biological matrix. The chemical and physical complexity of the proteome means that there is no single method currently existing that allows the extraction of all proteins from a sample or that can be used to extract proteins from all species (Wang et al. 2006). For plants, protein extraction is hindered by the extensive membrane structures and the rigidity of the cell walls, and also by the high concentrations of potentially interfering compounds (e.g., nucleic acids, phenols, lipids, carbohydrates, etc.). Furthermore, the protein content in plant tissues is relatively low compared to bacteria or animal tissues. Finally the concentration of interfering compounds can change enormously during a growing season or in response to environmental changes, as we recently observed when working with drought-exposed oak (Sergeant et al. 2011). From the first 1-step extraction protocol prior to 2D-PAGE (O'Farrell 1975) to the present, numerous protocols have been developed and found to be optimal for a certain sample, (for an overview see Table 1 in Carpentier et al. 2008). Currently, nearly all protocols used in plant proteomics include a precipitation step. Through the addition of salts, extreme pH values and/or high concentrations of organic solvents, proteins precipitate while compounds that can interfere with gel electrophoresis remain in solution. The most commonly used method for extraction of plant proteins is the TCA/acetone precipitation method (Damerval et al. 1986). Despite the dominance of this extraction protocol, it also comes with its problems, the most important one being the loss of proteins because of the difficult solubilization of TCA-precipitated proteins (Nandakumar et al. 2003). The second axis along which extraction protocols are developed and optimized is based on a phenol extraction. Using a two-phase phenol extraction scheme, the proteins are enriched in the phenolic phase while interfering carbohydrates and cellular debris can be eliminated with the aqueous phase.

In a few articles wherein the process leading to the optimal extraction method for specific experiments is described, most often the conclusion leads to a more or less equal quotation for TCA/acetone methods and methods based on phenol extraction (Saravanan and Rose 2004; Carpentier et al. 2005; Xu et al. 2008; Amalraj et al. 2010). Using the same criteria such as yield and quality of the gel images, some of these studies finally select a phenol-based protocol while others prefer a TCA-based one, thereby illustrating that, for optimal results the extraction procedure has to be optimized for each sample/project. An additional step can be the inclusion of a fractionation step on the fresh sample or on the protein sample after extraction. This fractionation step can be inserted at different points but always with the objective to decrease the complexity of a sample or to lower the dynamic range in protein abundance in the sample (Stasyk and Huber 2004; Yates et al. 2005). One of the problems when working with plants is the extremely high abundance of some proteins in specific tissues, RuBisCo in green tissues and storage proteins in seeds. Attempts to eliminate RuBisCo from leaf samples have been made (Hashimoto and Komatsu 2007), and commercial kits based on antibodies are now available (Widjaja et al. 2009). The Proteominer, developed by Righetti's team (Boschetti and Righetti 2008; Righetti and Boschetti 2008), is another tool that might be put to use for this aim (Fasoli et al. 2011). A second fractionation that is used is the limitation of the study to specific subcellular compartments. A first important group of methods was developed to focus on the different membrane fractions (Ephritikhine et al. 2004; Komatsu et al. 2007). By the isolation of the membrane fraction, low abundant integral membrane or membrane-associated proteins can be studied. Although some work was done using these techniques on poplar (Song et al. 2011) and banana (Vertommen et al. 2011), the study of membrane proteins is currently largely limited to *Arabidopsis* and some important crop plants. The isolation of cellular compartments (chloroplasts, proplastides, mitochondria, nuclei, Golgi apparatus, etc.), and the subsequent study of the proteome present in these compartments (van Wijk and Baginsky 2011; Taylor et al. 2011), might provide a more in-depth understanding of the molecular events occurring in plants. In general the different subcellular compartments are separated using differential centrifugation in sucrose or Percoll gradients although other approaches have been used (Bohler et al. 2011; Canut et al. 1999; Islinger et al. 2010). The first studies on the chloroplast proteome of poplar have been published (Yuan et al. 2010), and recently a study on the adaptations of the poplar chloroplast proteome during ozone exposure was done (Bohler et al. 2011).

2.1.2 2D-PAGE

The use of other techniques such as (2D)LC-MS/MS (Wolters et al. 2001; Pirondini et al. 2006; Vertommen et al. 2011), liquid phase isoelectric focusing (Vincent and

Solomon 2011), and techniques for the selective isolation of peptides containing specific properties such as COFRADIC (Gevaert et al. 2002) is now widespread in proteomics; however, the workhorse of proteomics studies on trees is still 2D-PAGE. This is illustrated by the fact that of the nearly 60 publications on proteomics in forest trees discussed by Abril et al. only a small minority use an alternative fractionation technique (Abril et al. 2011). The reasons for this near-monopoly situation are multiple, but the most important is without any doubt the lack of comprehensive sequence databases for most of the woody species (Carpentier et al. 2008).

The use of 2D-PAGE and its potential, but also problems associated with it, were the main topic of many recent publications (Rabilloud 2002; Gorg et al. 2004; Friedman et al. 2009; Rabilloud et al. 2010b). Given the limitations of 2D-PAGE it is not surprising that the search for new and improved methods is continuing, and some of these new developments have recently been reviewed (Rabilloud 2010; Miller et al. 2010). The current state of the art in 2D-gel-based proteomics is far from the initial technique used more than 35 years ago (O'Farrell 1975). Perhaps the most significant improvement was the introduction of commercially available immobilized gradient strips for the first dimensional separation (Bjellqvist et al. 1982, 1993). While the introduction of these strips made the approach user-friendlier and the commercial availability increased the reproducibility of the 2D gel images, these two properties remain to date the Achilles heel of 2D-PAGE. Other improvements were the creation of shorter pI gradients, allowing the cutting of the entire pI range into smaller portions, as small as 1–1.5 pI units, thereby increasing the number of spots that can be resolved (Gelfi et al. 1987; Gorg et al. 2000; Poznanovic et al. 2005). Another way to increase the number of spots that can be detected on gels is by increasing the detection limits during the scanning of them. Initially post-separation stains, such as CBB or silver nitrate, were used to visualize protein spots, but the introduction of fluorescent stains such as Sypro Ruby (Bio-Rad) or Lava Purple (Serva) has increased the sensitivity of the approach. The current state of the art however is the use of the DIGE (Differential in-Gel Electrophoresis, GE Healthcare) technology, a pre-separation fluorescent stain that exists in three different colors that can be separated on one gel and visualized separately (Unlu et al. 1997). Using this technology, two samples, control and treated or healthy and diseased, can be separated on the same gel, leaving the third Cy-dye that can be used to visualize a balanced mixture of both samples, the so-called internal standard. It is the possibility to add this internal standard, an identical mixture separated on all gels coming from an experiment, that has facilitated the matching of identical spots on numerous gels, thereby allowing the relative expression of proteins across different gels to be compared (Lilley and Friedman 2004; Bech-Serra et al. 2009). Despite the increased reproducibility and the ease-of-use offered, 2D-gel electrophoresis remains a critical step in each gel-based proteomics study. Good results are still only just within the grasp of skilled researchers and the exact method used, including the combination of extraction method and 2D-gel separation method, must be optimized for each experiment. Only when this is done can the full power of 2D-gel electrophoresis be put to use.

2.1.3 Protein Identification

Protein identification in proteomics is nearly exclusively done using mass spectrometry (Aebersold and Mann 2003), and although other techniques, such as ladder sequence analysis (Chait et al. 1993), are still used, the majority of identification is done using either peptide mass fingerprint analysis (PMF) or fragmentation of peptides (MS/MS). PMF was developed in 1993 and is based on the fact that the mass of a group of peptides derived from a protein by sequence-specific proteolysis can be used to identify the protein (Henzel et al. 1993; Thiede et al. 2005). Despite the introduction of new, more accurate, mass spectrometers, offering the possibility for routine analyses with an accuracy better than 1 ppm (Clauser et al. 1999), the use of PMF as the sole tool for protein identification is currently questioned (Carr et al. 2004). The relatively low specificity of the PMF used to be sufficient, but it is now overwhelmed by the amount of data available in sequence databases. More specific protein identification can be obtained using MS/MS. Selected peptides, precursors, are allowed to collide with neutral gas molecules, resulting in sequence-specific fragments (Paizs and Suhai 2005) with a defined nomenclature (Roepstorff and Fohlman 1984). These fragmentation spectra are then generally used in searches against amino acid and/or nucleotide sequence databases (Perkins et al. 1999; Aebersold and Goodlett 2001; Aebersold and Mann 2003), using dedicated algorithms such as MASCOT and Sequest, to name the two most commonly used. Although many studies have focused on the understanding of the mechanism by which peptides fragment during collisions (Burlet et al. 1992; Wysocki et al. 2000; Polce et al. 2000), influenced by the sequence of the precursor (Tabb et al. 2003, 2004), the presence or absence of peptide fragments in MS/MS spectra remains difficult to predict. Our increased understanding, although far from complete, of the factors that determine the intensity of fragments in MS/MS spectra is now encrypted in the code of newly launched algorithms for database searches (Narasimhan et al. 2005; Li et al. 2011). Despite the dominance of MS/MS as a tool for protein identification in proteomics, it is recognized that only a fraction of the information in gel-based but all the more so in LC-based proteomics experiments, even on fully sequenced species such as yeast and human, is actually used (Peng et al. 2003; Resing et al. 2004; Michalski et al. 2011). Compared to PMF, MS/MS-based identification in principle allows the identification of a protein based on one fragmentation spectrum corresponding to a single peptide sequence; one-hit wonders that if good enough are more and more considered to be valuable (Gupta and Pevzner 2009; Kilner et al. 2011). As for other poorly characterized species the major hurdle in the use of proteomics techniques in woody species stems from the fact that a limited number of genetic and/or protein sequences is available. Even for poplar, the most studied woody species, following the publication of its genome (Tuskan et al. 2006; Jansson and Douglas 2007; Quesada et al. 2008), the annotation is far from finished (Sjodin et al. 2009; Yang et al. 2009; Geraldes et al. 2011). For most other tree species that are commonly studied in relation to forestry, ecology and ecophysiology, the absolute lack of extensive

Table 1 The number of protein and EST sequences in the publicly available NCBI sequence database on the 15th March 2011

Scientific name	Number of EST	Number of proteins
Populus trichocarpa	89,943	87,536
Populus tremula	37,313	3,289
Quercus robur	81,646[a]	61
Quercus petraea	58,173[a]	41
Quercus suber	6,690	97
Fagus sylvatica	31,309[a]	284
Pinus sylvestris	2,355	3,671
Pinus taeda	328,662	2,350
Pinus pinaster	34,061	1,335
Eucalyptus gunnii	19,841	47
Eucalyptus globulus	14,026	402
Eucalyptus grandis	1,939	447
Salix spp (>40 species)	420	1,234
Alnus spp (24 species)	0	179
Betula pendula	2,549	410
Ulmus spp (6 species)	1,277	32
Malus x domestica	324,565	2,312
Citrus sinensis	213,830	671
Arabidopsis thaliana	1,529,700	183,955

[a]More than 95% of these database entries were added in the last 2 years

sequence libraries is striking. Despite huge efforts by different national and international consortia such as Evoltree (http://www.evoltree.eu/) and the Portuguese cork oak EST consortium, adding an enormous number of mainly EST sequences in recent years, most species remain very poorly represented (Table 1). This conclusion is even more apparent when it is considered that for instance pine has a genome 200 times larger than *Arabidopsis* (Wullschleger et al. 2002).

The problems associated with the identification of proteins from species poorly represented in accessible databases should, however, not be overstated. Too often, this is considered as being prohibitively difficult, despite the fact that numerous solutions to overcome a lack of accessible sequences have been proposed. Pivotal in this is a publication by Wilkins and Williams wherein the conservation of different sequence-related parameters, tryptic peptide masses, amino acid composition and molecular weight, of homologous proteins was compared (Wilkins and Williams 1997). For the two most important tools for protein identification, PMF, which requires that a number of peptide masses are conserved between the database entry and the protein that is studied, and MS/MS, in which the conservation of a single stretch of amino acids is sufficient to allow identification, the possibility of cross-species identification is illustrated in Fig. 2. Using MS/MS of the precursors LLDFTEK and MTKDLAL(I/L)HGSK, conserved in the three homologues indicated in blue, is sufficient for the identification of the oak (gi|311078729) and poplar protein (gi|75267781) should only the *Arabidopsis* homologue (gi|19171469) be included in the database. However using PMF only 2 of the 5

```
gi|19171469   NAKLLDFTEKLEAACVGTVESGKMTKDLALIIHGSKLSRDTYLNTEEFIDAVAAELK
gi|311078729  NARLLDFTEKLEAACVGTVESGKMTKDLALLIHGSKVTRSQYLNTEEFIDAVAEEL
gi|75267781   NAKLLDFTEKLEAACVGAVESGKMTKDLALLIHGSKVSRDQYLNTEEFIDAVAEELK
```

Fig. 2 Alignment of part of the sequence of isocitrate dehydrogenase of three plant species. Conserved residues are indicated in *black*, variable residues in *red*, tryptic peptides conserved between two of the three homologous in *green* and tryptic peptides conserved between the three proteins in *blue*

predicted trypsin cleavage products would be matched, limiting the possibility of significant identification. Although 70% sequence identity is generally accepted as the threshold above which cross-species identification is possible (Wilkins and Williams 1997; Lester and Hubbard 2002), Wright et al. present a figure to illustrate that in some cases proteins that have a sequence similarity of 90% can share no conserved tryptic peptide masses (Wright et al. 2010).

Although Fig. 2 shows that a protein can be identified based on a single conserved stretch of amino acids, it is obvious that a precise match is more probable when the databases contain sequences from closely related species. When database searches do not result in a significant match, only de novo sequence analysis, the determination of the sequence of a protein/peptide without using a database as basis, can result in identification. Older chemical techniques for de novo sequence analysis, Edman degradation and chemical C-terminal sequence analysis (Samyn et al. 2000), are now generally replaced by MS-based approaches. Sequences can be determined from MS/MS spectra either manually, based on native or derivatized peptides (Keough et al. 2000; Sergeant et al. 2005; Conrotto and Hellman 2008), or through the use of ever better software developed for this purpose. The techniques that can be used and the different possibilities for the identification of proteins have recently been reviewed (Gevaert and Vandekerckhove 2000; Forne et al. 2010; Seidler et al. 2010), as has cross-species protein identification (Liska and Shevchenko 2003). The use of these approaches are also discussed in some general reviews on plant proteomics (Canovas et al. 2004; Jorrin et al. 2007; Carpentier et al. 2008; Jorrin-Novo et al. 2009). Once de novo sequence data are obtained, proteins are identified based on homology preferably using specifically developed software tools optimized to do homology searches with short, partially incorrect stretches of sequence. An added difficulty when working with MS/MS-based de novo-determined sequences is that the order of the different short sequences in the protein is unknown. MSBlast developed by the group of A. Shevchenko, a pioneer in this field, allows to use short manually or in silico, de novo-determined sequences for the identification of proteins across species boundaries (Shevchenko et al. 2001; Liska and Shevchenko 2003; Waridel et al. 2007). Habermann et al. evaluated the applicability of MSBlast for the identification of proteins from species for which no sequences of closely related species are in the databases, and found that more than 80% of mammalian proteins can be identified by homology to the sequences that were publicly accessible in 2004 (Habermann et al. 2004). Today, different software tools are available online that are likewise optimized to do homology alignments using MS-derived de novo-determined sequences, including MSHomology and FASTS (Mackey et al. 2002; Kayser et al. 2004; Samyn et al. 2007). Because de novo sequence determination is relatively slow compared to database-dependent protein identification, most research groups working with proteins that are not represented in databases use a hybrid approach. Initially all acquired spectra are passed through a fast procedure of searches of the spectra against available protein and EST sequence databases. Those spectra that fail to be matched are subsequently passed via a de novo sequence approach (Wright et al. 2010). Since recent studies confirm that de novo sequence determination can be beneficial even for model organisms (Savitski et al. 2005), the authors also applied this scheme on proteins from fully characterized species to improve the significance of near-to-threshold identifications, for the discovery of PTMs and unknown forms of proteins. In recent studies on trees exposed to different abiotic constraints this has resulted in the discovery/experimental validation of transit peptides, isoforms and PTMs (Bohler et al. 2010; Durand et al. 2011; Sergeant et al. 2011).

2.2 Case Study: Proteome Study of Poplar and Cd Pollution

The number of studies using proteomics to study the responses and adaptations of trees growing on heavy metal-polluted substrates are limited to studies on poplar exposed to Cd (Kieffer et al. 2008, 2009; Durand et al. 2010b; Visioli et al. 2010). One of the things that must be considered when comparing results from proteomics studies is the time of exposure; two of the above-mentioned studies had a comparable exposure time respectively of 56 days for Kieffer et al. 2009 and 61 days for Durand et al. Both studies were done on young poplar cuttings in hydroponic culture or in pots, respectively (Kieffer et al. 2009; Durand et al. 2010b). Currently, no proteome studies on the effect of heavy metal pollution on mature trees growing on polluted industrial soils have been done. However, initial results of such a study on willow and alder growing on an old mining site in Belgium indicate the same tendencies, at least concerning the course of change for the major photosynthesis and carbohydrate related proteins (authors, unpublished results). There were several reasons to choose poplar in these studies: besides the fact that poplar is the model tree for molecular studies, the availability of different clones and the potential of poplar as tool for phytoremediation (recently reviewed in Yadav et al. 2010) are equally important. Although follow-up studies aimed at disentangling the biology behind the general but also the specific changes observed in these studies are needed, it is nonetheless clear from Table 2 that not only very general changes, such as the increase of glutathione S-transferase (GST) in all datasets, are observed but also very specific changes. For instance, the major quantitative changes due to Cd in bark were of a group of spots containing an aspartyl protease family protein. These two studies provide a view on the protein changes in poplar when exposed to a relatively high non-lethal concentration of Cd; three different tissues are included in these two studies with the leaf proteome being studied twice. Although most of

Table 2 Comparison of the differentially accumulated proteins in the different plant tissues studied

Pathway	Durand et al. 2010b				Kieffer et al. 2009			
	Leaf		Cambium		Leaf		Root	
Photosynthesis	Rubisco	–			Rubisco	–		
	Rubisco activase	–			Rubisco activase	–		
	Chlorophyll binding proteins	–						
	Oxygen evolving protein	–			Oxygen evolving protein	–		
	Ferredoxin-NADP oxidoreductase	–			Ferredoxin-NADP oxidoreductase	–		
Carbohydrate metabolism	Triosephosphate isomerase	+			Triosephosphate isomerase	–	Triosephosphate isomerase	–
			Enolase	+	Enolase	+	Enolase	–
	Glyceraldehyde-3-P dehydrogenase	+			Glyceraldehyde-3-P dehydrogenase	+	Glyceraldehyde-3-P dehydrogenase	–
			UDP-glucose dehydrogenase	–			UDP-glucose dehydrogenase	–
	Sedoheptulose bisphosphatase	–						
			α-D-xylosidase/mannosidase	+				
			Alcohol dehydrogenase	+				
					Aconitate hydratase	+		
					Malate dehydrogenase	+	Malate dehydrogenase	–
					Malic enzyme	+		
			2,3-bisphosphate-independent Phosphoglycerate mutase	–	2,3-bisphosphate-independent Phosphoglycerate mutase	+	2,3-bisphosphate-independent Phosphoglycerate mutase	–
Energy metabolism	Carbonic anhydrase	+			Carbonic anhydrase	–	Carbonic anhydrase	–
	ATPase β chain	–			ATPase β chain	+		

	ATPase α chain	+			ATPase α chain	+		
Protein metabolism/ folding	Nascent polypeptide associated complex	–						
			Methionine synthase	–				
	1-cys peroxiredoxin	++						
			S-adenosyl methionine synthase	–				
	Chaperonins	+/–			Chaperonins	+/–	Chaperonins	+
			Proteasome subunits	–	Proteasome subunits	+	Proteasome subunits	–
			Disulfide isomerase	–	Disulfide isomerase	+	Disulfide isomerase	+
					Glutamine synthase	+	Glutamine synthase	–
Oxidoreductase/ glutathione	Glutathione-S-transferase	+	Glutathione-S-transferase theta	+	Glutathione-S-transferase	+/–	Glutathione-S-transferase	+
			Peroxidase	+	Peroxidase	+	Peroxidase	–
			6-Phosphogluconate dehydrogenase	–				
							Monodehydroascorbate reductase	–
Cytoskeleton and cell wall			Tubulin (different forms)	–				
			Actin	+/–				
			Reversible glycosylated protein	–				
			Proline-rich protein	+/–				
			Phenylcoumaran benzylic ether reductase	–				
Lipis metabolism			Acetylcholinesterase	+				
			GDSL-motif lipase/hydrolase family protein	–				
DNA-binding aspartyl peptidase			Aspartyl protease family protein	+/–				
Pathogenesis-related proteins							MLP34	–
					Pathogenesis-related protein	+		
					β-1,3-glucanase	+		

'+' denotes an increase and '–' a decrease in abundance

the changes in leaf proteins are the same in both studies, for instance a decreased abundance of photosynthesis-related proteins (RuBisCo and RuBisCo activase), the two studies do not completely confirm each other. Some proteins change in the opposite direction depending on the study. Examples of this are triosephosphate isomerase, carbonic anhydrase and the ATPase β chain. The cause of these discrepancies is not clear, but may be due to the differences in the clones, the substrate, or the Cd conditions. However, the lesson that should be drawn from this is that comparing results from proteomics experiments is a task that can only be done with trepidation, even when the same species is used.

Furthermore, the presence in 2D-gel images of multiple spots containing a polypeptide with the same attributed function might make it necessary to completely characterize the polypeptide/protein in order to allow an in-depth biological description. As said previously, proteins can be identified based on a number of peptides, as was done in these two studies; however, in order to identify the specific regulation of different isoforms more sequence information might be needed. Similarly, in many plant proteomic studies, if RuBisCo large chain is identified at a lower molecular weight than expected, changes in the abundance of these spots can be misinterpreted as changes in the abundance of the intact, active, protein. In the datasets represented in Table 2, it is, for instance, imaginable that the same GSTs have an increased abundance in the leaves studied in Durand et al. (2010b) and Kieffer et al. (2009). However, in the latter study, some spots containing a GST were unexpectedly also of lower abundance; the biological significance of this can, however, not be determined without full characterization of the proteins.

3 Conclusion

Proteomics has a large potential to produce new biological insights into the adaptation of trees to environmental constraints. However, for the moment it is relatively little used in tree physiological studies, certainly when the adaptation of trees to growth in conditions with an excess or limitation of trace metals is studied. The lack of genetic data on the commonly studied tree species certainly accounts for a part of this, but other factors may be more important. Ideally, adult trees, living in a natural environment, should be studied, as the study of young cuttings or plantlets growing in the strictly controlled conditions of a greenhouse or incubator can only be an approximation. Nonetheless, this approximation that is likewise used in the study of non-woody species is often the only solution available to come to some understanding of the biology of plants. More specific for woody species is that the in-depth study and the biological validation of observations made during proteome studies are difficult. In annual plants, the creation of genetically modified plants, overexpressing or knock-out for a protein found to be involved in a physiological response, can provide these answers. The long life span of trees and the problems associated with the transformation of trees preclude the use of this approach. Even in poplar, the tree species used in this type of experiment, such studies encompassing the entire normal life span have not been carried. Given the few proteomics studies ever performed on trees and trace metals, we have tried to give a short overview of the potential and the different tools of current proteomics studies in trees. This overview is not comprehensive but it can be used as a starting point for newcomers to the field.

Acknowledgements Cofinancé par le Fonds européen de développement régional dans la cadre du programme Interreg IV A Grande Région and supported by the National Research Fund, Luxembourg; project ENERREM (FNR CORE 2008).

References

Abril N, Gion JM, Kerner R, Muller-Starck G, Cerrillo RM, Plomion C, Renaut J, Valledor L, Jorrin-Novo JV (2011) Proteomics research on forest trees, the most recalcitrant and orphan plant species. Phytochemistry. doi:10.1016/j.phytochem.2011.01.005

Aebersold R, Goodlett DR (2001) Mass spectrometry in proteomics. Chem Rev 101:269–295

Aebersold R, Mann M (2003) Mass spectrometry-based proteomics. Nature 422:198–207

Ahsan N, Renaut J, Komatsu S (2009) Recent developments in the application of proteomics to the analysis of plant responses to heavy metals. Proteomics 9:2602–2621

Alkorta I, Becerril JM, Garbisu C (2010) Phytostabilization of metal contaminated soils. Rev Environ Health 25:135–146

Amalraj RS, Selvaraj N, Veluswamy GK, Ramanujan RP, Muthurajan R, Palaniyandi M, Agrawal GK, Rakwal R, Viswanathan R (2010) Sugarcane proteomics: establishment of a protein extraction method for 2-DE in stalk tissues and initiation of sugarcane proteome reference map. Electrophoresis 31:1959–1974

Baryla A, Carrier P, Franck F, Coulomb C, Sahut C, Havaux M (2001) Leaf chlorosis in oilseed rape plants (*Brassica napus*) grown on cadmium-polluted soil: causes and consequences for photosynthesis and growth. Planta 212:696–709

Bech-Serra JJ, Borthwick A, Colome N, Albar JP, Wells M, del Sanchez PM, Canals F (2009) A multi-laboratory study assessing reproducibility of a 2D-DIGE differential proteomic experiment. J Biomol Tech 20:293–296

Bjellqvist B, Ek K, Righetti PG, Gianazza E, Gorg A, Westermeier R, Postel W (1982) Isoelectric focusing in immobilized pH gradients: principle, methodology and some applications. J Biochem Bioph Meth 6:317–339

Bjellqvist B, Pasquali C, Ravier F, Sanchez JC, Hochstrasser D (1993) A nonlinear wide-range immobilized pH gradient for two-dimensional electrophoresis and its definition in a relevant pH scale. Electrophoresis 14:1357–1365

Bohler S, Sergeant K, Lefevre I, Jolivet Y, Hoffmann L, Renaut J, Dizengremel P, Hausman JF (2010) Differential impact of chronic ozone exposure on expanding and fully expanded poplar leaves. Tree Physiol 30:1415–1432

Bohler S, Sergeant K, Hoffmann L, Dizengremel P, Hausman JF, Renaut J, Jolivet Y (2011) A difference gel electrophoresis study on thylakoids isolated from poplar leaves reveals a negative impact of ozone exposure on membrane proteins. J Proteome Res 10:3003–3011

Boschetti E, Righetti PG (2008) The ProteoMiner in the proteomic arena: a non-depleting tool for discovering low-abundance species. J Proteomics 71:255–264

Brune A, Dietz KJ (1995) A comparative analysis of element composition of roots and leaves of barley seedlings grown in the presence of toxic cadmium, molybdenum, nickel, and zinc concentrations. J Plant Nutr 18:853–868

Burlet O, Yang C, Gaskell SJ (1992) Influence of cysteine to cysteic acid oxidation on the collision-activated decomposition of protonated peptides: evidence for intraionic interactions. J Am Soc Mass Spectrom 3:337–344
Canovas FM, Dumas-Gaudot E, Recorbet G, Jorrin J, Mock HP, Rossignol M (2004) Plant proteome analysis. Proteomics 4:285–298
Canut H, Bauer J, Weber G (1999) Separation of plant membranes by electromigration techniques. J Chromatogr B 722:121–139
Capuana M (2011) Heavy metals and woody plants-biotechnologies for phytoremediation. iForest 4:7–15
Carpentier SC, Witters E, Laukens K, Deckers P, Swennen R, Panis B (2005) Preparation of protein extracts from recalcitrant plant tissues: an evaluation of different methods for two-dimensional gel electrophoresis analysis. Proteomics 5:2497–2507
Carpentier SC, Panis B, Vertommen A, Swennen R, Sergeant K, Renaut J, Laukens K, Witters E, Samyn B, Devreese B (2008) Proteome analysis of non-model plants: a challenging but powerful approach. Mass Spectrom Rev 27:354–377
Carr S, Aebersold R, Baldwin M, Burlingame A, Clauser K, Nesvizhskii A (2004) The need for guidelines in publication of peptide and protein identification data. Mol Cell Proteomics 3:531–533
Chait BT, Wang R, Beavis RC, Kent SB (1993) Protein ladder sequencing. Science 262:89–92
Chen S, Harmon AC (2006) Advances in plant proteomics. Proteomics 6:5504–5516
Chen CY, Stemberger RS, Klaue B, Blum JD, Pickhardt PC, Flot CL (2000) Accumulation of heavy metals in food web components across a gradient of lakes. Limnol Oceanogr 45:1525–1536
Cheng S (2002) Effects of heavy metals on plants and resistance mechanisms. Environ Sci Pollut Res Int 10:256–264
Clauser KR, Baker P, Burlingame AL (1999) Role of accurate mass measurement (+/− 10 ppm) in protein identification strategies employing MS or MS/MS and database searching. Anal Chem 71:2871–2882
Clemens S (2006) Toxic metal accumulation, responses to exposure and mechanisms of tolerance in plants. Biochimie 88:1707–1719
Conrotto P, Hellman U (2008) Lys Tag: an easy and robust chemical modification for improved de novo sequencing with a matrix-assisted laser desorption/ionization tandem time-of-flight mass spectrometer. Rapid Commun Mass Spectrom 22:1823–1833
Cooke CA, Balcom PH, Biester H, Wolfe AP (2009) Over three millennia of mercury pollution in the Peruvian Andes. Proc Natl Acad Sci U S A 106:8830–8834
Cosio C, Martinoia E, Keller C (2004) Hyperaccumulation of cadmium and zinc in *Thlaspi caerulescens* and *Arabidopsis halleri* at the leaf cellular level. Plant Physiol 134:716–725
Cox J, Mann M (2007) Is proteomics the new genomics? Cell 130:395–398
Cravatt BF, Simon GM, Yates JR III (2007) The biological impact of mass-spectrometry-based proteomics. Nature 450:991–1000
Cunningham SD, Ow DW (1996) Promises and prospects of phytoremediation. Plant Physiol 110:715–719
Dalcorso G, Farinati S, Furini A (2010) Regulatory networks of cadmium stress in plants. Plant Signal Behav 5:663–667
Damerval C, De Vienne D, Zivy M, Thiellement H (1986) Technical improvements in two-dimensional electrophoresis increase the level of genetic variation detected in wheat-seedling proteins. Electrophoresis 7:52–54
Devkota B, Schmidt GH (2000) Accumulation of heavy metals in food plants and grasshoppers from the Taigetos mountains, Greece. Agric Ecosyst Environ 78:85–91
Dickinson NM, Pulford ID (2005) Cadmium phytoextraction using short-rotation coppice Salix: the evidence trail. Environ Int 31:609–613
Dickinson NM, Baker AJM, Doronila A, Laidlaw S, Reeves RD (2008) Phytoremediation of inorganics: realism and synergies. Int J Phytoremediation 11:97–114
Dixon RK, Solomon AM, Brown S, Houghton RA, Trexier MC, Wisniewski J (1994) Carbon pools and flux of global forest ecosystems. Science 263:185–190

Durand TC, Hausman JF, Carpin S, Alberic P, Baillif P, Label P, Morabito D (2010a) Zinc and cadmium effects on growth and ion distribution in *Populus tremula x Populus alba*. Biol Planta 54:191–194

Durand TC, Sergeant K, Planchon S, Carpin S, Label P, Morabito D, Hausman JF, Renaut J (2010b) Acute metal stress in *Populus tremula* x *P. alba* (717-1B4 genotype): leaf and cambial proteome changes induced by cadmium(2+). Proteomics 10:349–368

Durand TC, Sergeant K, Renaut J, Carpin S, Label P, Morabito D, Hausman JF (2011) Poplar under drought: comparison of leaf and cambial proteomic responses. J Proteomics. doi:10.1016/j.jprot.2011.03.013

Ephritikhine G, Ferro M, Rolland N (2004) Plant membrane proteomics. Plant Physiol Bioch 42: 943–962

Ernst WHO (2005) Phytoextraction of mine wastes-Options and impossibilities. Chemie Erde Geochem 65:29–42

FAO (2009) State of the world's forests 2009. Electronic Publishing Policy and Support Branch, Communication Division, FAO, Rome

Fasoli E, D'Amato A, Kravchuk AV, Boschetti E, Bachi A, Righetti PG (2011) Popeye strikes again: the deep proteome of spinach leaves. J Proteomics 74:127–136

Forne I, Abian J, Cerda J (2010) Fish proteome analysis: model organisms and non-sequenced species. Proteomics 10:858–872

French CJ, Dickinson NM, Putwain PD (2006) Woody biomass phytoremediation of contaminated brownfield land. Environ Pollut 141:387–395

Friedman DB, Hoving S, Westermeier R (2009) Isoelectric focusing and two-dimensional gel electrophoresis. Methods Enzymol 463:515–540

Garbisu C, Alkorta I (2003) Basic concepts on heavy metal soil bioremediation. Eur J Min Process Environ Protect 3:58–66

Geebelen W, Sappin-Didier V, Ruttens A, Carleer R, Yperman J, Bongue-Boma K, Mench M, van der Lelie N, Vangronsveld J (2006) Evaluation of cyclonic ash, commercial Na-silicates, lime and phosphoric acid for metal immobilisation purposes in contaminated soils in Flanders (Belgium). Environ Pollut 144:32–39

Gelfi C, Bossi ML, Bjellqvist B, Righetti PG (1987) Isoelectric focusing in immobilized pH gradients in the pH 10–11 range. J Biochem Biophys Methods 15:41–48

Geraldes A, Pang J, Thiessen N, Cezard T, Moore R, Zhao Y, Tam A, Wang S, Friedmann M, Birol I, Jones SJ, Cronk QC, Douglas CJ (2011) SNP discovery in black cottonwood (*Populus trichocarpa*) by population transcriptome resequencing. Mol Ecol Resour 11(Suppl 1):81–92

Gevaert K, Vandekerckhove J (2000) Protein identification methods in proteomics. Electrophoresis 21:1145–1154

Gevaert K, Van Damme J, Goethals M, Thomas GR, Hoorelbeke B, Demol H, Martens L, Puype M, Staes A, Vandekerckhove J (2002) Chromatographic isolation of methionine-containing peptides for gel-free proteome analysis: Identification of more than 800 *Escherichia coli* proteins. Mol Cell Proteomics 1:896–903

Gorg A, Obermaier C, Boguth G, Harder A, Scheibe B, Wildgruber R, Weiss W (2000) The current state of two-dimensional electrophoresis with immobilized pH gradients. Electrophoresis 21:1037–1053

Gorg A, Weiss W, Dunn MJ (2004) Current two-dimensional electrophoresis technology for proteomics. Proteomics 4:3665–3685

Gunthardt-Goerg MS, Vollenweider P (2007) Linking stress with macroscopic and microscopic leaf response in trees: new diagnostic perspectives. Environ Pollut 147:467–488

Gupta N, Pevzner PA (2009) False discovery rates of protein identifications: a strike against the two-peptide rule. J Proteome Res 8:4173–4181

Habermann B, Oegema J, Sunyaev S, Shevchenko A (2004) The power and the limitations of cross-species protein identification by mass spectrometry-driven sequence similarity searches. Mol Cell Proteomics 3:238–249

Han X, Aslanian A, Yates JR III (2008) Mass spectrometry for proteomics. Curr Opin Chem Biol 12: 483–490
Hashimoto M, Komatsu S (2007) Proteomic analysis of rice seedlings during cold stress. Proteomics 7:1293–1302
Henzel WJ, Billeci TM, Stults JT, Wong SC, Grimley C, Watanabe C (1993) Identifying proteins from two-dimensional gels by molecular mass searching of peptide fragments in protein sequence databases. Proc Natl Acad Sci U S A 90:5011–5015
Islinger M, Eckerskorn C, Volkl A (2010) Free-flow electrophoresis in the proteomic era: a technique in flux. Electrophoresis 31:1754–1763
Jansson S, Douglas CJ (2007) Populus: a model system for plant biology. Annu Rev Plant Biol 58: 435–458
Järup L (2003) Hazards of heavy metal pollution. Braz Med Bull 68:167–182
Jorrin JV, Maldonado AM, Castillejo MA (2007) Plant proteome analysis: a 2006 update. Proteomics 7:2947–2962
Jorrin-Novo JV, Maldonado AM, Echevarria-Zomeno S, Valledor L, Castillejo MA, Curto M, Valero J, Sghaier B, Donoso G, Redondo I (2009) Plant proteomics update (2007–2008): second-generation proteomic techniques, an appropriate experimental design, and data analysis to fulfill MIAPE standards, increase plant proteome coverage and expand biological knowledge. J Proteomics 72:285–314
Kayser JP, Vallet JL, Cerny RL (2004) Defining parameters for homology-tolerant database searching. J Biomol Tech 15:285–295
Keough T, Lacey MP, Youngquist RS (2000) Derivatization procedures to facilitate de novo sequencing of lysine-terminated tryptic peptides using postsource decay matrix-assisted laser desorption/ionization mass spectrometry. Rapid Commun Mass Spectrom 14:2348–2356
Kidd P, Barcelo J, Pilar Bernal M, Navari-Izzo F, Poschenrieder C, Shilev S, Clemente R, Monterroso C (2009) Trace element behaviour at the root-soil interface: implications in phytoremediation. Environ Exp Bot 67:243–259
Kieffer P, Dommes J, Hoffmann L, Hausman JF, Renaut J (2008) Quantitative changes in protein expression of cadmium-exposed poplar plants. Proteomics 8:2514–2530
Kieffer P, Schröder P, Dommes J, Hoffmann L, Renaut J, Hausman JF (2009) Proteomic and enzymatic responses of poplar to cadmium stress. J Proteomics 72:379–396
Kilner J, Zhu L, OW SY, Evans C, Corfe BM (2011) Assessing the loss of information through application of the "two-hit rule" in iTRAQ datasets. J Integr Omics 1:124–134
Komatsu S, Konishi H, Hashimoto M (2007) The proteomics of plant cell membranes. J Exp Bot 58:103–112
Lane TW, Morel FM (2000) A biological function for cadmium in marine diatoms. Proc Natl Acad Sci U S A 97:4627–4631
Laureysens I, Blust R, De Temmerman L, Lemmens C, Ceulemans R (2004) Clonal variation in heavy metal accumulation and biomass production in a poplar coppice culture: I. Seasonal variation in leaf, wood and bark concentrations. Environ Pollut 131:485–494
Lepp NW, Eardley GT (1978) Growth and trace metal content of European sycamore seedlings grown in soil amended with sewage sludge. J Environ Qual 7:413–416
Lester PJ, Hubbard SJ (2002) Comparative bioinformatic analysis of complete proteomes and protein parameters for cross-species identification in proteomics. Proteomics 2:1392–1405
Lewandowski I, Schmidt U, Londo M, Faaij A (2006) The economic value of the phytoremediation function- assessed by the example of cadmium remediation by willow (*Salix* ssp). Agric Syst 89:68–89
Li W, Ji L, Goya J, Tan G, Wysocki VH (2011) SQID: an intensity-incorporated protein identification algorithm for tandem mass spectrometry. J Proteome Res 10:1593–1602
Lilley KS, Friedman DB (2004) All about DIGE: quantification technology for differential-display 2D-gel proteomics. Expert Rev Proteomic 1:401–409
Liska AJ, Shevchenko A (2003) Expanding the organismal scope of proteomics: Cross-species protein identification by mass spectrometry and its implications. Proteomics 3:19–28

Lombi E, Zhao FJ, Dunham SJ, McGrath SP (2001) Phytoremediation of heavy metal-contaminated soils: natural hyperaccumulation versus chemically enhanced phytoextraction. J Environ Qual 30:1919–1926

Mackey AJ, Haystead TA, Pearson WR (2002) Getting more from less: algorithms for rapid protein identification with multiple short peptide sequences. Mol Cell Proteomics 1:139–147

Mann M (2009) Comparative analysis to guide quality improvements in proteomics. Nat Methods 6:717–719

Mench M, Renella G, Gelsomino A, Landi L, Nannipieri P (2006) Biochemical parameters and bacterial species richness in soils contaminated by sludge-borne metals and remediated with inorganic soil amendments. Environ Pollut 144:24–31

Mertens J, Vervaeke P, De Schrijver A, Luyssaert S (2004) Metal uptake by young trees from dredged brackish sediment: limitations and possibilities for phytoextraction and phytostabilisation. Sci Total Environ 326:209–215

Michalski A, Cox J, Mann M (2011) More than 100,000 detectable peptide species elute in single shotgun proteomics runs but the majority is inaccessible to data-dependent LC-MS/MS. J Proteome Res 10:1785–1793

Migeon A, Richaud P, Guinet F, Chalot M, Blaudez D (2009) Metal accumulation by woody species on contaminated sites in the North of France. Water Air Soil Poll 204:89–101

Miller I, Eberini I, Gianazza E (2010) Other than IPG-DALT: 2-DE variants. Proteomics 10: 586–610

Mirck J, Isebrands JG, Verwijst T, Ledin S (2005) Development of short-rotation willow coppice systems for environmental purposes in Sweden. Biomass Bioenerg 28:219–228

Nandakumar MP, Shen J, Raman B, Marten MR (2003) Solubilization of trichloroacetic acid (TCA) precipitated microbial proteins via NaOH for two-dimensional electrophoresis. J Proteome Res 2:89–93

Narasimhan C, Tabb DL, Verberkmoes NC, Thompson MR, Hettich RL, Uberbacher EC (2005) MASPIC: intensity-based tandem mass spectrometry scoring scheme that improves peptide identification at high confidence. Anal Chem 77:7581–7593

O'Farrell PH (1975) High resolution two-dimensional electrophoresis of proteins. J Biol Chem 250:4007–4021

Paizs B, Suhai S (2005) Fragmentation pathways of protonated peptides. Mass Spectrom Rev 24: 508–548

Pawlowski TA (2010) Proteomic approach to analyze dormancy breaking of tree seeds. Plant Mol Biol 73:15–25

Peng J, Elias JE, Thoreen CC, Licklider LJ, Gygi SP (2003) Evaluation of multidimensional chromatography coupled with tandem mass spectrometry (LC/LC-MS/MS) for large-scale protein analysis: the yeast proteome. J Proteome Res 2:43–50

Perfus-Barbeoch L, Leonhardt N, Vavasseur A, Forestier C (2002) Heavy metal toxicity: cadmium permeates through calcium channels and disturbs the plant water status. Plant J 32:539–548

Perkins DN, Pappin DJ, Creasy DM, Cottrell JS (1999) Probability-based protein identification by searching sequence databases using mass spectrometry data. Electrophoresis 20:3551–3567

Pirondini A, Visioli G, Malcevschi A, Marmiroli N (2006) A 2-D liquid-phase chromatography for proteomic analysis in plant tissues. J Chromatogr B 833:91–100

Polce MJ, Ren D, Wesdemiotis C (2000) Dissociation of the peptide bond in protonated peptides. J Mass Spectrom 35:1391–1398

Poznanovic S, Schwall G, Zengerling H, Cahill MA (2005) Isoelectric focusing in serial immobilized pH gradient gels to improve protein separation in proteomic analysis. Electrophoresis 26:3185–3190

Pulford ID, Watson C (2003) Phytoremediation of heavy metal-contaminated land by trees–a review. Environ Int 29:529–540

Quesada T, Li Z, Dervinis C, Li Y, Bocock PN, Tuskan GA, Casella G, Davis JM, Kirst M (2008) Comparative analysis of the transcriptomes of *Populus trichocarpa* and *Arabidopsis thaliana*

suggests extensive evolution of gene expression regulation in angiosperms. New Phytol 180:408–420
Rabilloud T (2002) Two-dimensional gel electrophoresis in proteomics: old, old fashioned, but it still climbs up the mountains. Proteomics 2:3–10
Rabilloud T (2010) Variations on a theme: changes to electrophoretic separations that can make a difference. J Proteomics 73:1562–1572
Rabilloud T, Chevallet M, Luche S, Lelong C (2010) Two-dimensional gel electrophoresis in proteomics: past, present and future. J Proteomics 73:2064–2077
Resing KA, Meyer-Arendt K, Mendoza AM, Aveline-Wolf LD, Jonscher KR, Pierce KG, Old WM, Cheung HT, Russell S, Wattawa JL, Goehle GR, Knight RD, Ahn NG (2004) Improving reproducibility and sensitivity in identifying human proteins by shotgun proteomics. Anal Chem 76:3556–3568
Riddell-Black D (1994) Heavy metal uptake by fast growing willow species. In: Aronsson P, Perttu K (eds) Willow vegetation filters for municipal wastewaters and sludges: a biological purification system. Swedish University of Agricultural Sciences, Uppsala, pp 145–151
Righetti PG, Boschetti E (2008) The ProteoMiner and the FortyNiners: searching for gold nuggets in the proteomic arena. Mass Spectrom Rev 27:596–608
Robinson B, Mills T, Clothier B, Green S, Fung L (2002) Cadmium accumulation by willow clones used for soil conservation, stock fodder, and phytoremediation. Aust J Soil Res 40: 1331–1337
Roepstorff P, Fohlman J (1984) Proposal for a common nomenclature for sequence ions in mass spectra of peptides. Biomed Mass Spectrom 11:601
Rossignol M, Peltier JB, Mock HP, Matros A, Maldonado AM, Jorrin JV (2006) Plant proteome analysis: a 2004–2006 update. Proteomics 6:5529–5548
Ruttens A, Adriaensen K, Meers E, De Vocht A, Geebelen W, Carleer R, Mench M, Vangronsveld J (2010) Long-term sustainability of metal immobilization by soil amendments: cyclonic ashes versus lime addition. Environ Pollut 158:1428–1434
Samyn B, Hardeman K, Van der EJ, Van BJ (2000) Applicability of the alkylation chemistry for chemical C-terminal protein sequence analysis. Anal Chem 72:1389–1399
Samyn B, Sergeant K, Carpentier S, Debyser G, Panis B, Swennen R, Beeumen JV (2007) Functional proteome analysis of the banana plant (*Musa* spp.) using *de novo* sequence analysis of derivatized peptides. J Proteome Res 6:70–80
Saravanan RS, Rose JK (2004) A critical evaluation of sample extraction techniques for enhanced proteomic analysis of recalcitrant plant tissues. Proteomics 4:2522–2532
Savitski MM, Nielsen ML, Kjeldsen F, Zubarev RA (2005) Proteomics-grade de novo sequencing approach. J Proteome Res 4:2348–2354
Scott Angle J, Linacre NA (2005) Metal phytoextraction-a survey of potential risks. Int J Phytoremediation 7:241–254
Seidler J, Zinn N, Boehm ME, Lehmann WD (2010) De novo sequencing of peptides by MS/MS. Proteomics 10:634–649
Sergeant K, Samyn B, Debyser G, Van Beeumen J (2005) De novo sequence analysis of N-terminal sulfonated peptides after in-gel guanidination. Proteomics 5:2369–2380
Sergeant K, Spiess N, Renaut J, Wilhelm E, Hausman JF (2011) One dry summer: a leaf proteome study on the response of oak to drought exposure. J Proteomics. doi:10.1016/j.jprot.2011.03.011
Shevchenko A, Sunyaev S, Loboda A, Shevchenko A, Bork P, Ens W, Standing KG (2001) Charting the proteomes of organisms with unsequenced genomes by MALDI-quadrupole time-of-flight mass spectrometry and BLAST homology searching. Anal Chem 73:1917–1926
Sjodin A, Street NR, Sandberg G, Gustafsson P, Jansson S (2009) The Populus genome integrative explorer (PopGenIE): a new resource for exploring the Populus genome. New Phytol 182: 1013–1025

Song D, Xi W, Shen J, Bi T, Li L (2011) Characterization of the plasma membrane proteins and receptor-like kinases associated with secondary vascular differentiation in poplar. Plant Mol Biol. doi:10.1007/s11103-011-9771-3

Stasyk T, Huber LA (2004) Zooming in: fractionation strategies in proteomics. Proteomics 4: 3704–3716

Tabb DL, Smith LL, Breci LA, Wysocki VH, Lin D, Yates JR III (2003) Statistical characterization of ion trap tandem mass spectra from doubly charged tryptic peptides. Anal Chem 75: 1155–1163

Tabb DL, Huang Y, Wysocki VH, Yates JR III (2004) Influence of basic residue content on fragment ion peak intensities in low-energy collision-induced dissociation spectra of peptides. Anal Chem 76:1243–1248

Taylor NL, Heazlewood JL, Millar AH (2011) The *Arabidopsis thaliana* 2-D gel mitochondrial proteome: Refining the value of reference maps for assessing protein abundance, contaminants and post-translational modifications. Proteomics. doi:10.1002/pmic.201000620

Thiede B, Hohenwarter W, Krah A, Mattow J, Schmid M, Schmidt F, Jungblut PR (2005) Peptide mass fingerprinting. Methods 35:237–247

Turner AP, Dickinson NM (1993) Survival of *Acer pseudoplatanus* L. (sycamore) seedlings on metalliferous soils. New Phytol 123:509–521

Tuskan GA, Difazio S, Jansson S, Bohlmann J, Grigoriev I, Hellsten U et al (2006) The genome of black cottonwood, *Populus trichocarpa* (Torr. & Gray). Science 313:1596–1604

Unlu M, Morgan ME, Minden JS (1997) Difference gel electrophoresis: a single gel method for detecting changes in protein extracts. Electrophoresis 18:2071–2077

Unterbrunner R, Puschenreiter M, Sommer P, Wieshammer G, Tlustos P, Zupan M, Wenzel WW (2007) Heavy metal accumulation in trees growing on contaminated sites in Central Europe. Environ Pollut 148:107–114

Vallee BL, Auld DS (1990) Zinc coordination, function, and structure of zinc enzymes and other proteins. Biochemistry 29:5647–5659

Van Nevel L, Mertens J, Oorts K, Verheyen K (2007) Phytoextraction of metals from soils: how far from practice? Environ Pollut 150:34–40

van Wijk KJ, Baginsky S (2011) Plastid proteomics in higher plants: current state and future goals. Plant Physiol 155:1578–1588

Vandecasteele B, Meers E, Vervaeke P, De Vos B, Quataert P, Tack FM (2005) Growth and trace metal accumulation of two *Salix* clones on sediment-derived soils with increasing contamination levels. Chemosphere 58:995–1002

Vangronsveld J, Van Assche F, Clijsters H (1995) Reclamation of a bare industrial area contaminated by non-ferrous metals: in situ metal immobilization and revegetation. Environ Pollut 87:51–59

Verbruggen N, Hermans C, Schat H (2009) Mechanisms to cope with arsenic or cadmium excess in plants. Curr Opin Plant Biol 12:364–372

Vertommen A, Moller AL, Cordewener JH, Swennen R, Panis B, Finnie C, America AH, Carpentier SC (2011) A workflow for peptide-based proteomics in a poorly sequenced plant: a case study on the plasma membrane proteome of banana. J Proteomics. doi:10.1016/j.jprot.2011.02.008

Vincent D, Solomon PS (2011) Development of an in-house protocol for the OFFGEL fractionation of plant proteins. J Integr Omics. doi:10.5584/jiomics.v2011i2011.59

Visioli G, Marmiroli M, Marmiroli N (2010) Two-dimensional liquid chromatography technique coupled with mass spectrometry analysis to compare the proteomic response to cadmium stress in plants. J Biomed Biotechnol 2010. doi:10.155/2010/567510

Wang W, Vignani R, Scali M, Cresti M (2006) A universal and rapid protocol for protein extraction from recalcitrant plant tissues for proteomic analysis. Electrophoresis 27:2782–2786

Waridel P, Frank A, Thomas H, Surendranath V, Sunyaev S, Pevzner P, Shevchenko A (2007) Sequence similarity-driven proteomics in organisms with unknown genomes by LC-MS/MS and automated de novo sequencing. Proteomics 7:2318–2329

Widjaja I, Naumann K, Roth U, Wolf N, Mackey D, Dangl JL, Scheel D, Lee J (2009) Combining subproteome enrichment and Rubisco depletion enables identification of low abundance proteins differentially regulated during plant defense. Proteomics 9:138–147

Wilkins MR, Williams KL (1997) Cross-species protein identification using amino acid composition, peptide mass fingerprinting, isoelectric point and molecular mass: a theoretical evaluation. J Theor Biol 186:7–15

Witters N, Van Slycken S, Ruttens A, Adriaensen K, Meers E, Meiresonne L, Tack FMG, Laes E, Vangronsveld J (2009) Short-rotation coppice of willow for phytoremediation of a metal-contaminated agricultural area: a sustainability assessment. Bioenerg Res 2:144–152

Wolters DA, Washburn MP, Yates JR III (2001) An automated multidimensional protein identification technology for shotgun proteomics. Anal Chem 73:5683–5690

Wright JC, Beynon RJ, Hubbard SJ (2010) Cross species proteomics. In: Hubbard SJ, Jones AR (eds) Proteome bioinformatics. Humana, London, pp 123–135

Wullschleger S, Jansson S, Taylor G (2002) Genomics and forest biology: *Populus* emerges as the perennial favourite. Plant Cell 14:2651–2655

Wysocki VH, Tsaprailis G, Smith LL, Breci LA (2000) Mobile and localized protons: a framework for understanding peptide dissociation. J Mass Spectrom 35:1399–1406

Xu C, Xu Y, Huang B (2008) Protein extraction for two-dimensional gel electrophoresis of proteomic profiling in Turfgrass. Crop Sci 48:1608–1614

Yadav R, Arora P, Kumar S, Chaudhury A (2010) Perspectives for genetic engineering of poplars for enhanced phytoremediation abilities. Ecotoxicology 19:1574–1588

Yang X, Kalluri U, Difazio S, Wullschleger S, Tschaplinski TJ, Cheng M, Tuskan GA (2009) Poplar genomics: State of the science. Crit Rev Plant Sci 28:285–308

Yates JR III, Gilchrist A, Howell KE, Bergeron JJ (2005) Proteomics of organelles and large cellular structures. Nat Rev Mol Cell Bio 6:702–714

Yuan HM, Li KL, Ni RJ, Guo WD, Shen Z, Yang CP, Wang BC, Liu GF, Guo CH, Jiang J (2010) A systemic proteomic analysis of *Populus* chloroplast by using shotgun method. Mol Biol Rep. doi:10.1007/s11033-010-9971-y

Proteomics of Plant Hyperaccumulators

Giovanna Visioli and Nelson Marmiroli

Abstract Plant metal hyperaccumulators take up and detoxify high concentrations of metal ions in their roots and shoots. They constitute an exceptional biological material for understanding mechanisms regulating plant metal homeostasis and plant adaptation to extreme environments. Hyperaccumulation physiology has recently also been studied with molecular tools. Indeed making use of transcriptome analysis it has been demonstrated that different expression patterns of genes accompanied different responses to metals between hyperaccumulator and non-hyperaccumulator plant species. The proteomic approach can also be powerful in dissecting the hyperaccumulator phenotype and the complex involvement of the protein regulation in this phenomenon. This chapter focuses on the recent developments in the application of proteomics to the analysis of hyperaccumulators providing a comprehensive review of key literature data of plant–metal – in particular Cd, Ni and Zn – hyperaccumulation.

Abbreviations

2D-LC	Two-dimensional liquid chromatography
2D-PAGE	Two-dimensional polyacrylamide gel electrophoresis
APX	Ascorbate peroxidase
ESI	Electrospray ionization tandem mass spectrometry
GAPDH	Glyceraldehyde-3-phosphate dehydrogenase
GST	Glutathione S transferase
HS	Heat-shock
IEF	Isoelectric focusing
iTRAQs	Isobaric Tag for Relative and Absolute Quantification

G. Visioli • N. Marmiroli (✉)
Department of Environmental Sciences, University of Parma, Parma, Italy
e-mail: giovanna.visioli@unipr.it; nelson.marmiroli@unipr.it

D.K. Gupta and L.M. Sandalio (eds.), *Metal Toxicity in Plants: Perception, Signaling and Remediation*, DOI 10.1007/978-3-642-22081-4_8,

LC	La Calamine
LE	Lenninger
MALDI-TOF	Matrix-assisted laser desorption/ionization time-of-flight
MP	Monte Prinzera
MS	Mass spectrometry
MudPIT	Multidimensional protein identification technology
ROS	Reactive oxygen species
RpR	(Rožžnov pod Radhštěm-Moravia)
RuBisCo	Ribulose 1,5-bisphosphate carboxylase/oxybenase
SOD	Superoxide dismutase

1 Plant Hyperaccumulators

High tolerance to trace elements, including metals or metalloids, has evolved in a number of plant species. Excluders, e.g. plants in which both the entry and the root-to-shoot translocation of trace metals are limited, are often metal tolerant. However, many plant species are capable of accumulating metals in the above-ground tissues at concentrations which are significantly higher than those present in the soil without showing visible toxicity symptoms (Baker and Brooks 1989). Plants are termed hyperaccumulators when metal concentrations were from 50 to 100 times higher than in non-hyperaccumulator plants (Baker and Brooks 1989). The criterion for defining metal hyperaccumulation in plants depends on the abundance and toxicity of the element involved. Conventionally, the threshold concentrations are taken as 1% of shoot dry mass for Zn and Mn, 0.1% for Ni, Co, Cu and Se and 0.01% for As and Cd (Baker and Brooks 1989; Reeves and Baker 2000). Hyperaccumulation is a trait comparatively rare in the plant kingdom, being detected in about 450 species (0.2% of angiosperms). Hyperaccumulation involves metals, metalloids and non-metals, but the majority of hyperaccumulator species counted are Ni hyperaccumulators (75%) (Brooks et al. 1974, 1977; Baker and Brooks 1989; Reeves and Baker 2000; Ellis and Salt 2003; Reeves 2003, 2006; Sors et al. 2005; Milner and Kochian 2008; Baker et al. 2010; Krämer 2010). Different taxa of hyperaccumulators are known, the largest being *Alyssum* (family Brassicaceae), a genus native to southern Europe and Asia Minor, which contains 48 species capable of accumulating Ni up to 3% of dry shoot mass (Brooks et al. 1979). The wide distribution of hyperaccumulators has prompted for some ecological explanation of their role. A function as an elemental defence against herbivore or pathogen attack is presently one of the favored hypotheses (Pollard et al. 2002; Poschenrieder et al. 2006; Boyd 2007). Hyperaccumulators have also gained interest because of their potential use for practical environmental remediation such as phytoremediation (Chaney et al. 1997; Marmiroli and McCutcheon 2003; Pilon-Smits 2005), in plant-mediated milling or phytomining (Li et al. 2003) and in crop food bio-fortification to potentiate organoleptic proprieties (Broadley et al. 2007; Palmgren et al. 2008).

In the last 20 years, intensive research has been carried on about the biochemical, physiological and molecular features of the hyperaccumulators, in particular plant species widely distributed in Europe: *Thlaspi caerulescens* J. et C. Presl. hyperaccumulator of Pb, Zn, Cd, Ni, *Alyssum bertolonii* Desv. hyperaccumulator of Ni, Co and *Arabidopsis halleri* (L.) O'Kane and Al Shehbaz growing in soils with high levels of Zn, Cd and Pb (Bert et al. 2002). These species all belong to the Brassicaceae family, which provides good genetic syntheny and genomic correlation with *Arabidopsis thaliana* (L.) Heynh (www.arabidopsis.org). In particular, *T. careulescens* and *A. halleri* have become model plants for the genetic of hyperaccumulation.

Reviews and research papers describe some features of the genetic and physiological mechanisms of metal hyperaccumulation in *T. caerulescens* and in *A. halleri* (Macnair 2003; Ramos-Onsins et al. 2004; Yang et al. 2005; Basic et al. 2006; Clemens 2006; Krämer et al. 2007; Willems et al. 2007; Milner and Kochian 2008; Memon and Schroeder 2009; Verbruggen et al. 2009; Maestri et al. 2010). High throughput analyses at mRNA level combined with functional tests and complementation analysis in yeast have enhanced our understanding of the role of some genes in determining the hyperaccumulation phenotype (Weber et al. 2004; Becher et al. 2004; Rigola et al. 2006; van de Mortel et al. 2006, 2008). However, transcriptome analysis has some methodological limitations and sometimes there is a poor correlation between levels of mRNAs and those of the corresponding proteins. Indeed, RNA level does not reflect a possible protein regulation at translational and post-translational level; this on the proteins rather than transcripts better reflect, the dynamic condition of the organism which achieve an environmental adaptation (Gygi et al. 1999). A recent review on plant proteomic studies on proteomic and metal stresses has shown a gap for metal hyperaccumulators which claim to identify protein biomarkers for specific metals (Ahsan et al. 2010).

2 Methods in Plant Proteomics

Three steps are required to achieve a comprehensive picture of a plant proteome: (1) protein extraction, (2) protein separation and (3) protein identification (Fig. 1).

2.1 Protein Extraction

In developing new technologies for high-resolution protein separation and rapid automated protein identification, a critical step is protein extraction (Rose et al. 2004). Plant proteome analysis has some limitations in the extraction and resolution of proteins: a comparatively low concentration of protein per dry weight of tissue (Jacobs et al. 2001), a high protease activities (Gegenhemier 1990) and the presence of various interfering substances such as polysaccharides, lipids, and polyphenols

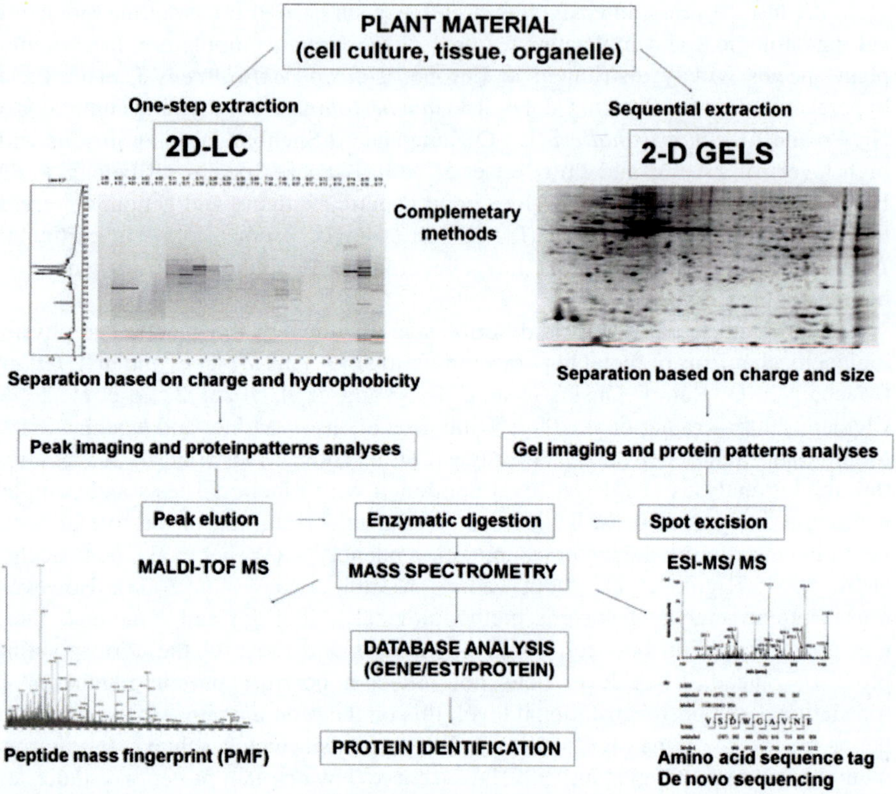

Fig. 1 Common steps in comparative proteomic analyses. These typically include protein "in gel" separation by two-dimensional polyacrylamide gel electrophoresis (2D-PAGE) or "gel free" separation by two-dimensional liquid chromatography (2D-LC), followed by protein pattern analysis and identification using spectra generated by MALDI-TOF MS or ESI MS/MS

(Michaud and Asselin 1995; Jacobs et al. 2001; Giavalisco et al. 2003; Carpentier et al. 2005). Different protocols for protein extraction from plant tissues have been reported in the literature; these are based on one-step extractions usually by means of different solvents followed by TCA/acetone or phenol-based precipitation, but the majority involve two or more steps (Rabilloud 1996). Sequential extractions are usually performed with a series of solvents followed by treatment with water-miscible organic solvents such as acetone and other acids (Michaud and Asselin 1995; Giavalisco et al. 2003). These multi-step extraction procedures are effective in dividing the proteome into distinct protein populations, thereby decreasing protein complexity, but on the other hand increase the loss of proteins due to either incomplete precipitation or resolubilization. In addition, protein instability and solubility in a cellular extract obtained with a multi-step treatment increase from the moment of extraction to the point of protein separation and detection, limiting reproducibility (Rose et al. 2004). Chemical extraction procedures involving a minimum number of steps, with stringent solvent cocktails capable of disrupting protein aggregates and of denaturing the constituent proteins are a valid alternative for comparative proteomic strategies (Lambert et al. 2005). In both cases, a drawback in plant protein extraction procedures is the exceeding high abundance of proteins like the large subunit of ribulose 1,5-bisphosphate carboxylase/ oxygenase (RuBisCo) which can foreshadow the identification of other less abundant proteins. Indeed no protein depletion system has been developed for the most abundant plant proteins (Vincent et al. 2006).

2.2 Protein Separation

Two different separation methodologies are currently used: two-dimensional polyacrylamide gel electrophoresis (2D-PAGE) and two-dimensional liquid chromatography (2D-LC) protein fractionation systems. The two methodologies are clearly complementary to obtain the most comprehensive picture of the entire plant proteome (Fig. 1). 2D-PAGE is a well-established technique that separates proteins in the first dimension based on their isoelectric point (p*I*) and in the second dimension according to their molecular mass (M*r*). Improvement of resolution and reproducibility has been obtained with the introduction of both broad p*I* range (typically p*I* 3–10) and narrow range (e.g., 1 p*I* unit) ampholine immobilized pH gradient (IPG) in the first dimension isoelectric focusing (IEF). This has increased both the resolution and the number of distinct proteins that can be separated in the second dimension. With this technique, which is still the most used for comparative proteome analyses, both the fractionation of specific classes of proteins as hydrophobic proteins and glycoproteins and the visualisation of low-abundance small peptides are usually limited. In addition, in 2D-PAGE the amount of proteins that can normally be loaded in a run is a maximum of up to 200 μg (Rose et al. 2004). A contribution to solve some of these problems comes from non-gel-based 2D-LC technologies such as the ProteomeLab™ PF2D (Eprogen, Darien, IL, USA), Multidimensional Protein Identification Technology (MudPIT) and isobaric Tag for Relative and Absolute Quantification (iTRAQs) These systems are based entirely on chromatographic separation and claim a high reproducibility which is important in comparative studies (Lambert et al. 2005; Pirondini et al. 2006; Visioli et al. 2010a, b).

2.3 Protein Patterns Analysis and Protein Identification

Once proteins have been separated, a comparative analysis can be performed (Fig. 1). For 2D-PAGE, commonly used software is PDQuest 2-D Analysis Software version 8.0 (Bio-Rad Laboratories, CA, USA) or DeCyder 2-D differential analysis Software v6.5 (GeHealthcare Bio-Sciences, Piscataway, NJ, USA); for

2D-LC analyses different software such as DeltaVue (Eprogen, Darien, IL, USA) or Progenesis LC-MS 3.1 (Nonlinear Dynamics, Newcastle upon Tyne, UK) can be utilised. Differentially abundant proteins are then identified by mass spectrometry (MS) analyses such as matrix-assisted laser desorption/ionization time-of-flight (MALDI-TOF) MS, that is typically used to measure the masses of the peptides derived from the trypsinized parent spot, thus generating a 'peptide mass fingerprint'. An alternative to this approach is de novo sequencing by electrospray ionization tandem mass spectrometry (ESI) MS/MS which yield amino acid sequences of selected tryptic digest peptides. Protein spectra are used with software packages (such as Mascot Matrix Science, http://www.matrixscience.com) for protein identification (for a review of MS methods, see Rose et al. 2004; Nesvizhskii 2007; Careri and Mangia 2011) (Fig. 1).

3 Proteomic Approaches for Identification of Key Functions in the Hyperaccumulators

Key steps involved in determining plant hyperaccumulation are: (1) interaction between soil and roots; (2) metal root uptake; (3) xylem loading and sequestration and detoxification of metal in shoots. We shall consider some of the relevant information obtained from recent proteomic approaches and the results are summarized in Table 1 and Fig. 2.

3.1 Proteins Involved in Plant-Soil Interaction

The hyperaccumulator phenotype evolves in specific environments in which limiting growth conditions allows for the adaptation of only a few plant species. An important aspect of this competition is the role played by the interaction between a plant's root and the soil. Plant access to trace elements is mediated by bioavailability and by root–plant interaction. There is a general lack of ecological knowledge about the hyperaccumulator and rhizosphere processes (Abou-Shanab et al. 2003a, b). The rhizosphere, defined as the root–soil interface, is a dynamic micro-system where microorganisms, roots and soil interact and where roots access soil trace elements (Alford et al. 2010). Root system development, root morphology, chemotropism are all recognised as equally important for establishing the hyperaccumulator phenotype, though the mechanisms are not fully understood (Comerford 2005; Ernts 1996; Moradi et al. 2009). Another important aspect of plant–soil interaction is the influence of roots on soil. Roots may change concentrations of chemicals in soil, pH and redox conditions; they can form organic acid complexes with nutrients and chelates metal (Hinsinger 1998). Root exudates can also play a role in metal mobilization.

Table 1 A summary of plant proteome analyses in hyperaccumulators published from 2005 to 2011

Metals	Plant species	Plant material	Proteomic technologies	Key results	Number of protein identified	References
Cd/Zn	*Thlaspi caerulescens* accessions (tolerant vs. hyperaccumulators Cd/Zn-treated vs control plants)	Root, shoot	2D-PAGE MALDI TOF-MS	Photosynthesis, nitrogen assimilation, carbohydrate metabolism and oxidative stress response proteins showed to be differentially regulated between accession and treatments. Difference in protein abundance were strong between accession also in absence of metal stress.	48	Tuomainen et al. (2006)
	Thlaspi caerulescens accessions (tolerant vs. hyperaccumulator)	Root, shoot	2D-PAGE, LC nano-ESI MS/MS and LTQ MS	Oxidative stress response and detoxification proteins co-segregate with Zn accumulating phenotypes.	46	Tuomainen et al. (2010)
	Arabidopsis halleri (Treated or not with rhizosphere microorganisms and Cd)	Shoot	2D-PAGE, nano HPLC ESI-Q-TOF MS	Interaction between plant and microorganisms: up-regulation of photosynthesis-related proteins and down-regulation of plant defence proteins	57	Farinati et al. (2009)
	Arabis paniculata Franch (Cd/Zn-treated vs. control plants)	Shoot	2D-PAGE, MALDI-TOF/TOF MS	Energy metabolism proteins are induced by Zn, xenobiotic/antioxidant defence and cellular metabolism proteins are induced by Cd	37	Zeng et al. (2011)
Ni	*Alyssum lesbiacum* (Ni-treated vs control plants)	Root, shoot	2D-PAGE, LC MS/MS	Sulfur metabolism (consistent with a reallocation of sulfur toward cystein and glutathione), protection against ROS and HS response proteins showed a modulation in presence of Ni	18	Ingle et al. (2005)

(continued)

Table 1 (continued)

Metals	Plant species	Plant material	Proteomic technologies	Key results	Number of protein identified	References
	Thlaspi caerulescens (metallicolous vs non-metallicolous Ni-treated vs. control plants)	Shoot	2D-LC MALDI TOF/MS	Sulfur metabolism, protection against oxidative stress, biotic and abiotic stress response and regulatory proteins differentiate between metallicolous and non-metallicolous phenotypes	28	Visioli et al. (2010b)
As	*Agrostis tenuis* (Tolerant, pseudometallophyte accumulator As-treated vs. control)	Shoot	2D-PAGE, MALDI-TOF/MS	Proteins linked to photosynthis were down-regulated in presence of As. Defence and proteolysis proteins increase in abundance and could play a role in As tolerance	30	Duquesnoy et al. (2009)
	Pteris vittata (inoculated with arbuscular mycorrhizae with or without As)	Shoot	2D-PAGE, and Nanospray LC-MS/MS	An up-regulation of GAPDH, phosphoglycerate kinase and enolase suggests a central role of glycolitic enzymes in arsenic metabolism	112	Bona et al. (2010)
Pb	*Heliantus annuus* (Pb-treated vs control plants)	Shoot	2D-PAGE ESI MS/MS	Overexpression of abiotic and biotic stress proteins	5	Walliwalagedara et al. (2010)

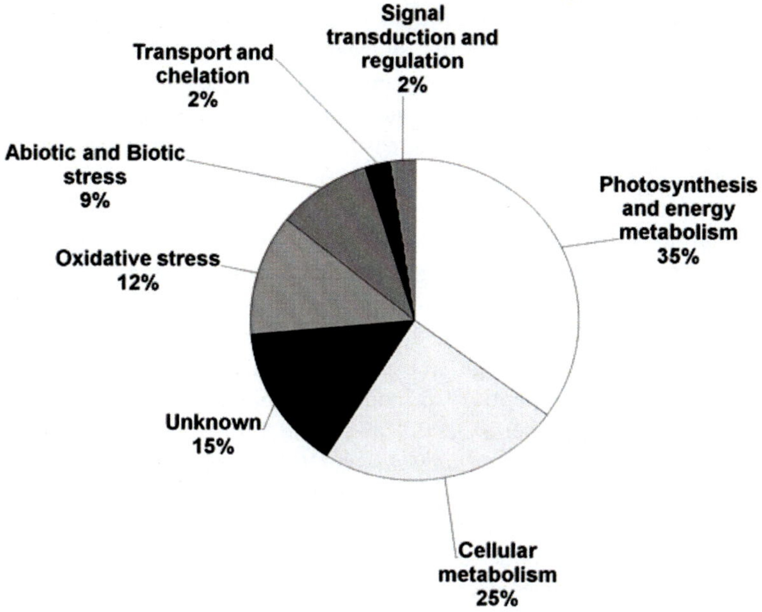

Fig. 2 Percentage of various classes of proteins identified in hyperaccumulators as up and down-regulated in response to different metals (Data refer to publications reported in Table 1)

Soil microbial communities are also very important in determining the interaction between plant and soil: plant growth promoting rhizobacteria, endophytic bacteria, and arbuscular mycorrhizae all seem to be very important in increasing root surface area by hairy root production, element solubility, and increasing the soluble elements in the rhizhosphere, thus contributing to the hyperaccumulator phenotype (Whiting et al. 2001). Contrasting results were obtained by comparing hyperaccumulators grown with inoculated microorganisms or non-sterile soils with hyperaccumulators grown on non-inoculated or sterile soils. In some cases, plants accumulate more trace elements when they were inoculated, in other cases no differences were found (Whiting et al. 2001; Abou-Shanab et al. 2003a, b; Abou-Shanab et al. 2006; Turnau and Mesjasz-Przybylowicz 2003; Liu et al. 2005, 2009; Trotta et al. 2006; Vogel-Mikuš et al. 2006; Li et al. 2007; Farinati et al. 2009). Recent proteomic works showed how the rhizosphere microorganisms influence plant metal accumulation and correlate specific proteins with the hyperaccumulation process (Farinati et al. 2009; Bona et al. 2010).

Farinati and collaborators (2009) described a microbial rhizosphere community in the Cd/Zn hyperaccumulator *A. halleri* collected on a metalliferous soil in northern France (Table 1). A positive correlation was found between the presence of microorganisms and the accumulation of Cd and Zn in the shoots. In plants treated with microorganisms shoots' proteomic analyses showed: (1) a down regulation of proteins linked to plant defence responses against herbivores, insects and pathogen attacks; (2) a down regulation of glutathione-S-transferase (GST) proteins; (3) an up-regulation of proteins involved in photosynthetic processes and in the Calvin cycle. These results suggest a general correlation between the presence of microorganisms and the plant fitness: the presence of microorganisms increases the hyperaccumulation capacity and takes the role of specific plant defence functions in another way increasing the energy requirement of the plant (Farinati et al. 2009).

Arbuscular mycorrhizae between the roots and soil fungi improve mineral nutrition, protect against pathogens and enhance resistance or tolerance to biotic and abiotic stresses. It has been suggested that arbuscular mycorrhizae enhances metal tolerance in plants grown on polluted soils by increasing phosphourus bioavailability and reducing Cd uptake, and thus arbuscular mycorrhizae leads to metal tolerance in plants through an exclusion mechanism. Proteomic approaches have been performed to identify Cd-induced proteins in peas roots colonized by *Glomus mosseae* (Repetto et al. 2003) and to analyse root proteins in *Medicago truncatula* colonised by *Glomus intraradices* (Valot et al. 2006; Aloui et al. 2009). A proteomic analysis has also been performed in the As hyperacccumulator species *Pteris vittata* in order to understand the role of arbuscular mycorrhizae in the protein synthesis during As detoxification (Bona et al. 2010). An up-regulation of glyceraldehyde-3-phosphate dehydrogenase (GAPDH), phosphoglycerate kinase and enolase suggests a central role of glycolitic enzymes in As metabolism (Table 1) (Bona et al. 2010). In considering the correlation between specific proteins and the presence of arbuscular mycorrhizae, it should be considered that this symbiosis reduces uptake and increases tolerance in plants, thus reducing As toxicity due to its accumulation.

Identification of all the proteins present in soil, also termed metaproteomic (Bastida et al. 2009), can also contribute to explain the phenomenon of plant hyperaccumulation. The lower amount of proteins in soil or macrocosm as compared to purified microbial cultures, the association of soil proteins with compounds which interfere with their identification, such as humic acids, and the lack of genome databases for protein matches for soil microorganisms are all gaps that wait to be closed to obtain a more accurate picture of the soil proteome (Bastida et al. 2009). Metaproteomic is a science still in its infancy but future technical developments and more precise standardization of methods will allow this science to be used more proficiently. Presence or absence of different soil proteins will become a process indicator, such as for the interaction between root and soil, in determining the plant hyperaccumulator phenotype.

3.2 Root Proteome

The root system represents the gate a metal had to cross to enter into the plant. Enhanced metal uptake combined to a reduced sequestration in the root vacuoles and the high translocation diversify hyperaccumulators and non-hyperaccumulator species (Verbruggen et al. 2009; Maestri et al. 2010). In roots, many Zn-transporters including ZNT1 are Zn-regulated in non-hyperaccumulators, e.g. they are only detectable under Zn deficiency, but in hyperaccumulators, they are constitutively expressed, independently of Zn supply (Pence et al. 2000; Assunção et al. 2001). The iron transporter IRT1 has shown to also be correlated with the transport of Cd and Zn in the hyperaccumulator *T. caerulescens* (Lombi et al. 2002). Proteomic studies on root systems have been performed in the Ni hyperaccumulator *Alyssum lesbiacum* (Ingle et al. 2005) and in the Cd/Zn hyperaccumulator *T. caerulescens* (Tuomainen et al. 2006, 2010) (Table 1). In the former work, proteins that increase in abundance after a short, high Ni treatment (0.3 mM) were involved in sulfur metabolism, protection against reactive oxygen species (ROS) and heat-shock (HS) response. Few modifications were observed in the root system after a prolonged treatment with a 0.03-mM Ni concentration which is considered optimum for growth of this species but enough to trigger hyperaccumulation. Under this latter Ni treatment, constitutively expressed genes and proteins in high Ni-tolerant species may allow for effective chelation and sequestration of Ni without the need of further protein synthesis. In the work of Tuomainen and collaborators (2006) the roots proteome of three different *T. caerulescens* accessions [Zn/Cd tolerant La Calamine (LC), and Zn/Cd hyperaccumulator but less tolerant Lenningen (LE), and Monte Prinzera (MP)] was compared after 3-week treatments at various Zn and Cd concentrations. Different classes of proteins whose abundance changes between metal exposure but also between accessions were identified. Proteins related to removal of ROS were more abundant in the more hyperaccumulator but less tolerant accessions MP and LE while showing a lower level of abundance in the less hyperaccumulator but more tolerant LC accession. The enzyme superoxide dismutase (SOD), which is a Zn-requiring enzyme, decreases in Zn deprivation conditions in the LC and MP. A decrease in ascorbate peroxidase (APX) in root at high metal concentrations was also observed in the same accessions. The authors postulate that these modifications could be due to an increase in ROS leading to increased lignifications and in different metal binding and uptake capacity.

Root cell walls can also be influenced by metal stress. Synthesis of a protein related to cell wall structure, a putative glycosyl hydrolase family of 18 proteins, showed a modulation according to accession and treatment. This protein which participates in the re-assembling of cell wall and particularly in cell expansion was less abundant in the roots of MP accession which effectively accumulate Cd and Ni in the roots rather than in the LC accession. The root cytosolic glutamine synthetase, which is specifically expressed in root pericycle and is involved in ammonium assimilation, is known to be inhibited by Cd, causing ammonium accumulation and toxicity. The presence and abundance of this protein was correlated with Cd tolerance rather than to Cd accumulating capacity in the three accessions. Tuomainen and collaborators (2010), using a proteomic approach on a segregating population of the cross between *T. caerulescens* LC and LE accessions demonstrated that proteins whose abundance co-segregates with Zn accumulation in the F3 progeny are, for the greatest majority root proteins. Therefore, differences in Zn accumulation between the two accessions were mainly determined by the root proteome. In *A. halleri*, the transcript of HMA4 gene, one of the genetic determinants of the hyperaccumulator phenotype, is also preferentially expressed in the roots (Hanikenne et al. 2008). Furthermore in the hyperaccumulator *T. caerulescens* the concentration of histidine, which contributes to metal (Zn and Ni) loading, is enhanced in roots but not in shoots as compared with the non-hyperaccumulator *Thlaspi arvense* (Richau et al. 2009). Protein,s, transcript,s and metabolite,s concentrations were all more abundant in roots of hyperaccumulators suggesting a specific function according to the literature data. Proteome research regarding root tissues has failed to identify root transporters differentially abundant between treatments or between hyperaccumulator and non-hyperaccumulator plants. These data are in accordance with transcriptomics analyses which showed a constitutive expression of metal transporter genes in hyperaccumulators (Assunção et al. 2001; Weber et al. 2004; Hammond et al. 2006). On the other hand it cannot be excluded that a particular transporter protein cannot be identified after the two-dimensional separation phase, due to some technical limitation of the proteomic approach (Rose et al. 2004).

3.3 Shoot Proteome

The shoot proteome of some hyperaccumulators has been searched to find protein functions correlated with sequestration and detoxification of metals (Table 1). Different times of exposures, different plants and different metals have been considered, but some results can be compared. In the work of Ingle and collaborators (2005), only three proteins were induced upon a long-term exposure to a Ni concentration optimum for *Alyssum lesbiacum* growth (0.03 mM). One of these, a Fe-SOD, can be associated with defence against antioxidants in Ni hyperaccumulators even when Ni concentrations in soil were not inhibitory for growth (Freeman et al. 2004). The two other enzymes are involved in carbohydrate metabolism: a transketolase, and a chloroplastic phosphoglycerate kinase. A consistent up-regulation of proteins involved in photosynthesis such as a sub-unit of the complexes responsible for light harvesting (chlorophyll a/b binding protein) and a membrane extrinsic subunit of photosystem II (PSBP), was observed in the hyperaccumulator *A. halleri* treated with Cd and Zn (Farinati et al. 2009). This was interpreted as due to an enhanced energy demand of the entire cellular metabolism in this hyperaccumulator species (Farinati et al. 2009). On the other hand, enzymes also involved in the Calvin cycle, such as phosphoribulokinase and fructose-bisphosphate aldolase and RuBisCo large sub-unit were up-regulated in the same condition. Similar results were obtained in *Arabis paniculata* Franch., a Zn/Cd-tolerant and hyperaccumulator plant when exposed to both low and high Zn or Cd levels for 10 days (Zeng et al. 2011), and in *Helianthus annuus* an hyperaccumulator of Pb (Walliwalagedara et al. 2010) (Table 1). A consistent increase in proteins involved in energy metabolism, xenobiotic/antioxidant defense and HS response was found: at high metal concentration the induction of enzymes, such as aldo/keto reductases, GST, Cytochrome P450, SOD and APX able to detoxify peroxides deriving from lipids' breakdown, were shown. All these data suggest that hyperaccumulation of metals is also extremely costly at shoot level. The plants can afford it only by reinforcing the photosynthetic mechanism and by overexpressing proteins that act as scavengers of the unavoidable breakdown products that occur with such a massive energy buildup.

Proteins linked to energy metabolism seem to be important in also conferring metal tolerance. Tuomainen and collaborators (2006), comparing shoot proteomes of *T. caerulescens* accessions with different levels of tolerance to Zn and Cd, found higher levels of PSBP responsible for stabilizing the PSII in the tolerant *T. caerulescens* LC accession than in the other less tolerant accessions, MP and LE, in response to Zn. Several Calvin cycle enzymes, such as RuBisCo large subunit, GAPDH and sedoheptulose-1,7 bisphosphatase, were also significantly more abundant in LC accession (Tuomainen et al. 2006). In the same work, differences between accessions were also evidenced for proteins related to ROS removal. In the less Zn/Cd-tolerant MP accession, larger levels of APX can indicate a need to eliminate a larger amount of ROS. In *Agrostis tenuis* an As-tolerant monocotyledon, tolerance is correlated with the induction of a set of proteins including enzymes such as oxygen-evolving enhancer protein, RuBisCO small and large subunits, RuBisCO activase and ATP synthase involved in the Calvin and Krebs cycles (Duquesnoy et al. 2009) (Table 1). Two different *T. caerulescens* populations, the Ni hyperaccumulator *T. caerulescens* MP from a metalliferous soil and *T. caerulescens* RpR from the Czech Republic adapted to grow on a non-metalliferous soil, were compared for their shoots' proteome (Table 1) (Visioli et al. 2010b). Plants were treated with a 0.010 mM Ni for 3 weeks. In this condition, the Ni hyperaccumulator MP grows better than in the absence of Ni and accumulated Ni in the shoots. Comparison of shoot proteome evidenced that only 28 proteins for MP and 23 for RpR from a total of 500 were up- or down-regulated in the presence of Ni. Exposure for a longer time to a relatively low Ni concentration has certainly allowed plants to adjust metal homeostasis as shown by growth and by other morphological features. In the same work a different modulation of some classes of proteins was evidenced between the two populations in the presence of Ni. In particular a putative metal transporter ZRT/IRT like protein 8 and a metal chelator methallothionein protein 1B increase their abundance in the MP but not in RpR accession. Genes codifying for these classes of proteins were also revealed to be more expressed in metal-adapted than in non-metal-adapted *Thlaspi* populations at the transcriptomic level (Assunção et al. 2001; Rigola et al. 2006; van de Mortel et al. 2006) and could be considered biomarkers of the hyperaccumulation phenotype.

An increase in proteins, such as putative thioredoxin, GST, cytochrome P450 and aspartate aminotransferase, all linked to the detoxification of endogenous and exogenous xenobiotics such as metals, ROS and ROS-induced lipid peroxidation products were more abundant in MP population in condition of Ni treatment. In particular GSTs have also been identified and characterized in bacteria, algae and fungi in response to different Cd, Zn and Cu (Fraser et al. 2002; Waschke et al. 2006), suggesting that GST can provide cells with a mechanism that make them refractory to cytotoxic agents. A proteomic approach was also performed on the Cd hyperaccumulator lichen *Physca ascendes* and the presence of a higher amount of a GST protein was observed both during a short-term and a long-term Cd treatment revealing a possible role of this enzyme both in detoxification and hyperaccumulation also in lower organisms (Rustichelli et al. 2008). Few data are available up to now on levels of protein regulators in all proteomic works published on hyperaccumulators (Fig. 2). Identification of regulatory proteins in a proteomic approach is impeded by their low abundance. Though many efforts have been made to increase the sensitivity of protein detection, proteomic analyses are still far from the high resolution levels obtained with high-throughput transcriptomic analyses based either on PCR or microarrays (Weber et al. 2004; Becher et al. 2004; Rigola et al. 2006; van de Mortel et al. 2006, 2008). Using a one-step protein extraction procedure coupled with 2D-LC separation and MS analyses in shoots of *T. caerulescens* MP a putative Ras-related protein (Rab 7) and a Myb transcription factor were found to increase in their abundance in condition of Ni treatment whereas in the non-metalliferous RpR population, that increase was not observed (Visioli et al. 2010b).

4 Follow-Up in Proteomic of Hyperaccumulators

To better dissect some of the genetic features of the hyperaccumulator phenotype, comparative genetic and molecular analyses between hyperaccumulator and non-hyperaccumulator phenotypes are of primary interest (Fig. 3). It is essential that the genetic and molecular comparison involves species with identical or similar genetic background. In some cases, the analysis of the interspecies variability has allowed the detection of features related with hyperaccumulation. It was from the interspecific cross between the hyperaccumulator *A. halleri* and the non-hyperaccumulator *Arabidopsis lyrata* that some interesting QTLs for hyperaccumulation were detected (Bert et al. 2002, 2003). Transcriptomic studies were also performed at interspecies level with microarray and real-time PCR technologies comparing *A. thaliana* and phylogenetically related hyperaccumulator species such as *A. halleri* or *T. caerulescens*, leading to the identification of genes specifically modulated in hyperaccumulators (Weber et al. 2004; Becher et al. 2004; Rigola et al. 2006; van de Mortel et al. 2006, 2008). But the potential of these methodologies alone is not sufficient because of the different level of genomic information between *A. thaliana* and the hyperaccumulator species. Genomic sequences of hyperaccumulators are very limited and this impedes the identification of unique gene functions.

At the proteomic level, the comparison of two different species can also be difficult. Attempts were performed with the proteome of the Ni hyperaccumulator *A. lesbiacum* with that of a related non-accumulator species, *Alyssum montanum*, in the absence of Ni. Unfortunately, the 2D-PAGE protein patterns of the two species were not similar enough to allow a quantitative comparative analysis (Ingle et al. 2005). Therefore, comparative proteome analysis seems more suitable for

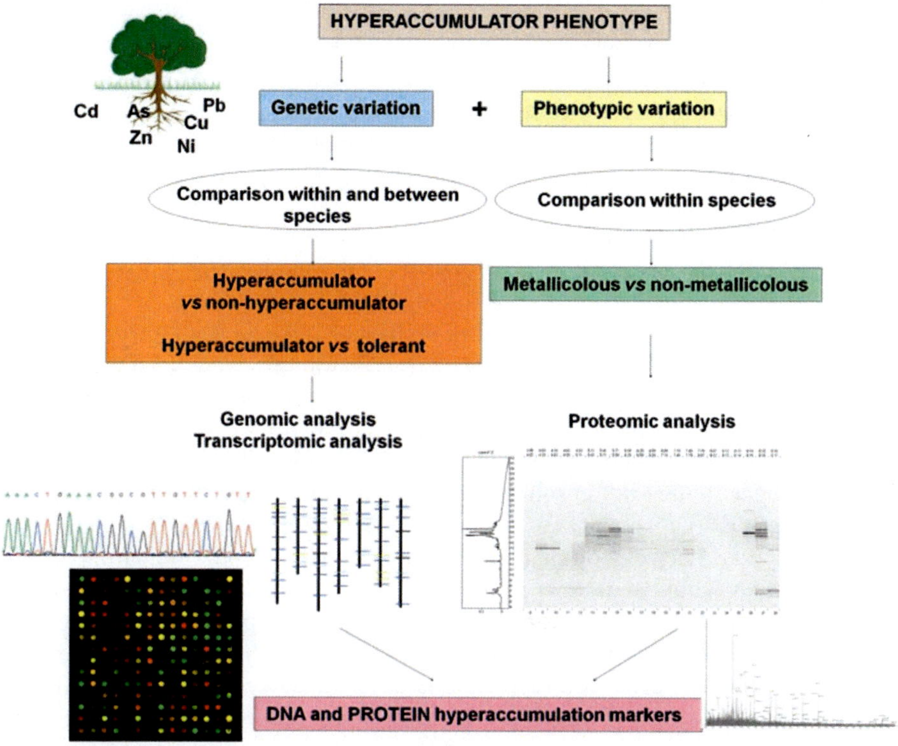

Fig. 3 Genomic and proteomic strategies proposed for molecular analyses in hyperaccumulators. Comparative proteomic approaches can be coupled with DNA and transcriptomic analyses to study the genotypic and phenotypic variation within and between species, thus providing useful DNA and protein markers for the elucidation of hyperaccumulator phenotypes

individuals within the same species adapted to grow in different environments, where it is possible, such as in the case of *T. caerulescens* where genetic variation within the species has been well documented (Assunção et al. 2003a, b; Yang et al. 2005; Richau and Schat 2009) (Fig. 3). Inter-population comparisons at transcriptomic and proteomic levels can be carried on avoiding the drawback due to inter-species variation, and could help in the identification of possible gene/protein biomarkers for hyperaccumulation of a specific metal. In the work of Toumainen and collaborators (2006), comparative proteomic analyses were performed on three *T. caerulescens* populations LC, LE and MP considering not only different metal stress conditions but also plant accessions differing in tolerance and hyperaccumulation for Zn and Cd. Many differences were also seen among the *Thlaspi* accessions in absence of metal treatments. Proteomic variations between populations of the same hyperaccumulator species *T. caerulescens* were found in another work by comparing the metallicolous MP population and the non-metallicolous RpR population (Visioli et al. 2010b). The inter-population variation observed in both cases was not unexpected, because the different *Thlaspi* accessions grow naturally in different environments, and have other phenotypic differences. Selection by environmental conditions can lead to local adaptation and to differentiation of sub-populations in quite a short time, which can eventually also result in changes in protein patterns (David et al. 1997). In addition, in the specific metalliferous environment, the plant proteome is not only affected by metal concentration but by multiple environmental factors like: light, heat and water which all interfere with the metal stress response, the metal accumulation and the metal tolerance. The thesis in this case is that for any primary stress there are also both a secondary and a tertiary stress, and all contribute to establish the global phenotype (Levitt 1980).

Until now, most of the proteomic work performed refers to laboratory conditions in which the metal is the only stressor. Differences were found in the accumulation capacity when comparing the same plant accession grow in laboratory conditions and in natural soil. *T. caerulescens* LC which is a Cd hyperaccumulator in its soil of origin, in laboratory conditions shows a great Cd tolerance but a lower hyperaccumulation capacity than in the soil of origin (Tuomainen et al. 2006). Studying the relationships between plant and environment can help in elucidating important protein functions putative for the hyperaccumulator phenotype. Comparing phenotypic variation within the Ni hyperaccumulator *T. caerulescens* MP population grown in its peculiar environment, it was found that the hyperaccumulation capacity is affected not only by soil composition but also by different micro-environmental conditions. These different accumulation capacities within population is reflected in abundance of specific sets of proteins related to metals, as metal transporters, and proteins related to defence against biotic and abiotic stresses and proteins of general metabolism (Visioli et al. unpublished data).

5 Conclusion

Environmental stress acts as selective agent that alter gene frequencies in a population and causes genetic divergence within and among populations, but it also changes the gene expression and regulation modifying both transcription and translation. The importance of studying how the protein patterns change in hyperaccumulators in metalliferous or non-metalliferous soils as compared with non-hyperaccumulator or tolerant species has been paramount to explain this phenotype (Fig. 3). Comparison of hyperaccumulator and closely related species differing in other environmental proprieties showed how the threshold of protein pattern perturbation can be raised or lowered according to the need of adaptation. Most of the studies reported and discussed in this chapter on proteome's variation in hyperaccumulators are evidence that, among all the species considered (though only a few), there is not a behavior that can be considered as a major one. Rather there are many different responses which may have something in common. A modulation of protein levels associated with the energy metabolism and oxidative stress response reveals a tendency to a new trade-off between the genotype and the environment in the hyperaccumulators (Fig. 2). Because the cost of adaptation increases this is paid by a surplus in energy production. Another common feature in many hyperaccumulators is the enhanced accumulation of stress responsive proteins (for both biotic and abiotic stress) and of detoxifying proteins against ROS or reactive toxic intermediates of the hyperaccumulation response (Fig. 2). This is also explained in term of the new trade-off because the signal transduction chain, that is triggered to permit the hyperaccumulation, also has elements in common with the general stress response mechanism and because the hyperaccumulation is certainly more exposed to the deleterious effect of other stresses, especially the primary ones (Levitt 1980). Finally, hyperaccumulation in terms of specific proteome variation so far detected has been below the expectations in the sense that few specific proteins were identified, whose presence or absence could be clearly correlated with the hyperaccumulation of a specific metal respect to non-hyperaccumulator or tolerant species. Indeed there is a certain percentage (15%) of proteins of un-known functions that can contribute to explain the hyperaccumulation behavior (Fig. 2).

In conclusion the hyperaccumulator is a very complex phenotype affected by many genetic and environmental causes. From the proteome point of view, it seems more the result of phenotypic plasticity. For this reason the proteome analysis is interesting behind the specific curiosity of these plants, because it can be a paradigmatic case to discuss the relative roles that genetic and phenotypic plasticity play in plant evolution and adaptation.

References

Abou-Shanab RA, Angle JS, Delorme TA, Chaney RL, van Berkum P, Moawad H, Ghanem K, Ghozlan HA (2003a) Rhizobacterial effects on nickel extraction from soil and uptake by *Alyssum murale*. New Phytol 158:219–224

Abou-Shanab RI, Delorme TA, Angle JS, Chaney RL, Ghanem K, Moawad H, Ghozlan HA (2003b) Phenotypic characterization of microbes in the rhizosphere of *Alyssum murale*. Int J Phytoremediation 5:367–379

Abou-Shanab RAI, Angle JS, Chaney RL (2006) Bacterial inoculants affecting nickel uptake by *Alyssum murale* from low, moderate and high Ni soils. Soil Biol Biochem 38:2882–2889

Ahsan N, Renaut J, Komatsu S (2010) Recent development in the application of proteomics to the analysis of plant responses to heavy metals. Proteomics 9:2602–2621

Alford ER, Pilon-Smits EAH, Paschke MW (2010) Metallophytes- A view from the rhizosphere. Plant Soil 337:33–50

Aloui A, Recorbet G, Gollotte A, Robert F, Valot B, Gianinazzi-Pearson V, Aschi-Smiti S, Dumas-Gaudot E (2009) On the mechanisms of cadmium stress alleviation in *Medicago truncatula* by Arbuscular mycorrhizal symbiosis: a root proteomic study. Proteomics 9:420–433

Assunção AGL, Da Costa MP, De Folter S, Vooijs R, Schat H, Aarts MGM (2001) Elevated expression of metal transporter genes in three accessions of metal hyperaccumulator *Thlaspi caerulescens*. Plant Cell Environ 24:217–226

Assunção AGL, Schat H, Aarts MGM (2003a) *Thlaspi caerulescens*, an attractive model species to study heavy metal hyperaccumulation in plants. New Phytol 159:411–419

Assunção AGL, Ten Bookum WM, Nelissen HJM, Vooijs R, Schat H, Ernst WHO (2003b) Differential metal-specific tolerance and accumulation patterns among *Thlaspi caerulescens* populations originating from different soil types. New Phytol 159:411–419
Baker AJM, Brooks RR (1989) Terrestrial higher plants which hyperaccumulate metallic elements: a review of their distribution, ecology and phytochemistry. Biorecovery 1:81–126
Baker AJM, Ernst WHO, van der Ent A, Malaisse F, Ginocchio R (2010) Metallophytes: the unique biological resource, its ecology and conservational status in Europe, Central Africa and Latin America. In: Batty LC, Hallberg KB (eds) Ecology of industrial pollution. Cambridge University Press, Cambridge
Basic N, Salamin N, Keller C, Gallard N, Besnard G (2006) Cadmium hyperaccumulatin and genetic differentiation of *Thlaspi caerulescens* populations. Biochem Syst Ecol 34:667–677
Bastida F, Moreno JL, Nicolas C, Hernandez T, Garcia C (2009) Soil metaproteomics: a review of an emerging environmental science. Significance, methodology and perspectives. Eur J Soil Sci 60:845–859
Becher M, Talke IN, Krall L, Krämer U (2004) Cross-species miroarray transcript profiling reveals high constitutive expression of metal homeostasis genes in shoots of the zinc hyperaccumulator *Arabidopsis halleri*. Plant J 37:251–268
Bert V, Bonnin I, Saumitou-Laprade P, de Laguérie P, Petit D (2002) Do *Arabidopsis halleri* from nonmetallicolous populations accumulate zinc and cadmium more effectively than those from metallicolous populations? New Phytol 155:47–57
Bert V, Meerts P, Samitou-Laprade P, Salis P, Gruber W, Verbruggen N (2003) Genetic basis of Cd tolerance and hyperaccumulation in *Arabidopsis halleri*. Plant Soil 249:9–18
Bona E, Cattaneo C, Cesaro P, Marsano F, Lingua G, Cavalletto M, Berta G (2010) Proteomic analysis of *Pteris vittata* fronds: two arbuscular fungi differentially modulate protein expression under arsenic contamination. Proteomics 10:811–3834
Boyd RS (2007) The defense hypothesis of elemental hyperaccumulation: status, challenges and new directions. Plant Soil 293:153–176
Broadley MR, White PJ, Hammond JP, Zelko I, Lux A (2007) Zinc in plants. New Phytol 173:677–702
Brooks RR, Lee J, Jaffré T (1974) Some New Zealand and New Caledonian plant accumulators of nickel. J Ecol 62:493–499
Brooks RR, Lee J, Reeves RD, Jaffré T (1977) Detection of nickeliferous rocks by analysis of herbarium specimens of indicator plants. J Geochem Explor 7:49–57
Brooks RR, Morrison RS, Reeves RD, Dudley TR, Akman Y (1979) Hyperaccumulation of nickel by *Alyssum Linnaeus* (Cruciferae). Proc R Soc Lond B 203:387–403
Careri M, Mangia A (2011) Trends in analytical atomic and molecular mass spectrometry in biology and the life sciences. Anal Bioanal Chem 399:2585–2595
Carpentier SC, Witters E, Laukens K, Deckers P, Swennen R, Panis B (2005) Preparation of protein extracts from recalcitrant plant tissues: an evaluation of different methods for two dimensional gel electrophoresis analysis. Proteomics 5:2497–2507
Chaney R, Malik M, Li YM, Brown SL, Brewer EP, Angle JS, Baker AJM (1997) Phytoremediation of soil metals. Curr Opin Biotech 8:279–284
Clemens S (2006) Toxic metal accumulation, responses to exposure and mechanisms of tolerance in plants. Biochimie 88:1707–1719
Comerford NB (2005) Soil factors affecting nutrient bioavailability. In: BassiriRad H (ed) Nutrient acquisition by plants an ecological perspective. Springer, Heidelberg
David JL, Zivy M, Cardin ML, Brabant P (1997) Protein evolution in dynamically managed populations of wheat: adaptative responses to macro-environmental conditions. Theor Appl Genet 95:932–941
Duquesnoy I, Goupil P, Nadaud I, Branlard G, Piquet-Pissaloux A, Ledoigt G (2009) Identification of *Agrostis tenuis* leaf proteins in response to As(V) and As(III) induced stress using a proteomics approach. Plant Sci 176:206–213
Ellis DR, Salt DE (2003) Plants, selenium and human health. Curr Opin Plant Biol 6:273–279

Ernst WHO (1996) Bioavailability of heavy metals and decontamination of soils by plants. Appl Geochem 11:163–167

Farinati S, DalCorso G, Bona E, Corbella M, Lampis S, Cecconi D, Polati R, Berta G, Vallini G (2009) A Proteomic analysis of *Arabidopsis halleri* shoots in response to the heavy metals cadmium and zinc and rhizosphere microorganisms. Proteomics 9:4837–4850

Fraser JA, Davis MA, Hynes MJ (2002) A gene from *Aspergillus nidulans* with similarity to URE2 of *Saccharomyces cerevisiae* encodes a glutathione S-transferase which contributes to heavy metal and xenobiotic resistance. Appl Environ Microb 68:2802–2808

Freeman JL, Persans MW, Nieman K, Albrect C, Peer W, Pickering IJ, Salt DE (2004) Increased glutathione biosynthesis plays a role in nickel tolerance in *Thlaspi* nickel hyperaccumulators. Plant Cell 16:2176–2191

Gegenhemier P (1990) Preparation of extracts from plants. Methods Enzymol 182:174–193

Giavalisco P, Nordhoff E, Lehrach H, Gobom J, Klose J (2003) Extraction of proteins from plant tissues for two-dimensional electrophoresis analysis. Electrophoresis 24:207–216

Gygi SP, Rochon Y, Franza BR, Aebersold R (1999) Correlation between protein and mRNA abundance in yeast. Mol Cell Biol 19:1720–1730

Hammond JP, Bowen HC, White PJ, Mills V, Pyke KA, Baker AJ, Whiting SN, May ST, Broadley MR (2006) A comparison of the *Thlaspi caerulescens* and *Thlaspi arvense* shoot transcriptomes. New Phytol 170:239–260

Hanikenne M, Talke IN, Haydon MJ, Lanz C, Nolte A, Motte P, Kroymann J, Weigel D, Krämer U (2008) Evolution of metal hyperaccumulation required cis regulatory changes and triplication of HMA4. Nature 453:391–395

Hinsinger P (1998) How do plant roots acquire mineral nutrients? Chemical processes involved in the rhizosphere. In: Sparks DL (ed) Advances in agronomy, vol 64. Academic, San Diego

Ingle RA, Smith JA, Sweetlove LJ (2005) Responses to nickel in the proteome of the hyperaccumulator plant *Alyssum lesbiacum*. Biometals 18:627–641

Jacobs DI, van Rijssen MS, Heijden RVD, Verpoorte R (2001) Sequential solubilisation of proteins precipitated with trichloroacetic acid in acetone from cultured *Catharanthus roseus* cell yields 52% more spots after two dimensional electrophoresis. Proteomics 1:1345–1350

Krämer U (2010) Metal hyperaccumulation in plants. Annu Rev Plant Biol 61:517–534

Krämer U, Talke I, Hanikenne M (2007) Transition metal transport. FEBS Lett 581:2263–2272

Lambert JP, Ethier M, Smith JC, Figeys D (2005) Proteomics: from gel based to gel free. Anal Chem 77:3771–3788

Levitt J (1980) Responses of plants to environmental stresses. Academic, New York

Li YM, Chaney R, Brewer E, Roseberg R, Angle JS, Baker A, Reeves R, Nelkin J (2003) Development of a technology for commercial phytoextraction of nickel: economic and technical considerations. Plant Soil 249:107–115

Li WC, Ye ZH, Wong MH (2007) Effects of bacteria an enhanced metal uptake of the Cd/Zn-hyperaccumulating plant, *Sedum alfredii*. J Exp Bot 58:4173–4182

Liu Y, Zhu YG, Chen BD, Christie P, Li XL (2005) Influence of the arbuscular mycorrhizal fungus *Glomus mosseae* on uptake of arsenate by the as hyperaccumulator fern *Pteris vittata* L. Mycorrhiza 15:187–192

Liu Y, Christie P, Zhang JL, Li XL (2009) Growth and arsenic uptake by Chinese brake fern inoculated with an Arbuscular mycorrhizal fungus. Environ Exp Bot 66:435–441

Lombi E, Tearall KL, Howarth JR, Zhao FJ, Hawkesford MJ, McGrath SP (2002) Influence of iron status on cadmium and zinc uptake by different ecotypes of the hyperaccumulator *Thlaspi caerulescens*. Plant Physiol 128:1359–1367

Macnair MR (2003) The hyperaccumulation of metals by plants. Adv Bot Res 40:63–105

Maestri E, Marmiroli M, Visioli G, Marmiroli N (2010) Metal tolerance and hyperaccumulation: costs and trade-offs between traits and environment. Environ Exp Bot 68:1–13

Marmiroli N, McCutcheon SC (2003) Making phytoremediation a successful technology. In: McCutcheon SC, Schnoor JL (eds) Phytoremediation: transformation and control of contaminants. Wiley, Hoboken

Memon AR, Schroeder P (2009) Implications of metal accumulation mechanisms to phytoremediation. Environ Sci Pollut R 16:162–175

Michaud D, Asselin A (1995) Application to plant proteins of gel electrophoretic methods. J Chromatogr A 698:263–279

Milner MJ, Kochian LV (2008) Investigating heavy-metal hyperaccumulation using *Thlaspi caerulescens* as a model system. Ann Bot (Lond) 102:3–13

Moradi AB, Conesa HM, Robinson BH, Lehmann E, Kaestner A, Schulin R (2009) Root responses to soil Ni heterogeneity in a hyperaccumulator and a non-accumulator species. Environ Pollut 157:2189–2196

Nesvizhskii AI (2007) Protein identification by tandem mass spectrometry and sequence database searching. Methods Mol Biol 367:87–119

Palmgren MG, Clemens S, Williams LE, Krämer U, Borg S, Schjørring JK, Sanders D (2008) Zinc biofortification of cereals: problems and solutions. Trends Plant Sci 13:464–473

Pence NS, Larsen PB, Ebbs SD, Letham DLD, Lasat MM, Garvin DF, Eide D, Kochian LV (2000) The molecular physiology of heavy metal transport in the Zn/Cd hyperaccumulator *Thlaspi caerulescens*. Proc Natl Acad Sci U S A 97:4956–4960

Pilon-Smits E (2005) Phytoremediation. Annu Rev Plant Biol 56:5–39

Pirondini A, Visioli G, Malcevschi A, Marmiroli N (2006) A 2-D liquid–phase chromatography analysis in plant tissues. J Chromatogr B 833:91–100

Pollard AJ, Powell KD, Harper FA, Smith JAC (2002) The genetic basis of metal hyperaccumulation in plants. Crit Rev Plant Sci 21:539–566

Poschenrieder C, Tolrà R, Barceló J (2006) Can metals defend plants against biotic stress? Trends Plant Sci 11:288–295

Rabilloud T (1996) Solubilization of proteins for electrophoretic analyses. Electrophoresis 17:813–829

Ramos-Onsins SE, Stranger BE, Mitchell-Olds T, Aguadé M (2004) Multilocus analysis of variation and speciation in the closely related species *Arabidopsis halleri* and *A. lyrata*. Genetics 166:373–388

Reeves RD (2003) Tropical hyperaccumulators of metals and their potential for phytoextraction. Plant Soil 249:57–65

Reeves RD (2006) Hyperaccumulation of trace elements by plants. In: Morel JL, Echevarria G, Goncharova N (eds) Phytoremediation of metal-contaminated soils. NATO science series: IV: earth and environmental sciences, Vol. 68 New York, NY, USA, Springer 1–25

Reeves RD, Baker AJM (2000) Phytoremediation of toxic metals: using plants to clean up the environment. In: Raskin I, Ensley BD (eds) Metal accumulating plants. Wiley, New York

Repetto O, Bestel-Corre G, Dumas-Gaudot E, Berta G, Gianinazzi-Pearson V, Gianinazzi S (2003) Targeted proteomics to identify cadmium-induced protein modifications in *Glomus mosseae*-inoculated pea roots. New Phytol 157:555–568

Richau KM, Schat H (2009) Intraspecific variation of nickel and zinc accumulation and tolerance in the hyperaccumulator *Thlapsi caerulescens*. Plant Soil 293:107–119

Richau KH, Kozhevnikova AD, Seregin IV, Vooijs R, Koevoets PLM, Smith JAC, Ivanov VB, Schat H (2009) Chelation by histidine inhibits the vacuolar sequestration of nickel in roots of the hyperaccumulator *Thlaspi caerulescens*. New Phytol 183:106–116

Rigola D, Fiers M, Vurro E, Aarts MGM (2006) The heavy metal hyperaccumulator *Thlaspi caerulescens* expresses many species-specific genes, as identified by comparative expressed sequence tag analysis. New Phytol 170:753–766

Rose JKC, Bashir S, Giovannoni JJ, Jahn MM, Saravanan RS (2004) Tackling the plant proteome: practical approaches, hurdles and experimental tools. Plant J 39:715–733

Rustichelli C, Visioli G, Kostecka D, Vurro E, Sanità di Toppi L, Marmiroli N (2008) Proteomic analysis in the lichen *Physcia adscendens* exposed to cadmium stress. Environ Pollut 156:1121–1127

Sors TG, Ellis DR, Na GN, Lahner B, Lee S, Leustek T, Pickering IJ, Salt DE (2005) Analysis of sulfur and selenium assimilation in *Astragalus* plants with varying capacities to accumulate selenium. Plant J 42:785–797

Trotta A, Falaschi P, Cornara L, Minganti V, Fusconi A, Drava G, Berta G (2006) Arbuscular mycorrhiza increase the arsenic translocation factor in the As hyperaccumulating fern *Pteris vittata* L. Chemosphere 65:74–81

Tuomainen MH, Nunan N, Lehesranta SJ, Tervahauta AI, Hassinen VH, Schat H, Koistinen KM, Auriola S, McNicol J, Karenlampi S (2006) Multivariate analysis of protein profiles of metal hyperaccumulator *Thlaspi caerulescens* accessions. Proteomics 6:3696–3706

Tuomainen M, Tervahauta A, Hassinen V, Schat H, Koistinen KM, Lehesranta S, Rantalainen K, Häyrinen J, Auriola S, Anttonen M, Kärenlampi S (2010) Proteomics of *Thlapsi caerulescens* accessions and an interaccession cross segregating for zinc accumulation. J Exp Bot 61:1075–1087

Turnau K, Mesjasz-Przybylowicz J (2003) *Arbuscular mycorrhiza* of *Berkheya coddii* and other Ni-hyperaccumulating members of Asteraceae from ultramafic soils in South Africa. Mycorrhiza 13:185–190

Valot B, Negroni L, Zivy M, Gianinazzi S, Dumas-Gaudot E (2006) A mass spectrometric approach to identify Arbuscular mycorrhiza-related proteins in root plasma membrane fractions. Proteomics 6:S145–S155

van de Mortel JE, Almar Villanueva L, Schat H, Kwekkeboom J, Coughlan S, Perry D, Moerland PD, van Themaat EVL, Koornneef M, Aarts MGM (2006) Large expression differences in genes for iron and zinc homeostasis, stress response, and lignin biosynthesis distinguish roots of *Arabidopsis thaliana* and the related metal hyperaccumulator *Thlaspi caerulescens*. Plant Physiol 142:1127–1147

van de Mortel JE, Schat H, Moerland PD, van Themaat EVL, van der Ent S, Blankestijn H, Ghandilyan A, Tsiatsiani S, Aarts MGM (2008) Expression differences for genes involved in lignin, glutathione and sulphate metabolism in response to cadmium in *Arabidopsis thaliana* and the related Zn/Cd hyperaccumulator *Thlaspi caerulescens*. Plant Cell Environ 31:301–324

Verbruggen N, Hermans C, Schat H (2009) Molecular mechanisms of metal hyperaccumulation in plants. New Phytol 181:759–776

Vincent D, Wheatley DM, Cramer RG (2006) Optimization of protein extraction and solubilization for mature grape berry clusters. Electrophoresis 27:1853–1865

Visioli G, Marmiroli M, Marmiroli N (2010a) Two dimensional liquid chromatography (2D-LC) technique coupled with mass spectrometry analysis to compare the proteomic response to Cd stress in plants. J Biomed Biotechnol 2010:1–10

Visioli G, Pirondini A, Malchevshi A, Marmiroli N (2010b) Comparison of proteomic variation in *Thlaspi caerulescens* populations from metalliferous and non metalliferous soils. Int J Phytoremediat 12:805–819

Vogel-Mikuš K, Pongrac P, Kump P, Nečemer M, Regvar M (2006) Colonisation of a Zn, Cd and Pb hyperaccumulator *Thlaspi praecox* Wulfen with indigenous Arbuscular mycorrhiza fungal mixture induces changes in heavy metal and nutrient uptake. Environ Pollut 139:362–371

Walliwalagedara C, Atkinson I, van Keulen H, Cutright T, Wei R (2010) Differential expression of proteins induced by lead in the dwarf sunflower *Helianthus annuus*. Phytochemistry 71:1460–1465

Waschke A, Sieh D, Tamasloukht M, Fischer K, Mann P, Franken P (2006) Identification of heavy metal-induced genes encoding glutathione S-transferases in the Arbuscular mycorrhira fungus *Glomus intraradices*. Mycorrhiza 17:1–10

Weber M, Harada E, Vess C, Roepenack-Lahaye EV, Clemens S (2004) Comparative microarray analysis of *Arabidopsis thaliana* and *Arabidopsis halleri* roots identifies nicotinamine synthase, a ZIP transporter and other genes as potential hyperaccumulation factors. Plant J 37:269–281

Whiting SN, De Souza MP, Terry N (2001) Rhizosphere bacteria mobilize Zn for hyperaccumulation by *Thlaspi caerulescens*. Environ Sci Technol 35:3144–3150

Willems G, Drager DB, Courbot M, Gode C, Verbruggen N, Saumitou-Laprade P (2007) The genetic basis of zinc tolerance in the metallophyte *Arabidopsis halleri* ssp. *halleri* (*Brassicaceae*): an analysis of quantitative trait loci. Genetics 176:659–674

Yang XE, Feng Y, He ZL, Stoffella PJ (2005) Molecular mechanisms of heavy metal hyperaccumulation and phytoremediation. J Trace Elem Med Bio 18:339–353

Zeng XW, Qiu RL, Ying RR, Tang YT, Tang L, Fang XH (2011) The differentially-expressed proteome in Zn/Cd hyperaccumulator *Arabis paniculata* Franch. in response to Zn and Cd. Chemosphere 82:321–328

Heavy Metal Toxicity: Oxidative Stress Parameters and DNA Repair

Dinara Jaqueline Moura, Valéria Flores Péres, Rosangela Assis Jacques, and Jenifer Saffi

Abstract Plant growth and productivity are adversely affected by frequent exposure to various abiotic and biotic stress factors, such as heavy metals. Heavy metals are elements with a relatively high density and are toxic or poisonous. Different molecular mechanisms for heavy metal toxicity have been described, and, among these, the production of reactive oxygen species (ROS) deserves special attention. ROS are highly reactive atoms or molecules naturally produced in plants and predominantly formed in the electron transport chain of cellular respiration (chloroplasts) and in photoreactivation. Oxidative attack on DNA generates both altered bases and damaged sugar residues that undergo fragmentation and lead to strand breaks. DNA damage caused by exposure to ROS is one of the primary causes of DNA decay in most organisms. The irreversible DNA damage can interfere with plant development and affect crop productivity. To protect the cells, a complex network of proteins is activated for damage control and repair that includes both antioxidant enzymes removing ROS and DNA repair proteins. The DNA repair response includes different pathways, such as mismatch repair (MMR), base excision repair (BER), nucleotide excision repair (NER), nonhomologous end joining (NHEJ), and homologous recombination (HR). Recent advances in the study of DNA repair in higher plants show that they use similar mechanisms to those present in other eukaryotes to remove and/or tolerate oxidized bases and other oxidative DNA lesions. The aim of this chapter is to review the latest data regarding the induction of oxidative stress by heavy metals in higher plants,

D.J. Moura • V.F. Péres • J. Saffi (✉)
Department of Basic Health Sciences, Federal University of Health Sciences of Porto Alegre, Porto Alegre, RS, Brazil
e-mail: dinjamoura@gmail.com; valefp@hotmail.com; jenifers@ufcspa.edu.br

R.A. Jacques,
Department of Chemistry, Federal University of Rio Grande do Sul, Porto Alegre, RS, Brazil
e-mail: rosangela.j@globo.com

D.K. Gupta and L.M. Sandalio (eds.), *Metal Toxicity in Plants: Perception, Signaling and Remediation*, DOI 10.1007/978-3-642-22081-4_9,

introducing some basic concepts and seeking a relationship with DNA damage repair mechanisms.

Abbreviations

Ag	Silver
AP	Site apurinic and/or apyrimidinic site
APX	Ascorbate peroxidase
As	Arsenic
BER	Base excision repair
CAT	Catalase
Cd	Cadmium
Co	Cobalt
Cr	Chromium
Cu	Copper
Cu/Zn-SOD	Cooper/zinc superoxide dismutase
Cys	Cysteine
DMA^V	Dimethylarsinic acid
DMDSe	Dimethyl diselenide
DMSe	Dimethyl selenide
DSB	Double strand breaks
Fe	Iron
Fe-SOD	Iron superoxide dismutase
gEC	γ-Glutamylcystein
gECS	γ-Glutamylcystein synthetase
GGR	Global genome repair
GPX	Glutathione peroxidase
GSH	Glutathione
H_2O_2	Hydrogen peroxide
Hg	Mercury
HR	Homologous recombination
LMWT	Low-molecular-weight thiols
MMR	Mismatch repair
Mn	Manganese
Mn-SOD	Manganese superoxide dismutase
Mo	Molybdenum
NER	Nucleotide excision repair
NHEJ	Nonhomologous end joining
Ni	Nickel
$O_2^{\bullet -}$	Superoxide radicals
$OH^{\bullet}$	Hydroxyl radicals
Pb	Lead
PC	Phytochelatins

RAPD	Random amplification of polymorphic DNA
SCE	Sister chromatid exchanges
SOD	Superoxide dismutase
TCR	Transcription-coupled repair
Zn	Zinc

1 Introduction

An increase in world population and urbanization has created environmental problems such as air pollution, soil sealing, and water quality. Heavy metals are elements commonly present in particulate matter, soils, and wastewater. These may be toxic when present in high concentrations and can react with macromolecules and ligands in molecular membranes to promote bioaccumulation (Giampaoli 2010). Based on the chemical and physical properties of the various heavy metals, three different molecular mechanisms for heavy metal toxicity have been recognized: (1) production of reactive oxygen species (ROS) by autoxidation and Fenton reactions, (2) blocking essential functional groups in biomolecules, and (3) displacement of essential metal ions from biomolecules (Cuypers et al. 2002; Schützendübel and Polle 2002). The most common result of this abiotic stress is an increased production of ROS. ROS may lead to the nonspecific attack of vital molecules, such as lipids, proteins, and nucleic acids, which can result in structural, metabolic, and physiological disorders in cells and may lead to cell death (Halliwell and Gutteridge 1989; Bray et al. 2000).

Heavy metals in the environment can originate from both natural (e.g., volcanoes and continental dust) and anthropogenic (e.g., mining, fossil fuels, phosphate fertilizers, and metal-working) sources (Nagajyoti et al. 2010). Heavy metals, including copper (Cu), cadmium (Cd), lead (Pb), zinc (Zn), and chromium (Cr), that have come from activities such as mining and smelting have contaminated extensive regions of the world. This has resulted in significant environmental changes that are likely to override the adaptive potential of plants. This chapter reviews the latest data regarding the induction of oxidative stress by heavy metals in higher plants, introducing some basic concepts and seeking a relationship with DNA damage repair mechanisms.

2 Oxidative Stress and Cell Defenses

Reactive oxygen species are highly reactive atoms or molecules found in all aerobic biological organisms. They are naturally produced in plants and are predominantly formed in the electron transport chain of chloroplasts (cellular respiration) and in photoreactivation. The main reactive species formed from these processes are superoxide radicals ($O_2^{\bullet-}$), hydroxyl radicals ($OH^{\bullet}$), which are highly reactive molecules with a short half-life, and hydrogen peroxide (H_2O_2), a signaling molecule that can cross membranes and accumulate within cells (Giampaoli 2010; Pinto et al. 2003; Apel and Hirt 2004).

Plants have developed a variety of strategies to prevent excessive accumulation of nonessential metals within cells and/or transform these metals into less toxic forms (Cobbett 2000). Some plants produce metabolites that bind to heavy metals in the cytosol, such as glutathione (GSH), polypeptides and proteins (e.g., metallothioneins and phytochelatins) and proline (Hall 2002). However, when these defense mechanisms are not sufficient, ROS overproduction occurs, which causes oxidative stress and activates other mechanisms (Patra et al. 2004). The defense system includes enzymatic and nonenzymatic antioxidants. Representatives of the enzymatic antioxidant defense system include superoxide dismutase, ascorbate peroxidase, and catalase for ROS removal from cells (Halliwell and Gutteridge 1989). The main nonenzymatic water-soluble defenses include ascorbic acid (AA), thiols (predominantly glutathione), α-tocopherol, carotenoids, flavonoids, and polyamines, which have also been shown to promote reparation of injuries caused by ROS (Bray et al. 2000). The ROS produced in young leaf cells, for example, are removed by nonenzymatic systems (e.g., ascorbate, glutathione, α-tocopherol) and enzymatic (e.g., catalase (CAT), ascorbate peroxidase (APX), glutathione peroxidase (GPX), and superoxide dismutase (SOD)) antioxidant systems (Panda et al. 2003; Giampaoli 2010).

Superoxide dismutase is a metalloenzyme classified into three groups based on the specific metal cofactor used. Typically, plants have different SOD isoforms located in the various compartments of the cell. Cu/Zn-SOD is found in the cytosol, Cu/Zn and/or Fe-SOD are located in the chloroplasts, and Mn-SOD is in the mitochondria (Alscher et al. 2002; Apel and Hirt 2004; Giampaoli 2010). These enzymes act as the first line of defense against ROS by catalyzing the dismutation of $O_2^{\bullet-}$ to hydrogen H_2O_2. Lower concentrations of H_2O_2 formed from $O_2^{\bullet-}$ dismutation can be a signal for the activation of the antioxidant system of plants (Gratão et al. 2005). Either H_2O_2 addition or its endogenous induction into leaf tissues promote stomata closing and consequently reduce the transpiration rate. This acts as a signal for inducing the expression of genes related to the activation of CAT, APX, GPX, and guaiacol peroxidase (Deuner et al. 2008; Giampaoli 2010). However, high concentrations of H_2O_2 can be damaging to tissues because it can cross cell membranes and promote lipid peroxidation (Resende et al. 2003). H_2O_2 is decomposed by catalase or by the ascorbate/glutathione cycle, which is the primary system for ROS removal in chloroplasts.

Ascorbate peroxidase is found in higher-order plants, chlorophytes, and red algae. This enzyme has a high specificity for ascorbate (or ascorbic acid) as the reducing substrate, and it subsequently catalyzes the reduction of hydrogen peroxide into water (Carvezan 2008). The ascorbic acid acts both as a cofactor and reducing agent for the enzyme, allowing the enzyme to detoxify various organic radicals, counteract $O_2^{\bullet-}$ and H_2O_2 and remove active oxygen (Halliwell and Gutteridge 1989).

As previously mentioned, ROS can readily oxidize most types of macromolecules, including lipids, protein, and nucleic acids. Major targets in living cells are the electron-rich bases of DNA, which can undergo oxidation to produce a diverse range of genotoxic modifications (Dizdaroglu et al. 2002). The irreversible DNA damage can interfere with plant development and affect crop productivity (Tuteja et al. 2009; Gill and Tuteja 2010). Oxidative attack to DNA generates both modified bases and damaged sugar residues that facilitate fragmentation and lead to strand breaks. Oxidative attack to DNA bases generally involves $OH^{\bullet}$ addition to double bonds, while sugar damage mainly results from hydrogen abstraction from deoxyribose (Roldán-Arjona and Ariza 2009). To protect the cells, a complex network of proteins is activated for damage control and repair that includes both antioxidant enzymes removing ROS and DNA repair proteins (Balestrazzi et al. 2011).

3 DNA Repair Mechanisms: A General Overview

The DNA repair response includes different pathways, such as photoreactivation or direct repair, mismatch repair (MMR), base excision repair (BER), nucleotide excision repair (NER), nonhomologous end joining (NHEJ), and homologous recombination (HR). Photoreactivation is a mechanism of UV-damage DNA repair that is performed by photolyases using the energy of visible light. In MMR, NER, and BER, damaged bases are excised from the genome and replaced by corrected nucleotides following the most common DNA repair rules. HR and NHEJ are responsible for repairing double strand breaks (DSB) (Kimura and Sakaguchi 2006; Balestrazzi et al. 2011).

3.1 Base Excision Repair (BER)

BER is a multistep process and is involved in the repair of the most frequent lesions, such as abasic sites (i.e., apurinic and/or apyrimidinic-AP site) and single strand breaks (SSB). Oxidized bases in DNA are frequently removed by BER. Several studies specifically performed in *Arabidopsis thaliana* have shown that BER proteins from plants share a common structure and/or homology of most animal BER proteins (Roldán-Arjona and Ariza 2009). BER is initiated by DNA glycosylases that, through hydrolysis, cleave the glycosyl bond between the deoxyribose and oxidized base, releasing the free damaged base and leaving an abasic site. The AP site formed from the removal reaction is then incised by an AP endonuclease. Some DNA glycosylases are known to have AP endonuclease activity. Ten types of DNA glycosylases and three types of AP endonucleases have been identified in the rice genome. In *A. thaliana*, *Saccharomyces cerevisiae* homologs of Ntg (*AtNTH1* and *AtNTH2*), Ogg1 (*AtOGG1*), and APN2 (*AtArp*,

AtAPE1 and *AtAPE2*) have been described (Roldán-Arjona and Ariza 2009). Further steps for complementation during BER require additional enzymes and accessory proteins. For example, following the cut made by AP endonucleases at positions 3′ and 5′ of DNA abasic sites, a phosphodiesterase then removes sugar-phosphate waste, and the gap is filled by polymerase, an enzyme commonly associated with replication factors PCNA and FEN. Finally, either DNA ligase I or a complex of DNA ligase III and XRCCI make the ligation. In mammalian cells, DNA polymerase β (pol β) fills the single nucleotide gap, preparing the strand for ligation by DNA ligase I or the DNA ligase III/XRCC1 complex. *Arabidopsis* lacks any clear pol β homolog, but it has been proposed that its role could be fulfilled by a homolog of another polymerase X family member, named pol λ (García-Díaz et al. 2000). Though XRCC1 is present in plants, DNA ligase III, which probably forms a complex with XRCC1 in animal systems, has not been found (Kimura and Sakaguchi 2006).

3.2 *Nucleotide Excision Repair (NER)*

Another mechanism of excision is NER, which is responsible for processing DNA lesions that generate major distortions in the helical DNA structure (Kunz et al. 2005). Although several NER-related functions have been characterized in plants, information concerning the regulatory aspects of NER is missing (Balestrazzi et al. 2011). Recently, Singh et al. (2011) highlighted the fact that NER proteins were relatively more conserved in amino acid sequence between *Arabidopsis* and rice than proteins required for other repair pathways. NER mechanisms rely on a series of reactions that include recognition of DNA damage, unwinding double-stranded DNA in the neighborhood of the damage, excising the damaged nucleotides, and filling the single-stranded gap by DNA synthesis. This DNA repair process can be classified as either global genome repair (GGR), which repairs DNA damage anywhere in the genome, or transcription-coupled repair (TCR), which specifically restores DNA strands that are being transcribed (Singh et al. 2011).

NER-related genes have been isolated by analysis of UV-sensitive mutants (*UVH*) in *Arabidopsis*, such as *AtRad1* (*XPF*), *AtRad2* (*XPG*), and *AtRad3* (*XPD*). In addition, the NER-related genes *XPB* (*AtRad25*), *AtUVDDB2*, *AtRad23*, and *ERCC1* (*AtRad10*) have also been isolated from *Arabidopsis*. The involvement of *XPF* (*AtRad1*) in the repair of oxidatively damaged DNA was demonstrated by an in vitro DNA repair assay using cell extracts of *Arabidopsis* (Kimura and Sakaguchi 2006).

3.3 *Mismatch Repair (MMR)*

MMR restores correct matches between mismatched base pairs formed by the incorporation of an incorrect base by DNA polymerase or during recombination.

The MMR system in animals and plants functions as a combination of at least five proteins that recognize mismatches and repair DNA via excision and replacement with correct nucleotides. This system occurs in four main steps: (1) incorrectly paired base recognition and association of enzymes to form the repair complex; (2) identification of the leader strand that contains the correct genetic information and mismatch base elimination, together with a small nucleotide sequence; (3) synthesis through the action of a DNA polymerase; and (4) restoration of the normal structure of DNA with DNA ligase (Lario et al. 2011). *MSH* genes, which are homologs of *Escherichia coli MutS*, have been isolated from plants but predominantly from *Arabidopsis*. The *MSH2/MSH6* and *MSH2/MSH7* complexes in *Arabidopsis* recognize different nucleotide mismatches. Additionally, *Mus*1 and *Mus*2, homolog of MutS, have been isolated from *Zea mays*. *MLH*1, a homolog of MutL, has also been isolated from *Arabidopsis* (Horwath et al. 2002; Gomez et al. 2011).

3.4 Double-Strand Break Repair

Oxidative attack on DNA generates both altered bases and damaged sugars that undergo fragmentation and lead to DNA strand breaks. DSB are often a consequence of ROS attack on DNA and are a very severe type of DNA damage that can be rapidly repaired in cells by two mechanisms, HR and NHEJ. While HR creates covalent linkages between DNA in regions of highly similar or identical sequence, the DNA ends of severed chains are joined directly in NHEJ, and deletions or insertions may then occur (Puchta 2005). It has been suggested that higher plants possess a more efficient repair of DSB than mammalian cells due to a higher tolerance for ionizing radiation (Yokota et al. 2005). Again, *Arabidopsis* has been the main plant studied. *AtRad50* and *AtMre11*, which are involved in homologous and nonhomologous DNA repair processes, as well as in DNA damage sensing and signaling, have been isolated and investigated (Gallego et al. 2001; Daoudal-Cotterell et al. 2002; Puizina et al. 2004). In rice and in *Arabidopsis*, the HR proteins *AtRad51*, *OsRad51*, *AtRadA*, and *OsRadA* showed a high degree of homology with mammalian Rad51 and RadA/Sms, a eubacterial homolog of RecA. *AtKu70*, *AtKu80*, DNA ligase IV, and *AtXRCC4* are all involved in NHEJ and have been isolated and analyzed in plants (West et al. 2002). Interestingly two NER-related genes, *AtXPF* and *AtERCC1*, have also been shown to be involved in plant DSB repair (Hefner et al. 2003; Dubest et al. 2004). As demonstrated by Cuypers et al. (1999), each heavy metal induces different oxidative stress mechanisms and different responses to the exposure that are determined by the chemical properties of each metal. As a transition element, Ni (nickel) can participate in the Haber–Weiss reactions, generating free radicals directly. Metals such as Cd and aluminum (Al) have only one oxidation state (2^+ and 3^+, respectively) and are not transition elements. Cd is often treated as one, at least with regard to the criterion of having an unpaired electron in its valence orbital. These properties influence the mechanisms by which each metal produces oxidative stress.

4 Heavy Metals

4.1 Arsenic

Arsenic (As) is a naturally occurring toxic metal, and its presence in food represents a potential risk to the health of both humans and animals (Al Rmalli et al. 2005). Inorganic arsenic occurs naturally in soil, air, and water in addition to anthropogenic sources, such as mining, agriculture, and nonagricultural activities. Arsenic is used industrially in the manufacture of numerous products, including glass, ceramics, electronics, cosmetics, and fireworks (Smith et al. 2009). In the latter half of the twentieth century, As was also widely used in pesticide and herbicide formulations and in wood preservation, although such use is now declining (Xie et al. 2009). More than 300 As minerals are known to occur in nature. Of these, approximately 60% are arsenates, 20% are sulfides and sulfosalts, 10% are oxides, and the rest are arsenites, arsenides, native elements, and metal alloys (Bowell and Parshley 2001). The most important primary As-bearing minerals are those in which As occurs as the anion (arsenide), dianion (diarsenide), or as the sulfarsenide anion(s); these anions are bonded to metals such as Fe (e.g., löllingite or arsenopyrite), Co (cobaltite), and Ni (gersdorffite). The simple As sulfide minerals realgar and orpiment can also commonly be found (Drahota and Filippi 2009). These primary As-bearing minerals are listed in Table 1. Arsenic exists in either organic or inorganic form but is normally not encountered in its elemental state. Typically, inorganic fractions contain oxygenated As anions or more complex As salts with, for example, sulfur and iron, of which arsenopyrite (FeAsS) is the most abundant (Smedley and Kinniburgh 2002). However, the most prevalent inorganic As species are the pentavalent As(V) (occurring as $H_2AsO_4^-$ and $HAsSO_4^{2-}$ in most environments) and the trivalent As(III) (As_2O_3), which dissolves in water as As$(OH)_3$ (Ali et al. 2009). In addition, large number of organic As compounds can be found. Living organisms can contain arsenolipids and arsenosugars that may be incorporated into cellular metabolism (Ali et al. 2009). The phytotoxicity of arsenic depends on its oxidation state. Arsenite As(III) and arsenate As(V) are both

Table 1 Empirical formula of the most common primary As-bearing minerals (Drahota and Filippi 2009)

Mineral	Formula	System
Arsenopyrite	FeAsS	Monoclinic
Cobaltite	CoAsS	Orthorhombic
Enargite	Cu_3AsS_4	Orthorhombic
Gersdorffite	NiAsS	Cubic
Löllingite	$FeAs_2$	Orthorhombic
Orpiment	As_2S_3	Monoclinic
Pyrite	FeS_2	Cubic
Realgar	AsS	Monoclinic
Tennantite	$(Cu,Ag,Fe,Zn)_{12}As_4S_{13}$	Cubic

inorganic, phytoavailable forms of arsenic and are highly toxic to plants. These inorganic forms are interconvertible, depending on the redox conditions of the aquatic ecosystem (Ozturk et al. 2010). Generally, the toxicity of these compounds is lower than that of inorganic arsenic species. However, there may be exceptions to this rule. For example, in the plants *Spartina patens* and radish, dimethylarsinic acid (DMA^V) appeared to be more toxic than inorganic arsenic, whereas in human hepatocytes, cytotoxicity decreased in the following order: MMAIII > As^{III} > As^V > MMA^V = DMA^V (Sakai et al. 2010).

In higher plants, absorbed As(V) is rapidly reduced to As(III) either enzymatically by As(V) reductase or nonenzymatically by glutathione (GSH). As(III) reacts with the thiol groups (−SH) of proteins in plant tissues, leading to the inhibition of cellular function and ultimately death. In addition to these direct toxicities, the role of As-induced oxidative stress in plants has also been suggested (Sakai et al. 2010). It is known that higher plants exposed to As show a substantial increase in low-molecular-weight thiols (LMWTs), such as cysteine (Cys), γ-glutamylcysteine (gEC), GSH, and phytochelatins (PCs), all of which form complexes with As(III) and result in the inactivation of As in the plant (Srivastava et al. 2005). The thiol groups of these LMWTs are responsible for binding to As(III). Among these LMWTs, GSH and PCs, which are simply oligomers of GSH, are regarded as the major ligands for As(III). Furthermore, LMWTs also contribute to alleviating the oxidative stress induced by As (Shri et al. 2009). Thus, LMWT synthesis plays a crucial role in mechanisms of As tolerance. Many studies have focused on the LMWT synthesis pathway to investigate the mechanisms of As tolerance in higher plants. Previous attempts to enhance As tolerance by overexpressing the genes involved in the production of enzymes responsible for LMWT synthesis, thus contributing to an increase in the LMWT pool, are an example. In particular, gEC synthetase (gECS) is often targeted because gEC synthesis is a rate-limiting step in the LMWT synthesis pathway (Rao et al. 2011). An in vitro study in *Vicia faba* with arsenate showed that arsenate induced an increase of SOD and CAT activities in both leaves and roots. Additionally, DNA damage increased, indicating the genotoxicity of arsenate. These results indicated that arsenate toxicity causes oxidative stress in *V. faba*, which might be one of the mechanisms through which arsenic induces DNA damage (Lin et al. 2008).

4.2 Cadmium

Cadmium (Cd), a nonessential element, is one of the most toxic environmental and industrial pollutants. It causes DNA damage, elevates lipid peroxidation, has a long biological half-life, and represents a serious environmental pollutant for plants. It negatively affects plant growth and development (Benavides et al. 2005). Several environmental factors influence Cd absorption by the plant, such as the average metal concentration, pH and plant species. Cd is easily absorbed by plant roots and translocated via the xylem to the shoot through transpiration (Prasad 1995). Cd is strongly phytotoxic, causing growth inhibition and even plant death (Schützendübel and Polle 2002). Cd alters the functionality of cell membranes by inducing changes in enzymatic activities (e.g., H^+-ATPase activity) associated with membranes. Moreover, Cd accumulation causes reductions in photosynthesis, diminishes the uptake of both water and nutrients, and results in visible symptoms of injury in plants, such as chlorosis and growth inhibition (Drazkiewicz et al. 2003). Cd was recognized as a cause of oxidative stress in plants by either inducing ROS production or by decreasing enzymatic and nonenzymatic defenses (Fornazier et al. 2002; Cho and Seo 2005). As a consequence of oxidative stress, Cd induces changes to the functionality of membranes through lipid peroxidation (Fodor et al. 1995; Gratão 2003). In several studies, it has been shown that Cd damaged DNA, RNA, and the nucleolar structure in both animal and plant cells (Hartwig and Schwerdtle 2002; Jonak et al. 2004; Cvjetko et al. 2010). Plant species respond differently to exposure of the same metal, depending on genetic structures. *Allium sativum*, for example, has the considerable capability of removing Cd from solutions and accumulating it (Ma et al. 1995; Zhang et al. 2005). The Cd content found in the roots of *A. sativum* has been shown to increase with increasing concentrations of Cd. It has been demonstrated that Cd exposure decreases the expression of MMR genes and that this parameter could be used as a potential biomarker indicating Cd exposure in plants for environmental studies (Liu et al. 2008, 2009). Cd can have an adverse effect on the mismatch recognition process of the MutSα and MutSβ complex and nucleotide excision process of the MMR systems in *Arabidopsis* (Liu et al. 2008). Therefore, evidence suggests that proteins participating in DNA repair systems, especially in excision and mismatch repair, are sensitive targets of Cd toxicity. Cd might contribute to an increased risk for tumor formation by interfering and inhibiting these DNA repair processes, and these actions could possibly explain the indirect contribution of Cd to mutagenic effects and/or carcinogenicity (Giaginis et al. 2006).

4.3 Chromium

Chromium (Cr) is considered to be one of the most detrimental elements to the environment. The most stable and common oxidation states of Cr are trivalent Cr(III), and hexavalent Cr(VI), species (Peralta-Videa et al. 2009). Hexavalent Cr is the most toxic species because it has a high oxidizing potential, solubility, and mobility across membranes in living organisms and through the environment (Peralta-Videa et al. 2009). It has been shown to retard growth, reduce the number of palisade and spongy parenchyma cells in leaves, and increase the number of vacuoles and electron dense material along the walls of xylem and phloem (Panda and Choudhury 2005). Trivalent Cr is relatively insoluble in water and tends to form hydroxide precipitates with Fe at typical ground water pH values. However, at high concentrations of oxygen or Mn oxides, Cr(III) can be oxidized to Cr(VI) (Peralta-Videa et al. 2009). Cr-induced oxidative stress involves induction of lipid peroxidation in plants that results in severe damage to cell membranes. Oxidative stress induced by chromium initiates the degradation of photosynthetic pigments, causing a decline in plant growth. High concentrations of chromium can disturb the chloroplast ultrastructure, thereby disturbing the photosynthetic process. Like copper and iron, chromium is also a redox metal and its redox behavior exceeds that of other metals, such as Co (cobalt), Fe (iron), Zn (zinc), and Ni. Its redox behavior can thus be attributed to the direct involvement of chromium in inducing oxidative stress in plants. Chromium affects antioxidant metabolism in plants. Antioxidant enzymes like SOD, CAT, POX, and GR are found to be susceptible to chromium, resulting in a decline of their catalytic activities. This decline in antioxidant efficiency is an important factor in generating oxidative stress in plants under chromium stress. However, both metallothioneins and organic acids are important to plants as components of tolerance mechanisms, and they are also involved in the detoxification of chromium (Panda and Choudhury 2005).

The chromium genotoxic effect in mammalian cells has also been shown. Chromium interacts with DNA breaks induced in the molecule; however, little is known about possible DNA interactions in plants. In this sense, Labra et al. (2004) demonstrated that potassium dichromate induced extensive methylation changes in CCGG-sequences with the net result being genome-wide hypermethylation, which suggested clear DNA modification in plants as a response to chromium exposure. Cr (III) binds to DNA molecules or can be complexed with hydrophobic ligands, such as 1,10-phenanthroline, 2,2-bipyridine, or picolinic acid, with which it is transported across plasma membranes to produce gene mutations (Peralta-Videa et al. 2009).

4.4 Copper

Copper (Cu) is an essential micronutrient and a component of several enzymatic systems, participating primarily in electron flow and catalysis of redox reactions, but excess Cu is toxic to organisms (Schützendübel and Polle 2002). At high concentrations, Cu is involved in the formation of OH^- from H_2O_2 via Haber–Weiss and Fenton reactions, and it initiates nonspecific lipid peroxidation.

Haber–Weiss reaction

$$\underbrace{\left(Fe^{3+}/Cu^{2+}\right)}_{\text{Metal oxidized}} + O_2^{\bullet -} \rightarrow \underbrace{\left(Fe^{2+}/Cu^{+}\right)}_{\text{Metal reduced}} + O_2 \quad (1)$$

Fenton reaction

$$\underbrace{\left(Fe^{2+}/Cu^{+}\right)}_{\text{Metal reduced}} + H_2O_2 \rightarrow \underbrace{\left(Fe^{3+}/Cu^{2+}\right)}_{\text{Metal oxidized}} + OH^{\bullet} + OH^- \quad (2)$$

Cumulative results in Haber–Weiss/Fenton reaction

$$O_2^{\bullet -} + H_2O_2 \rightarrow O_2 + OH^{\bullet} + OH^- \quad (3)$$

The genotoxic effects of increased amounts of Cu in plants have been observed. In *Vicia faba* and *Pisum sativum*, Cu induces chromosomal alterations and increases micronucleus induction at low concentrations (Marcato-Romain et al. 2009). Additionally, various concentrations of Cu(II) were shown to induce changes in the genome template stability of *Solanum melongena* L. using random amplification of polymorphic DNA (RAPD) analyses (Körpe and Aras 2011).

4.5 Lead

Lead (Pb) is a pollutant that readily accumulates in soil and sediments. Pb environmental contamination results from mining and smelting activities, as well as from gasoline and explosives (Sharma and Dubey 2005). As Pb is not an essential element, plants do not have channels for Pb uptake. Instead, this element binds to carboxylic groups of mucilage uronic acids on root surfaces, but it is still unknown how this element progresses into the root tissue (Peralta-Videa et al. 2009). High levels of Pb in soil cause abnormal morphology in many plants, e.g., the irregular radial thickening of pea roots and cell walls of the endodermis as well as lignification of cortical parenchyma. Pb inhibits the activity of enzymes by reacting with their sulfhydryl groups; moreover, Pb induces oxidative stress by increasing the production of ROS in plants (Sharma and Dubey 2005). Pb increases DNA damage in the root but not the leaf of *Nicotiana tabacum L.* Tobacco plants exposed to Pb^{2+} presented with increased levels of DNA damage, were severely injured, and showed stunted growth and distorted leaves, and brown root significantly increased (Gichner et al. 2008).

4.6 Mercury

Mercury (Hg) is a unique metal due to its different forms (e.g., HgS, Hg^{2+}, Hg^0, and methyl-Hg), but in agricultural soils it is predominantly found in the ionic form (Hg^{2+}) (Han et al. 2006). Hg is relatively uncommon in the earth's crust, but it is released by natural processes (e.g., erosion and volcanic activity) or mining. Anthropogenic activities, both intentional (e.g., manufacture and distribution of mercury-based products) and unintentional (e.g., fossil fuel burning), represent the main sources of environmental contamination (Nagajyoti et al. 2010). However, the toxic effects of mercury in plants, which absorb the heavy metal, depend on factors such as plant species and soil properties. The main concern regarding mercury in plants is its incorporation into the food chain through the consumption of contaminated plants (Loredo et al. 2010). Increasing evidence has shown that Hg^{2+} can readily accumulate in higher plants. In low levels it may not significantly affect plant growth, but Hg is strongly phytotoxic to plant cells at high concentrations because of its ability to bind water channel proteins, interfere with mitochondrial activity, and induce oxidative stress, which triggers ROS generation. The genotoxic effects of mercury are related to an increase of ROS (Collin-Hansen et al. 2005; Kuzmick et al. 2007).

4.7 Selenium

Selenium (Se) is a nonmetallic element that resembles sulfur (S) and occurs naturally as a trace element in most soils, rocks, and water (Djanaguiraman et al. 2010). Selenium has long been recognized as an essential micronutrient for animal and human nutrition because it is an integral part of the enzyme glutathione peroxidase, a selenoenzyme that prevents oxidative damage to body tissues (Akbulut and Cakir 2010). Recent studies have shown that Se, at low concentrations, can protect plants from several types of abiotic stresses (Hawrylak-Nowak et al. 2010). Many kinds of biotic and abiotic stresses, such as drought, extreme temperature, soil salinity, and heavy metals, are known to cause oxidative damage to plants, either directly or indirectly, by triggering an increase of ROS. To resist oxidation damage, antioxidant enzymes and certain metabolites play important roles, leading to adaptation and ultimately the survival of plants during periods of stress (Hartikainen et al. 2000). Plants play a crucial role in this respect. For example, Se-accumulating plants can be used to supply this element in the human or animal diet in many areas that are Se-deficient. Because of its antioxidative and defense gene expression role, it is possible that Se can delay leaf senescence and increase the carbon supply to developing grain under stress (Prasad et al. 2006).

Some studies suggest that Se toxicity leads to the production of lipid peroxides and induces some of the crucial enzymes in the antioxidant defense system in barley seedlings. On the other hand, the capacities of plants to absorb and sequester Se can also be utilized to manage environmental Se contamination by phytoremediation. Selenium becomes toxic at higher levels due to the incorporation of Se into S-containing molecules, especially the nonspecific replacement of Cys with SeCys in proteins (Xie et al. 2009). This replacement of sulfur with Se in molecules is due to the chemical similarity between these two elements. Additionally, most enzymes involved in sulfur metabolism can also catalyze the analogous reaction with the corresponding Se substrates with similar affinities. Se-hyperaccumulating plant species, such as *Astragalus bisulcatus*, and secondary accumulators, such as Indian mustard (*Brassica juncea*), have attracted significant interest for their ability to accumulate and volatilize Se in the phytoremediation of Se-contaminated soils (Bañuelos et al. 2007). Selenium volatilization converts highly toxic selenate and selenite into volatile dimethyl selenide (DMSe) and dimethyl diselenide (DMDSe), which are 500–700 times less toxic that Se (Wilber 1980). This process provides a low-cost, environmentally friendly, highly efficient approach for the cleanup of Se-contaminated environments. The conversion of inorganic forms of Se into volatile Se in plants is believed to occur via the sulfur metabolic pathway (Ellis and Salt

2003). Se is predominantly present in soils as selenate (SeO_4^{-2}) and selenite (SeO_3^{-2}). While selenate is actively taken up into plants through sulfur transporters, selenite enters plant cells passively. The reduction of these oxidized forms of Se results in the production of selenoamino acids, such as SeCys and SeMet.

In Se-nonaccumulating plants, SeCys and SeMet are readily incorporated into proteins nonspecifically. In Se-accumulating plants, they are metabolized primarily into various nonproteinogenic selenoamino acids. These selenoamino acids can be further metabolized into the volatile Se compounds DMSe and DMDSe. While Se nonaccumulators mainly volatilize DMSe, accumulators primarily emit DMDSe. The genotoxicity of selenium is related to the perturbation of the cell redox homeostasis as a consequence of ROS generation through its reaction with biological thiols. In other words, Se depletes the thiol content through adduct formation without affecting the rate of oxidation.

4.8 Zinc

Zinc (Zn) is an essential micronutrient that has a role in several metabolic processes of plants. Zinc deficiency inhibits plant growth and development. However, an excess of zinc results in a decrease in growth, development, and metabolism, as well as an oxidative damage in various plant species. In many plants, excess Zn generates ROS and displaces other metals from active sites in proteins. Chronic exposure to Zn in the aquatic plant *Lemna minor* L. resulted in elevated levels of lipid peroxidation, protein oxidation, and enhanced DNA damage (Balen et al. 2011). In *Vicia faba* roots, Zn also induced a genotoxic effect measured as significant micronucleus induction (Marcato-Romain et al. 2009). Several enzymes contain Zn, such as carbonic anhydrase, alcohol dehydrogenase, SOD, and RNA polymerase. Zinc is required to maintain the integrity of ribosome. It takes part in the formation of carbohydrates and catalyzes oxidation processes in plants. Zinc also provides a structural role in many transcription factors (Nagajyoti et al. 2010).

5 Heavy Metal Hyperaccumulator Phenotypes in Plants

Some plants, in the same way as other organisms, have developed complex homeostasis mechanisms to minimize the deleterious effects of heavy metals by controlling their absorption, accumulation, and translocation. These mechanisms protect the cell by avoiding the accumulation of excess free ions in the cytosol, which result in various degrees of tolerance to certain heavy metals. Some plants not only tolerate high concentrations of heavy metals, but also hyperaccumulate the metals. The ability to acquire a tolerance against heavy metals and to accumulate metals to very high concentrations has evolved both independently and together in a number of different plant species (Memon and Schroder 2009). Approximately 400 plant species are described as hyperaccumulators of heavy metals, which are defined as plants that can accumulate more than 0.1% of its dry weight as Ni, Co, and Pb, more than 1% Zn and 0.01% of its dry weight in Cd (Baker and Brooks 1989). Plants have different tolerance mechanisms in response to excess of heavy metals, such as cell wall binding, active transport of ions into the vacuole, formation of thiol-rich peptides (e.g., phytochelatins and metallothioneins), and formation of complexes with organic acids and peptides (Memon and Schroder 2009).

Hyperaccumulators are potential species for use in the cleanup of soil contaminated with heavy metals (i.e., phytoextraction) because their use is a low-cost process that does not harm the environment. However, this potential is limited by factors, such as their specificity for a limited number of elements, most hyperaccumulators grow slowly and produce low biomass and they are often endemic species and little is known about these plants, such as their agronomic cultivation and physiology (Hall 2002; Schröder et al. 2007). However, the potential use of plants for phytoremediation of polluted environments is very promising. Better understanding of the physiological genetic and biochemical bases of metal hyperaccumulation in plants is key to the success of phytoremediation. Although heavy metals have been shown to inhibit the DNA repair process in plants (Liu et al. 2009), the exact connection between the adaptive response, heavy metal hyperaccumulation phenotype, and DNA repair need still to be evaluated. This would provide additional information for the comprehension of the phytoremediation process.

6 Conclusion

DNA damage caused by exposure to reactive oxygen species is one of the primary causes of DNA decay in most organisms. In plants, endogenous ROS are generated not only by respiration and photosynthesis, but also by active responses to certain environmental challenges, such as heavy metal intoxication. Oxidative attack on DNA generates both altered bases and damaged sugar residues that undergo fragmentation and lead to strand breaks. Plants are well equipped to cope with oxidative damage to cellular macromolecules, including DNA. Although biochemical and genetic analyses of DNA repair pathways in plants are still crawling, recent advances in the study of DNA repair in higher plants show that they use mechanisms similar to those present in other eukaryotes to remove and/or tolerate oxidized bases and other oxidative DNA lesions. They are not only interesting from a purely biological viewpoint, but they are also likely to make significant contributions to the solution of environmental problems, such as heavy metal contamination.

References

Akbulut M, Cakir S (2010) The effects of Se phytotoxicity on the antioxidant systems of leaf tissues in barley (*Hordeum vulgare* L.) seedlings. Plant Physiol Biochem 48(2–3):160–166

Al Rmalli SW, Haris PI, Harrington CF, Ayub M (2005) A survey of arsenic in foodstuffs on sale in the United Kingdom and imported from Bangladesh. Sci Total Environ 337:23–30

Ali W, Isayenkov SV, Zhao FJ, Maathuis FJ (2009) Arsenite transport in plants. Cell Mol Life Sci 66:2329–2339

Alscher RG, Erturk N, Heath LS (2002) Role of superoxide dismutases (SODs) in controlling oxidative stress in plants. J Exp Bot 53:1331–1341

Apel K, Hirt H (2004) Reactive oxygen species: metabolism, oxidative stress, and signal transduction. Annu Rev Plant Biol 55:373–399

Baker AJM, Brooks RR (1989) Terrestrial higher plants which hyperaccumulate metallic elements–a review of their distribution, ecology and phytochemistry. Biorecovery 1:81–126

Balen B, Tkalec M, Sikić S, Tolić S, Cvjetko P, Pavlica M, Vidaković-Cifrek Z (2011) Biochemical responses of *Lemna minor* experimentally exposed to cadmium and zinc. Ecotoxicology. doi:10.1007/s10646-011-0633-1

Balestrazzi A, Confalonieri M, Macovei A, Donà M, Carbonera D (2011) Genotoxic stress and DNA repair in plants: emerging functions and tools for improving crop productivity. Plant Cell Rep 30:287–295

Bañuelos G, LeDuc DL, Pilon-Smits EA, Terry N (2007) Transgenic Indian mustard overexpressing selenocysteine lyase or selenocysteine ethyltransferase exhibit enhanced potential for selenium phytoremediation under field conditions. Environ Sci Technol 41:599–605

Benavides MP, Gallego SM, Tomaro ML (2005) Cadmium toxicity in plants. Braz J Plant Physiol 17:21–34

Bowell RJ, Parshley J (2001) Arsenic cycling in the mining environment. Characterization of waste, chemistry, and treatment and disposal, proceedings and summary report on U.S. EPA workshop on managing arsenic risks to the environment, Denver, CO, USA, May 1–3 2001

Bray EA, Bailey-Serres J, Weretilnyk E (2000) Responses to abiotic stresses. In: Jones RL (ed) Biochemistry and molecular biology of plants. American Society of Plant Physilogysts Press, New York

Carvezan A (2008) Caracterização funcional dos gêneros de ascorbato peroxidase de arroz (*Oryza sativa* L.) nas interações entre estresse oxidativo e estresses abióticos. Dissertation, Universidade Federal do Rio Grande do Sul, Porto Alegre, Brazil

Cho UH, Seo NH (2005) Oxidative stress in *Arabidopsis thaliana* exposed to cadmium in due to hydrogen peroxide accumulation. Plant Sci 168:113–120

Cobbett CS (2000) Phytochelatin biosynthesis and function in heavy-metal detoxification. Curr Opin Plant Biol 3:211–216

Collin-Hansen C, Andersen RA, Steinnes E (2005) Damage to DNA and lipids in *Boletus edulis* exposed to heavy metals. Mycol Res 109(Pt 12):1386–1396

Cuypers A, Vangronsveld J, Clijsters H (1999) The chemical behaviour of heavy metals plays a prominent role in the induction of oxidative stress. Free Radic Res 31:S39–S43

Cuypers A, Vangronsveld J, Clijsters H (2002) Peroxidases in roots and primary leaves of *Phaseolus vulgaris* copper and zinc phototoxicity: a comparison. J Plant Physiol 189:869–876

Cvjetko P, Tolic S, Sikic S, Balen B, Tkalec M, Vidakovic-Cifrek Z, Pavlica M (2010) Effect of copper on the toxicity and genotoxicity of cadmium in duckweed (*Lemna minor* L.). Arh Hig Rada Toksikol 61:287–296

Daoudal-Cotterell S, Gallego ME, White CI (2002) The plant Rad50-Mre11 protein complex. FEBS Lett 516:164–166

Deuner S, Alves JD, Fries DD, Zanandrea I, Lima AA, Henrique PC, Goulart PFP (2008) Peróxido de hidrogênio e ácido ascórbico influenciando a atividade de enzimas antioxidantes de mudas de cafeeiro. Rev Ceres 55:135–140

Dizdaroglu M, Jaruga P, Birincioglu M, Rodriguez H (2002) Free radical-induced damage to DNA: mechanisms and measurement. Free Radic Biol Med 32:1102–1115

Djanaguiraman M, Prasad PV, Seppanen M (2010) Selenium protects sorghum leaves from oxidative damage under high temperature stress by enhancing antioxidant defense system. Plant Physiol Biochem 48:999–1007

Drahota P, Filippi M (2009) Secondary arsenic minerals in the environment: a review. Environ Int 35:1243–1255

Drazkiewicz M, Tukendorf A, Baszynski T (2003) Age-dependent response of maize leaf segments to cadmium treatment: effect on chlorophyll fluorescence and phytochelatin accumulation. J Plant Physiol 160:247–254

Dubest S, Gallego ME, White CI (2004) Roles of the AtErcc1 protein in recombination. Plant J 39:334–342

Ellis DR, Salt DE (2003) Plants, selenium and human health. Curr Opin Plant Biol 6:273–279

Fodor A, Szabo-Nagy A, Erdei L (1995) The effects of cadmium on the fluidity and H^{+}-ATPase activity of plasma membrane from sunflower and wheat roots. J Plant Physiol 14:787–792

Fornazier RF, Ferreira RR, Vitória AP, Molina SMG, Lea PJ, Azevedo RA (2002) Effects of cadmium on antioxidant enzyme activities in sugar cane. Biol Plant 45:91–97

Gallego ME, Jeanneau M, Granier F, Bouchez D, Bechtold N, White CI (2001) Disruption of the *Arabidopsis* RAD50 gene leads to plant sterility and MMS sensitivity. Plant J 25:31–41

García-Díaz M, Domínguez O, López-Fernández LA, de Ler LT, Saníger ML, Ruiz JF, Párraga M, García-Ortiz MJ, Kirchhoff T, del Mazo J, Bernad A, Blanco L (2000) DNA polymerase lambda (Pol lambda), a novel eukaryotic DNA polymerase with a potential role in meiosis. J Mol Biol 301:851–867

Giaginis C, Gatzidou E, Theocharis S (2006) DNA repair systems as targets of cadmium toxicity. Toxicol Appl Pharmacol 213:282–290

Giampaoli P (2010) Efeito de cobre e zinco no desenvolvimento in vitro de Aechmea blanchetiana (Baker) LB Smith. Dissertation, Instituto de Botânica da Secretaria de Estado do Meio Ambiente-São Paulo, Brazil

Gichner T, Znidar I, Száková J (2008) Evaluation of DNA damage and mutagenicity induced by lead in tobacco plants. Mutat Res 652:186–190

Gill SS, Tuteja N (2010) Polyamines and abiotic stress tolerance in plants. Plant Signal Behav 5:26–33

Gomez RL, Galles C, Spampinato CP (2011) High-level production of MSH2 from Arabidopsis thaliana: a DNA mismatch repair system key subunit. Mol Biotechnol 47:120–129

Gratão PL (2003) Análise da resposta antioxidante de células de Nicotiana tabacum cv BY-2 submetidas ao Cádmio. Dissertation, Universidade de São Paulo, Brazil

Gratão PL, Prasad MNV, Cardoso PF, Lea PJ, Azevedo RA (2005) Phytoremediation: green technology for the clean up of toxic metals in the environment. Braz J Plant Phys 7:53–64

Hall JL (2002) Cellular mechanisms for heavy metal detoxification and tolerance. J Exp Bot 53:1–11

Halliwell B, Gutteridge JMC (1989) Free radicals in biology and medicine. Clarendon, Oxford

Han FX, Su Y, Monts DL, Waggoner CA, Plodinec MJ (2006) Binding, distribution, and plant uptake of mercury in a soil from Oak Ridge, Tennessee, USA. Sci Total Environ 368:753–768

Hartikainen H, Xue T, Piironen V (2000) Se as an anti-oxidant and pro-oxidant in ryegrass. Plant Soil 225:193–200

Hartwig A, Schwerdtle T (2002) Interaction by carcinogenic metal compounds with DNA repair processes: toxicological implications. Toxicol Lett 127:47–54

Hawrylak-Nowak B, Matraszek R, Szymańska M (2010) Selenium modifies the effect of short-term chilling stress on cucumber plants. Biol Trace Elem Res 138:307–315

Hefner E, Preuss SB, Britt AB (2003) Arabidopsis mutants sensitive to gamma radiation includes the homologue of the human repair gene ERCC1. J Exp Bot 54:669–680

Horwath M, Kramer W, Kunze R (2002) Structure and expression of the *Zea mays* mutS-homologs Mus1 and Mus2. Theor Appl Genet 105:423–430

Jonak C, Nakagami H, Hirt H (2004) Heavy metal stress. Activation of distinct mitogen-activated protein kinase pathways by copper and cadmium. Plant Physiol 136:3276–3283

Kimura S, Sakaguchi K (2006) DNA repair in plants. Chem Rev 106:753–766

Körpe DA, Aras S (2011) Evaluation of copper-induced stress on eggplant (*Solanum melongena* L.) seedlings at the molecular and population levels by use of various biomarkers. Mutat Res 719:29–34

Kunz BA, Anderson HJ, Osmond MJ, Vonarx EJ (2005) Components of nucleotide excision repair and DNA damage tolerance in *Arabidopsis thaliana*. Environ Mol Mutagen 45:115–127

Kuzmick DM, Mitchelmore CL, Hopkins WA, Rowe CL (2007) Effects of coal combustion residues on survival, antioxidant potential, and genotoxicity resulting from full-lifecycle exposure of grass shrimp (*Palaemonetes pugio* Holthius). Sci Total Environ 373:420–430

Labra M, Grassi F, Imazio S, Di Fabio T, Citterio S, Sgorbati S, Agradi E (2004) Genetic and DNA-methylation changes induced by potassium dichromate in *Brassica napus* L. Chemosphere 54:1049–1058

Lario LD, Ramirez-Parra E, Gutierrez C, Casati P, Spampinato CP (2011) Regulation of plant MSH2 and MSH6 genes in the UV-B-induced DNA damage response. J Exp Bot. doi:10.1093/jxb/err001

Lin A, Zhang X, Zhu YG, Zhao FJ (2008) Arsenate-induced toxicity: effects on antioxidative enzymes and DNA damage in *Vicia faba*. Environ Toxicol Chem 27:413–419

Liu W, Yang YS, Francis D, Rogers HJ, Li P, Zhang Q (2008) Cadmium stress alters gene expression of DNA mismatch repair related genes in *Arabidopsis* seedlings. Chemosphere 73:1138–1144

Liu W, Zhou Q, Li P, Gao H, Han YP, Li XJ, Yang YS, Li Y (2009) DNA mismatch repair related gene expression as potential biomarkers to assess cadmium exposure in *Arabidopsis* seedlings. J Hazard Mater 167:1007–1013

Loredo J, Petit-Domínguez MD, Ordóñez A, Galán MP, Fernández-Martínez R, Alvarez R, Rucandio MI (2010) Surface water monitoring in the mercury mining district of Asturias (Spain). J Hazard Mater 176:323–332

Ma TH, Xu ZD, Xu CG, Mcconnell H, Rabago EV, Arreola GA, Zhang H (1995) The improved Allium/Vicia root tip micronucleus assay for clastogenicity of environmental pollutants. Mutat Res 334:185–195

Marcato-Romain CE, Pinelli E, Pourrut B, Silvestre J, Guiresse M (2009) Assessment of the genotoxicity of Cu and Zn in raw and anaerobically digested slurry with the *Vicia faba* micronucleus test. Mutat Res 672:113–118

Memon AR, Schroder P (2009) Implications of metal accumulation mechanisms to phytoremediation. Environ Sci Pollut Res Int 16:162–175

Nagajyoti PC, Lee KD, Sreekanth TVM (2010) Heavy metals, occurrence and toxicity for plants: a review. Environ Chem Lett 8:199–216

Ozturk F, Duman F, Leblebici Z, Temizgul R (2010) Arsenic accumulation and biological responses of watercress (*Nasturtium officinale* R. Br.) exposed to arsenite. Environ Exp Bot 69:167–174

Panda SK, Choudhury S (2005) Chromium stress in plants. Braz J Plant Physiol 17:95–102

Panda SK, Chaudhury I, Khan MH (2003) Heavy metals induce lipid peroxidation and affect antioxidants in wheat leaves. Biol Plant 46:289–294

Patra M, Bhowmik N, Bandopadhyay B, Sharma A (2004) Comparison of mercury, lead and arsenic with respect to genotoxic effects on plant systems and the development of genetic tolerance. Environ Exp Bot 52:199–223

Peralta-Videa JR, Lopez ML, Narayan M, Saupe G, Gardea-Torresdey J (2009) The biochemistry of environmental heavy metal uptake by plants: implications for the food chain. Int J Biochem Cell Biol 41:1665–1677

Pinto E, Sigaud-Kutner TCS, Leitão MAS, Okamoto OK, Morse D, Colepicolo P (2003) Heavy metal-induced oxidative stress in algae. J Phycol 39:1008–1018

Prasad MNV (1995) Cadmium toxicity and tolerance in vascular plants. Environ Exp Bot 35:525–545

Prasad PVV, Boote KJ, Allen LH Jr, USDA ARS (2006) Adverse of high temperature effects on pollen viability, seed-set, seed yield and harvest index of grain-sorghum [Sorghum bicolor (L). Moench] are more severe at elevated carbon dioxide due to higher tissue temperatures. Agric For Meteorol 139:237–251

Puchta H (2005) The repair of double-strand breaks in plants: mechanisms and consequences for genome evolution. J Exp Bot 56:1–14

Puizina J, Siroky J, Mokros P, Schweizer D, Riha K (2004) Mre11 deficiency in *Arabidopsis* is associated with chromosomal instability in somatic cells and Spo11-dependent genome fragmentation during meiosis. Plant Cell 16:1968–1978

Rao KP, Vani G, Kumar K, Wankhede DP, Misra M, Gupta M, Sinha AK (2011) Arsenic stress activates MAP kinase in rice roots and leaves. Arch Biochem Biophys 506:73–82

Resende MLV, Salgado SML, Chaves ZM (2003) Espécies ativas de oxigênio na resposta de defesa de plantas a patógenos. Fitopatol Bras 28:123–130

Roldán-Arjona T, Ariza RR (2009) Repair and tolerance of oxidative DNA damage in plants. Mutat Res 681:169–179

Sakai Y, Watanabe T, Wasaki J, Senoura T, Shinano T, Osaki M (2010) Influence of arsenic stress on synthesis and localization of lowmolecular-weight thiols in *Pteris vittata*. Environ Pollut 158:3663–3669

Schröder P, Navarro-Aviñó J, Azaizeh H, Goldhirsh AG, DiGregorio S, Komives T, Langergraber G, Lenz A, Maestri E, Memon AR, Ranalli A, Sebastiani L, Smrcek S, Vanek T, Vuilleumier S, Wissing F (2007) Position paper: using phytoremediation technologies to upgrade waste water treatment in Europe. Environ Sci Pollut Res Int 14:490–497

Schützendübel A, Polle A (2002) Plant responses to abiotic stresses: heavy metal-induced oxidative stress and protection by mycorrhization. J Exp Bot 53:1351–1365

Sharma P, Dubey RS (2005) Lead toxicity in plants. Braz J Plant Physiol 17:35–52

Shri M, Kumar S, Chakrabarty D, Trivedi PK, Mallick S, Misra P, Shukla D, Mishra S, Srivastava S, Tripathi RD, Tuli R (2009) Effect of arsenic on growth, oxidative stress, and antioxidant system in rice seedlings. Ecotoxicol Environ Saf 72:1102–1110

Singh SK, Choudhury SR, Roy S, Sengupta DN (2011) Understanding DNA repair and recombination in higher plant genome: Information from genome-wide. Plant Signal Behav 6:120–122

Smedley PL, Kinniburgh DG (2002) A review of the source, behaviour and distribution of arsenic in natural waters. Appl Geochem 17:517–568

Smith E, Juhasz AL, Weber J (2009) Arsenic uptake and speciation in vegetables grown under greenhouse conditions. Environ Geochem Health 31:125–132

Srivastava M, Ma LQ, Singh N, Singh S (2005) Antioxidant responses of hyper-accumulator and sensitive fern species to arsenic. J Exp Bot 56:1335–1342

Tuteja N, Ahmad P, Panda BB, Tuteja R (2009) Genotoxic stress in plants: shedding light on DNA damage, repair and DNA repair helicases. Mutat Res 681:134–149

West CE, Waterworth WM, Story GW, Sunderland PA, Jiang Q, Bray CM (2002) Disruption of the *Arabidopsis* AtKu80 gene demonstrates an essential role for AtKu80 protein in efficient repair of DNA double-strand breaks in vivo. Plant J 31:517–528

Wilber CG (1980) Toxicology of selenium: a review. Clin Toxicol 17:171–230

Xie QE, Yan XL, Liao XY, Li X (2009) The arsenic hyperaccumulator fern *Pteris vittata* L. Environ Sci Technol 43:8488–8495

Yokota Y, Shikazono N, Tanaka A, Hase Y, Funayama T, Wada S, Inoue M (2005) Comparative radiation tolerance based on the induction of DNA double-strand breaks in tobacco BY-2 cells and CHO-K1 cells irradiated with gamma rays. Radiat Res 163:520–525

Zhang H, Jiang Y, He Z, Ma M (2005) Cadmium accumulation and oxidative burst in garlic (*Allium sativum*). J Plant Physiol 162:977–984

Protein Oxidative Modifications

Liliana Beatriz Pena, Claudia Elsa Azpilicueta, María Patricia Benavides, and Susana Mabel Gallego

Abstract Proteins are molecules especially susceptible to oxidative modifications owing to their abundance and reactivity to radicals. Amongst the protein oxidative (redox) changes, carbonylation of the molecules represents an irreversible process that leads to the lost of protein functionality. The bulk of carbonylated proteins are produced as a result of metal oxidative stress induction in plants. In addition, metal ions-catalyzed oxidation (MCO) systems have been used especially for the introduction of carbonyl groups in the protein molecules in vitro. The mechanism underlying protein carbonylation for redox active metals is the direct catalysis of reactive oxygen species (ROS) generation, while metals considered redox inactive act in decreasing the antioxidant defence system. Despite the fact that protein carbonylation is associated with general and random processes; recent advances indicate a great degree of selectivity in the protein oxidation process. In turn, there are proteins, such as catalase, that respond to metal-induced oxidative stress by regulating the translation of isoforms and thus inducing the synthesis of new subunits less sensitive to oxidation. Further, the intracellular level of oxidized proteins is the product of a balance between the rate of oxidation and the rate of degradation of proteins. Metals can alter plant cell capacity for removing damaged proteins. As part of the proteolytic system, the 20S proteasome is responsible for the proteolysis of the carbonylated proteins. The 20S proteasome activity is regulated through oxidative modification of the proteasome itself, where a moderate 20S protein oxidation increases its activity, but a severe oxidative condition decreases it, concomitantly producing oxidized protein accumulation. The widespread occurrence of protein modifications and regulated proteolysis, as well as the existence of regenerative mechanisms of oxidative modifications, is presented.

L.B. Pena • C.E. Azpilicueta • M.P. Benavides • S.M. Gallego (✉)
Departamento de Química Biológica, Facultad de Farmacia y Bioquímica, Universidad de Buenos Aires, IQUIFIB-CONICET, Buenos Aires, Argentina
e-mail: lpena@ffyb.uba.ar; clazpili@ffyb.uba.ar; mbenavi@ffyb.uba.ar; sgallego@ffyb.uba.ar

D.K. Gupta and L.M. Sandalio (eds.), *Metal Toxicity in Plants: Perception, Signaling and Remediation*, DOI 10.1007/978-3-642-22081-4_10,

1 Proteins as Molecular Targets of Oxidative Reactions

Proteins are one of the main cellular macromolecules susceptible to oxidative modifications. One of the reasons for this consideration is the abundance of proteins in living organisms. Proteins comprise the major, non-water, component in biological systems, at the tissue, cellular or biological fluid levels. For example, in a typical plant cell, proteins represent about 30% of the total dry weight (DW), but they are between 60 and 70% of DW when the cell wall and the starch are excluded (Taiz and Zeiger 2010). Moreover, proteins are found ubiquitously in the cell, not only in soluble forms but attached to or forming part of biological membranes. On the other hand, a further indication of the importance of proteins as targets for oxidants is the rate constants for the reaction of a range of reactive radicals with proteins with respect to other biological macromolecules (Davies 2005; Xu and Chance 2007). Oxidised proteins accumulation has been considered a cause of cellular damage (Berlett and Stadtman 1997). However, considering that there are many types of protein oxidative modifications and proteins play a variety of functions in the cell, ranging from catalytic activities, structural features or regulation of several processes, it is possible to assume that protein oxidation can directly modify cell structure, signaling and metabolism.

1.1 Protein Oxidative Products

Protein oxidation is defined as the covalent modification of a polypeptide induced either directly by reactive oxygen species (ROS) or indirectly by reaction with secondary by-products of oxidative stress (Møller et al. 2007). One of the most common oxidative protein modifications is the formation of carbonyl derivatives, in which the amino-acid side chains, mainly histidine, arginine, lysine, proline, threonine, and tryptophan residues, are converted by the action of ROS to aldehyde or keto groups (Møller et al. 2007). The majority of the ROS-mediated reactions are hydrogen atom abstraction from α carbon position of the protein or radical transfer from side chains (Davies 2005; Xu and Chance 2007). In addition to the direct action of ROS on amino acids, protein carbonyl derivatives can be formed indirectly on lysine, cysteine, and histidine by forming adducts with reactive carbonyl compounds on carbohydrates (glycoxidation products), lipids, and advanced glycation/lipoxidation end products (Madian and Regnier 2010). The rise in the number of carbonyl group per protein molecule is called protein carbonylation. Protein carbonylation is considered an irreversible process because damaged proteins are unable to be enzymatically repaired in the cell (Nyström 2005). The classical method developed for the detection and quantification of protein carbonyl groups involves derivatization of the carbonyl group with 2,4-dinitrophenylhydrazine (2,4-DNPH) and the subsequent immunodetection of the resulting hydrazone using monoclonal or polyclonal antibodies (Levine et al. 1990, 1994; Yan et al. 1998).

Detectable protein carbonyl content by assessment with 2,4-DNPH is one of the most common assays used to quantify oxidative stress in vivo. Tryptophan oxidation is another apparently irreversible aminoacid modification that involves the formation of *N*-formylkynurenine, a breakdown product of tryptophan caused by dioxygenation and ring breakage (Finley et al. 1998; Møller and Kristensen 2006). Polypeptides with sulfur-containing residues suffer oxidative modifications that involve cross-linking or the occurrence of sulfur-containing derivatives like sulfenic, sulfinic, sulfonic, or sulfoxide forms (Davies 2005). The oxidation of cysteine thiol group to disulfide is considered a reversible modification where the reduced form can be regenerated by the thioredoxin (Trx) or glutaredoxin systems. In the former, the enzyme thioredoxin reductase transfers electrons from NADPH to Trx via a flavin carrier. Glutaredoxin is also able to reduce disulfide bonds, but using GSH as an electron donor (Møller et al. 2007; Rey et al. 2007; Rouhier et al. 2008). Further oxidation of cysteine to cysteic acid ($R\text{-}SO_3H$) appears to be irreversible and damaging to the protein (Ghezzi and Bonetto 2003). Like cysteine, methionine belongs to the most easily oxidized amino acids owing to the presence of sulfur. The first stage of oxidation leads to methionine sulfoxide ($R\text{-}SOCH_3$, abbreviated as MetSO) a biologically occurring product (Vogt 1995; Hong and Schöneich 2001). Oxidation of methionine to MetSO, which results in modification of activity and structure for many proteins, is reversed by an enzyme present in most organisms named methionine sulfoxide reductase (MSR; EC 1.8.4.11). This enzyme catalyzes the thioredoxin-dependent reduction of MetSO back to the correct Met residue. Two types of MSR has been isolated, MSRA specific to the MetSO S-enantiomer, and MSRB, which catalytically reduces the MetSO R-enantiomer. Both enzymes are required, since the cell oxidation of Met residues at the sulfur atom results in a racemic mixture of the two stereoisomers (Rouhier et al. 2008). *MSRA* and *MSRB* genes encode different MSR isoforms and are found to be relatively ubiquitous, with homologues found in many different organisms from bacteria and yeast to insects and mammals (Rouhier et al. 2006). Oxidation of Met residues can block phosphorylation-induced regulation of proteins (Hardin et al. 2009). The highest level of Met oxidation, $R\text{-}SO_2CH_3$ (sulfone) appears to be irreversible.

2 Metals as Responsible of Protein Oxidation

Different methods are used to generate ROS capable of producing protein oxidation in vitro (Xu and Chance 2007). Among them, one of the most common and potent mode of inducing protein oxidation derives from the metal ions-catalyzed oxidation (MCO) systems. In the same sense, the natural presence of metal in many protein structures can increase the molecule susceptibility to suffer oxidative modifications.

2.1 *Metals Ions-Catalyzed Oxidation Systems*

A metal ions-catalyzed oxidation (MCO) system comprises the presence of transition metals, such as Fe (III) or Cu (II), and H_2O_2. In a number of metal-catalyzed oxidation (MCO) systems, H_2O_2 is formed by the catalysis of different electron donors in the presence of O_2 and Fe (III) or Cu (II). Reducing agents include nonenzymatic autooxidizable substrates, like ascorbate, or sulfhydryl compounds, and enzymatic systems, e.g., NAD(P)H dehydrogenases, xanthine oxidase, and cytochrome P450 reductases (Stadtman 1993). Some metal–chelator complexes, e.g., ethylenediaminetetraacetate (EDTA)–Fe(II) are more effective than the unchelated metals in MCO systems, producing greater amounts of radicals and at fast rates (Stadtman and Berlett 1991). EDTA also increases the solubility of the metal ions thus allowing the reaction to be carried out at neutral pH. The action of MCO systems on amino acids like Arg, Pro, His, and Lys were reported to result in the formation of a carbonyl derivative which provides a means for monitoring the protein oxidation process (Berlett and Stadtman 1997; Schöneich 2000; Temple et al. 2006). The ability of MCO systems to catalyze protein carbonylation is attributable to the hydroxyl radicals ($HO^{\cdot}$) generation via Fenton-type chemistry. The Fenton reaction is comprised in the Haber–Weiss reaction (Fig. 1). Fe (II) and Cu (I) ions bind to a specific metal binding site within the protein and react with H_2O_2 to generate $HO^{\cdot}$, which then attacks the amino acid residues near the metal binding site and in turn leads to the production of a carbon radical (Stadtman 1993). The quantitatively most important carbonyl products of the metal-catalyzed oxidation reaction are glutamic semialdehyde from arginine and proline, and aminoadipic semialdehyde from lysine (Requena et al. 2001, 2003). Metal ions-catalyzed oxidation system also oxidizes tyrosine residues with the concomitant formation of dityrosine (Kato et al. 2001). In this case, since $Cu(I)/H_2O_2$ oxidative conditions did not lead to the formation of dityrosine, the MCO system dityrosine production, it is unlikely to occur via Fenton chemistry (Ali et al. 2004).

2.2 *Metalloproteins Susceptibility to Oxidative Stress*

Metals are natural components of many proteins. The main biological roles played by metals in living organisms are carried out in relationship to proteins. Almost half

$$O_2^{-} + Fe^{3+} \rightarrow Fe^{2+} + O_2$$

$$Fe^{2+} + H_2O_2 \rightarrow Fe^{3+} + HO\cdot + OH^{-} \quad \text{(Fenton Reaction)}$$

$$O_2^{-} + H_2O_2 \rightarrow O_2 + HO\cdot + OH^{-} \quad \text{(Haber-Weiss Reaction)}$$

Fig. 1 The Haber–Weiss reaction generates $HO^{\cdot}$ from H_2O_2 and O_2^{-}. In the presence of metals such as iron or copper, hydrogen peroxide (H_2O_2) can be converted into a highly reactive species hydroxyl radical ($HO^{\cdot}$) in a chemical reaction called Fenton reaction. The oxidized metal can undergo a re-reduction in a subsequent reaction with superoxide anion radical (O_2^{-})

of all proteins are associated with metal ions to perform their specific functions, and the majority of these metalloproteins contain transition metals as part of their structure (Dudev and Lim 2003). The natural selection of proteins metal cofactors has been based on both their unique physicochemical properties and their bioavailability in the Earth's crust (Williams 1997). The presence of the metal ion adds new functionality to proteins and helps proteins to catalyze some of the most difficult biological reactions. Protein reactivity is finely tuned by using different metal ions, different redox states of the same metal ion, or different ligands and geometric arrangements. For example, metalloenzymes perform functions such as redox reactions that cannot easily be carried out by the limited set of functional groups found in amino acids. Thus, metalloproteins participate in the most important biochemical processes including respiration, nitrogen fixation and oxygenic photosynthesis. The available data demonstrate that metal cofactors interact with the host protein mainly in two different ways: some bind to a well-structured cavity in a folded protein, in this case protein largely directed the final structure of the metalloprotein; others bind to the unfolded polypeptide and assist in folding, in this case metal cofactors largely directed the final structure of the metalloprotein (Kharenko and Ogawa 2004; Wilson et al. 2004). Nowadays, there are considerable interests in creating metal binding sites in designed proteins to understand the structural roles of metal ions and to design new metalloproteins with useful functions (Dudev and Lim 2008). Metal-catalyzed oxidation of proteins binds a cation capable of redox cycling to a metal-binding site on the protein. At present, the evidence suggests that metalloprotein oxidative modifications occur in functional groups of amino acid residues at or near the metal-binding site (Hong and Schöneich 2001; Sharp et al. 2003; Bridgewater et al. 2006). For example, amino acid residues nearby of the metal center(s) in Cu,Zn-superoxide dismutase (Cu,Zn-SOD; EC 1.15.1.1) can be selectively oxidized by addition of H_2O_2, which generates reactive oxygen species via a Fenton-like reaction upon reaction with Cu ions (Stadtman 1993; Kurahashi et al. 2001; Bridgewater and Vachet 2005). The combination of MCO system with mass spectrometry (MS) constitutes an interesting method for determining the coordination structure of metalloproteins (Bridgewater et al. 2006; Sadineni et al. 2006). Moreover, the amino acid of the metal coordination in enzymes active sites can be detected using as strategy the exchange of the metal by Cu (Miyazaki et al. 2009).

3 Metal Stress in Plants Is Associated to an Increase in Protein Carbonylation

The excess of essential metals, or the presence of those considered not essential for plant survival – even though in low concentration – has been associated with an imbalance between ROS production and antioxidant defence system capacity in the cell with the concomitant oxidative stress generation (Gallego et al. 1996, 2002;

Sharma and Dietz 2009). In this context, since carbonylation is an irreversible protein modification, it is widely used as oxidative stress biomarker. Protein oxidation constitutes a post-translational modification (PTM) observed in different parts of the plant as result of metal toxicity. The bulk of protein carbonylation is produced as a result of the metal-induced oxidative stress in plants as shown in Table 1, which lists those metals that produce protein carbonylation on different plant species. It is important to highlight that protein carbonylation can be produced to different extents in the same plant species subjected to metal stress. This effect

Table 1 Different plants species subjected to metal stress with marked increase on carbonyl group content

Plant specie	Part of the plant	Metal	Exposure time (d)	Concentration (μM)	Reference
Cucumis sativus	Leaves	Al^{3+}	10	1–2,000	Belmonte Pereira et al. (2010)
	Seedlings	Cd^{2+}	10	400–1,000	Gonçalves et al. (2007)
	Seedlings	Hg^{2+}	10–15	250–500	Cargnelutti et al. (2006)
Hordeum vulgare	Leaves	Cu^{2+}	5	150–1,500	Demirevska-Kepova et al. (2004)
	Leaves	Mn^{2+}	5	183–18,300	
Pisum sativum	Leaves	Cd^{2+}	14	50	Romero-Puertas et al. (2002)
Zea mays	Leaves	Cd^{2+}	1	100	Pena et al. (2007)
	Shoot/root	Cd^{2+}, Hg^{2+}	7	6–30	Rellán-Álvarez et al. (2006)
Helianthus annuus	Callus	Cd^{2+}	28	150	Gallego et al. (2005)
	Cotyledons	Cd^{2+}	4	100–200	Gallego et al. (1999)
	Cotyledons	Cd^{2+}, Cu^{2+}, Al^{3+}, Co^{2+}, Pb^{2+}, Cr^{3+}, Ni^{2+}, Hg^{2+}	4	100	Pena et al. (2006a)
	Leaves	Cd^{2+}	4	100–300	Pena et al. (2006b)
Medicago sativa	Seedlings	Cd^{2+}, Hg^{2+}	7	30	Ortega-Villasante et al. (2005)
Allium cepa	Roots	Al^{3+}	0.6	200	Murali Achary et al. (2008)
Triticum durum	Roots	Cd^{2+}	3–7	20–40	Paradiso et al. (2008)
Triticum aestivum	Shoots	Cu^{2+}, Cd^{2+}, Ni^{2+}	7	75	Gajewska and Skłodowska (2010)
Arabidopsis thaliana	Leaves	Cd^{2+}	6	50	Polge et al. (2009)
Solanum lycopersicum	Leaves	Cd^{2+}	3–10	300	Djebali et al. (2008)

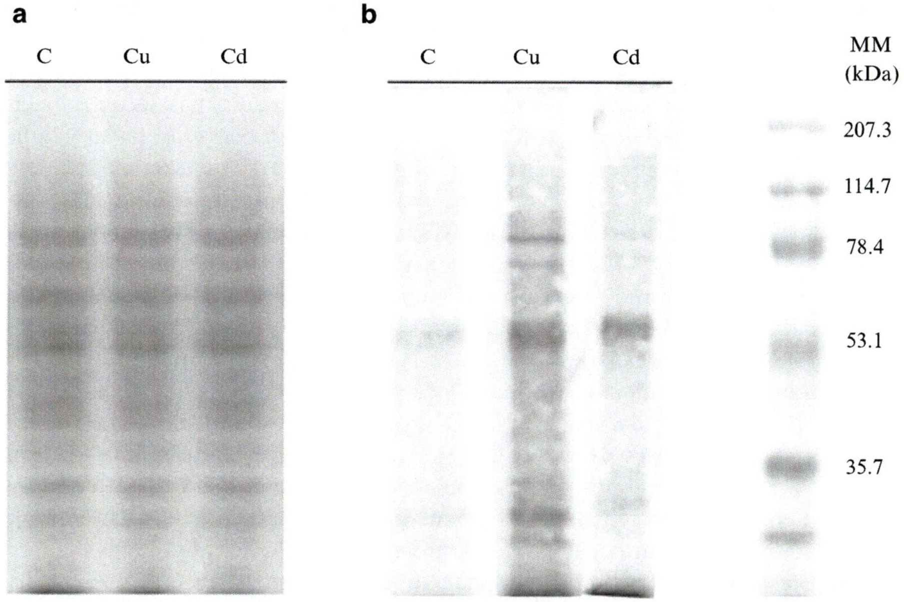

Fig. 2 One-dimensional PAGE of carbonylated proteins from wheat roots of control and metal-treated plants. The same protein amount was loaded per lane. (**a**) Coomassie Brilliant Blue (protein stain) and (**b**) anti-DNP immunoassay (carbonyl groups) are shown. *C* control, *MM* molecular mass marker

could be attributable to differences in the studied organs, the age of the plant, the metal concentration and the exposure time. The qualitative pattern obtained by immunodetection of carbonyl residues on plants treated with toxic metal concentrations shows a general protein oxidation profile, including proteins from low to high molecular weight (Romero-Puertas et al. 2002; Pena et al. 2008). As an example, Fig. 2 shows a typical immunoblot where proteins were tagged with 2,4-dinitrophenylhydrazine (DNPH) and detected with anti-DNP antibodies.

3.1 *Metals Catalyze Reactive Oxygen Species Generation Inside the Cell*

The mechanism involved on protein carbonylation process for redox active metals, like Fe and Cu ions, is related to the capacity of these metals to directly catalyze ROS generation (Stohs and Bagchi 1995). Other metals considered non-redox active, like Cd, are able to alter the redox cell status mainly by modifying the antioxidant defence system and thus increasing ROS cell level (Schützendübel et al. 2001; Romero-Puertas et al. 2004; Garnier et al. 2006). It could be inferred that ROS accumulation produced by metal stress is responsible for protein oxidation since the same targeted proteins were detected in pea (*Pisum sativum*) plants treated with Cd and treated with H_2O_2 (Romero-Puertas et al. 2002). Protein carbonylation is a non-enzymatic process and has been suggested to be rather non-specific and would be expected to be at random. However, new evidence has recently been found in relation to the selectivity of the protein oxidation process inside the cell. In this sense, a higher degree of protein oxidation linked to cellular compartments associated to ROS production would be expected. In plants, chloroplasts and peroxisomes are the main sources of ROS in autotrophic tissues under light conditions (Foyer and Noctor 2003; Gill and Tuteja 2010). On the other side, ROS come mostly from mitochondria in heterotrophic tissues or in green cells on darkness (Foyer and Noctor 2003; Gill and Tuteja 2010). In plants under metal stress, proteins from organelles appear to be particularly susceptible to oxidative modification. For example, in Cd-treated pea plants, the degree of protein carbonylation was proportionally higher in isolated peroxisomes compared to the whole-plant extracts (Romero-Puertas et al. 2002). Analysis of the soluble matrix fraction of rice (*Oryza sativa*) leaf mitochondria showed that proteins markedly affected by MCO system in vitro were particularly prone to oxidation in vivo (Kristensen et al. 2004). These authors further identified a group of mitochondrial proteins that are particularly susceptible to mild oxidation in vitro (Kristensen et al. 2004). The molecular basis for the apparent sensitivity of some proteins to carbonylation is not well understood, but it is likely that MCO is an intrinsic problem for proteins containing transition metals (Nyström 2005). In this sense, it has recently been demonstrated a strong correlation among the sets of immobilized metal affinity chromatography-interacting proteins, proteins predicted to contain metal-binding motifs, and protein sets known to be oxidized or degraded during abiotic stress in isolated *Arabidopsis thaliana* mitochondria (Tan et al. 2010).

4 Metals Can Alter Cell Metabolism by Mediating Protein Carbonylation

Proteins serve vital roles in the cells by catalyzing process, regulating structure or participating in signaling processes. Thus, protein oxidative modification can therefore rapidly affect plant metabolism. Analysis of the effect of MCO system on plant proteins allowed to study the oxidative modification of proteins in vitro, its implication on protein functionality and to assume the consequences of the oxidative stress in the whole plant. For example, the large subunit (LSU) of ribulose-1,5-bisphosphate carboxylase/oxygenase (rubisco) purified from wheat (*Triticum aestivum*) was broken down in the catalytic site by exposition to MCO system into two polypeptides (Ishida et al. 1999). Identical fragmentation of the protein was described in intact leaves of cucumber (*Cucumis sativus* L.) under chilling-light conditions (Nakano et al. 2006).

Moreover, different cell processes can be altered by metal oxidative inactivation of enzymes. For example, Kranner and Colville (2011)) proposed that metals can compromise seed germination by inducing oxidative damage to hydrolytic enzymes and storage proteins. Another metabolic process that has been greatly studied is the inactivation by MCO system of the enzymes involved in the nitrogen assimilation cycle. In the assimilation pathway, ammonium is incorporated into glutamine by glutamine synthetase (GS; EC 6.3.1.2), which is then converted to glutamate by glutamate synthase using 2-oxoglutarate (GOGAT; EC 1.4.1.13) (Lancien et al. 2000). Moreover, the GS/GOGAT cycle connects nitrogen and carbon metabolism in cells. Glutamine synthetase from *Escherichia coli* has been shown to be regulated by MCO, inducing an inactivation of the enzyme that precedes selective degradation by specific proteases (Kim et al. 1985). Besides GS from *E coli*, diverse glutamine synthetases have been described to undergo inactivation by some kind of oxidative modification produced by MCO system, as the enzymes of the green alga *Monoraphidium braunii* (Humanes et al. 1995) or the marine oxyphotobacterium *Prochlorococcus* (Gómez-Baena et al 2001). Ortega et al. (1999) demonstrated that GS from soybean (*Glycine max*) root extract subjected to MCO systems was not only inactive but more susceptible to degradation than non-oxidized GS. A decrease of both GS and GOGAT activities concomitant with the oxidative stress generation has been shown in the GS/GOGAT cycle in nodules and roots of soybean plants after Cd exposure (Balestrasse et al. 2001, 2003). In this case, the inactivation of the nitrogen assimilation cycle was closely related to the increase in the carbonylation level of GS and GOGAT proteins (Balestrasse et al. 2006).

4.1 *Regulation of the Translation of Isoforms: The Catalase*

In many cases, there is a lack of information not only about the effect of oxidation on protein activity or function but also in the form that plant metabolism copes with protein oxidative modification. In this regard, an interesting mechanism consisting of increasing the synthesis of protein subunits less sensible to oxidation has been described in plants for the enzyme catalase. Catalase (CAT; EC 1.11.1.6) is one of the main antioxidant enzymes that catalyzes the conversion of hydrogen peroxide to O_2 and H_2O, which in plants is localized inside peroxisomes. In plants, catalase activity shows a great degree of susceptibility to metal stress (Gallego et al. 1996, 2002; Balestrasse et al. 2001; Pandey and Sharma 2002; Singh et al. 2006). Moreover, catalase has been involved in Cd tolerance enhancement in plants, as was demonstrated in transgenic tobacco over-expressing a Cd-induced catalase cDNA from the hyperaccumulator species *Brassica juncea* (BjCAT3) (Guan et al. 2009). In contrast, catalase deficiency has been related to cadmium toxicity in tobacco plants (Iannone et al. 2010).

Catalase enzyme is encoded by a small unlinked nuclear gene family, and heterotetramers of CAT are formed when polypeptides encoded by distinct genes are simultaneously expressed in the same plant cell. In sunflower (*Helianthus annuus* L.), at least eight isoforms have been identified: CAT1 to CAT8 (Eising et al. 1990), where the biogenesis of the subunits is controlled by four different genes (*CATA1* to *CATA4*). Studies of CAT activity, protein expression, state of protein oxidation and CATA transcripts accumulation showed that inactivation of catalase produced by cadmium in sunflower was due to a mechanism involving the oxidation of CAT protein (Azpilicueta et al. 2007, 2008). Therefore, damaged CAT might cause the augmentation of free radical-mediated oxidative damage to other proteins, as has been observed under H_2O_2 treatment in peroxisomal enzymes of castor bean (*Ricinus communis*) endosperm (Anand et al. 2009). Interestingly, under redox stress conditions generated by metal ions, sunflower plant cells maintain catalase activity by regulating isoforms translation. The induction of transcription of *CATA3* and *CATA4* genes derived in CAT isoforms enriched in the subunit less sensitive to the oxidative damage (Grotjohann et al. 1997). Thus, differential expression of catalase genes under metal treatment in sunflower leads to the synthesis of enzyme isoforms less sensitive to oxidation, which prevent enzyme inactivation (Azpilicueta et al. 2007, 2008). Amino acid substitutions that occur in strictly conserved positions of *CATA* gene products can be expected to contribute to catalase ability to resist oxidative stress conditions (Engel et al. 2006).

5 Carbonylated Protein Degradation

Oxidative modifications of proteins result in physical changes in the protein structure that lead to dysfunction, inactivation, cross-linking of polypeptide chains or chemical fragmentation of the protein. The accumulated oxidized proteins tend to form insoluble high molecular weight aggregates, also known as inclusion bodies or plaques, all of which occur as a result of the increased level of hydrophobic bonds, ionic and covalent bonds that are potentially cytotoxic and can actively influence cellular metabolism (Grune et al. 2004; Davies 2005). Thus, oxidative stress conditions trigger an increase in protein turnover and degradation. The bulk of carbonylated proteins created must be degraded to prevent accumulation of the unfolded protein forms which are highly toxic for the cell. Proteolysis in plants not only regulates protein processing and intracellular protein levels, but removes abnormal or damaged proteins from the cell (Buchanan et al. 2000). So, the intracellular level of oxidized protein is the product of the balance between the rate of protein oxidation and the rate of oxidized protein degradation. The proteolytic process depends on many variables that determine the concentrations and/or activities of the proteases that degrade oxidatively damaged proteins, and many other factors like the presence of metal ions, endogenous inhibitors and regulatory proteins that modify the proteolytic activity (Berlett and Stadtman 1997). Moreover, a rise in the protein degradation could be due either to an increase in proteolytic enzyme activities or to modifications in the protein that make it a more suitable substrate for proteases already present. In this sense, protein carbonylation initially makes the molecule more susceptible to proteolytic attack, though, accumulated oxidized proteins tend to form high molecular weight aggregates with increasing resistance to proteolysis (Grune et al. 2004; Davies 2005). At the cellular level, protein breakdown is a complex process that includes proteases mainly localized in vacuole and other cell organelles, and the ubiquitin-proteasome system (UPS) active in both the cytoplasm and nucleus.

5.1 Role of Proteases

Proteolytic enzymes are classified as exopetidases or endopeptidases depending on the site of the hydrolytic cleavage of the peptide chain. Exopeptidases can be classified on the basis of the reaction that they catalyze. Depending on the hydrolyzed substrate, they are classified as dipeptidases (EC 3.4.13) and tripeptide aminopeptidase (EC 3.4.11.4), or can be characterized by the terminal group attacked as aminopeptidases (EC 3.4.11), carboxypeptidases (3.4.16–18) and omega-peptidase (EC 4.3.19). Exopeptidases are also classified according to the number of amino acid residues released as peptidyl-dipeptidase (EC 4.3.15), dipeptidyl-peptidase or tripeptidyl-peptidases (EC 3.4.14) (Dalling 1986). Endopeptidases, also known simply as proteases, are classified according to their catalytic mechanism. Different classes of proteases found in plants are serine endopeptidases (EC 3.4.21), cysteine endopeptidases (EC 3.4.22) and aspartic endopeptidases (EC 3.4.23) according to the amino acid in their active sites. On the other hand, there are the metalloendopeptidases (EC 4.3.24) which contain a metal such as Zn^{2+}, Co^{2+} or Mn^{2+} in the reaction center (Dalling 1986; Palma et al. 2002). The proteolytic system of plants comprises a large number of proteases. For example, in the genome of *Arabidopsis thaliana* more than 650 nucleotide sequences encoding proteases of different catalytic classes have been identified, but only some of them have a known function for the life of the plant. Moreover, 41 sequences have been described as known or putative protease inhibitors (Rawlings et al. 2010). The most common assays for determining total endoproteolytic activity use non-specific substrates such as gelatin, hemoglobin, casein or albumin, and then measure the amino acids released by hydrolysis. The use of chromogenic substrates, like azocasein, enables direct measurement of diazotized fragments that are soluble in trichloroacetic acid. Moreover, the immobilization of the protein substrate in SDS-polyacrylamide gels (SDS-PAGE) can be used to detect endoprotease activity and provide an estimated molecular weight. In this case, protease either has not to be sensitive to the presence of SDS, or has to be renatured after denaturation by SDS (Simpson 2001).

There are two important subjects to be highlighted with respect to the response of the proteolytic system to metal stress in plants: the effect of metals on protease activity and the relationship between the proteolytic activity and the level of carbonylated proteins. The action of metals on the proteolytic activity in sunflower has been shown to be independent of metal redox capacity or type of metal (Pena et al. 2008). Metal stress affected plant endoproteolytic activity but their effects could not be generalized. Moreover, the response of the proteolytic activity and the accumulation of carbonylated proteins lead to contrasting observations depending on the metal, plant or organ. In sunflower plants subjected to cadmium stress, oxidized protein accumulated in cotyledons and leaves even though protease activity increased (Pena et al 2006a, b, 2007). Carbonyl groups contents, global endopeptidase activity, specific protease activities and their transcript levels increased in the leaves of *Arabidopsis* plants exposed to cadmium (Polge et al. 2009). However, in axes and seeds of *Sorghum bicolor,* increasing Cd concentrations had a negative effect on protease activity, although additional new isozymes were induced (Kuriakose and Prasad 2008). Cadmium-treated pea plants did not show any significant change in the total endoproteolytic activity in leaf extracts, but increased protease activity in leaf peroxisomes as a consequence of the overall increase in the activity of the endopeptidases isozymes inside the organelle (Romero-Puertas et al. 2002). In spite of the results mentioned above for *S. bicolor* (Kuriakose and Prasad 2008) and pea plants (McCarthy et al. 2001; Romero-Puertas et al. 2002), cadmium produced an increase in protein carbonylation in both species. Cadmium treatment decreased protease activity without accumulation of carbonyl group content in roots of tomato (*Solanum lycopersicon*) plants, but both parameters increased in leaves of Cd-treated plants (Djebali et al. 2008). Copper-exposed wheat roots and shoots showed an enhanced protein carbonylation that corresponded to the induction of protease activity. While no activation of proteolysis was observed in wheat plants treated with Cd and Ni, accumulation of carbonylated proteins was only detected in shoots (Gajewska and Skłodowska 2010). The presence of putative proteases involved in degradation of the oxidatively damaged proteins was suggested in isolated mitochondria of *Arabidopsis* plants using H_2O_2 as a model stress (Sweetlove et al. 2002) whereas in root mitochondria of cucumber (*Cucumis sativus* L.) mutant lines (MSC16) with different mitochondrial genome rearrangement, low carbonyl group content was accompanied with a high protease activity (Juszczuk et al. 2008). Thus, different behaviors of the key elements of protein turnover machinery in response to metal stress has been observed, and although unspecific proteases could recycle oxidative proteins, their activities were not strong enough to diminish oxidized protein content in plant cells. Proteolysis susceptibility to metals should involve more than one mechanism, including direct effects on enzyme structure and/or functionality, as well as indirect mechanisms, such as increased levels of protein oxidation (Pena et al. 2007). Moreover, metals could have an impact on protease inhibitors that regulate endogenous proteolytic activity, a topic extensively surveyed under biotic stress conditions (van der Hoorn and Jones 2004; Farrokhi et al. 2008).

5.2 Role of 20S Proteasome

The selective removal of proteins in both the cytoplasm and the nucleus of eukaryotes involve the covalent attachment of multiple ubiquitin (Ub) molecules to protein substrates targeted for breakdown. The binding of the core protease 20S with the regulatory particles 19S forms the 26S proteasome, a 2-MDa ATP-dependent multi-subunit protease complex, which is responsible for the degradation of ubiquitinated proteins, releasing Ub molecules for recycling (Smalle and Vierstra 2004; Jung et al. 2009). The 20S proteasome core has a molecular weight of about 700 kDa and is arranged as a cylindrical stack of four heptameric rings with two central rings composed by of seven β-type subunits, and two peripheral rings composed of seven related α subunits. In the center of the barrel there is channel with an active site for protein degradation. Three of the β subunits are proteolytically active: β1 has cysteine protease with peptidylglutamyl peptide-hydrolyzing-like activity (PGPH), β2 presents serine protease with trypsin-like activity, and β5 is also a serine protease with chymotrypsin-like activity (Smalle and Vierstra 2004). The active sites of the plant 20S proteasome are very sensitive to the 26S proteasome inhibitors, MG115, MG132, lactacystin, and epoxomycin (Yang et al. 2004). Thus, the assay to determine proteasome peptidase activities is performed by monitoring cleavage of three different peptide substrates linked to the fluorescence reporters: Ala-Ala-Phe-7-amido-4-methylcoumarin, Boc-Leu-Ser-Thr-Arg-7-amido-4-methylcoumarin and N-Clz-Leu-Leu-Glu-β-nafthylamide for chymotrypsin-, trypsin-, and PGPH-like activities, respectively, either in the absence or the presence of a proteasome inhibitor. Besides its function as the proteolytic core of the 26S complex, 20S proteasome has a key role in degradation of mildly oxidatively modified proteins without non-ATP and ubiquitin requirement in mammalian cells (Shringarpure et al. 2003). The major recognition motif of the substrates by the 20S proteasome seems to be hydrophobic surface patches formed by partial unfolding and exposure of hydrophobic amino acid residues during oxidation. However, heavily oxidized proteins form covalent cross-links that decrease their susceptibility to proteolysis by the 20S proteasome (Grune et al. 2003). One of the effects exerted by metal stress condition is the inhibition of the 20S proteasome activity on sunflower leaves (Pena et al. 2008). But the mechanism underlying this inhibition seems to be dependent on the metal tested. Inactivation produced by Cd^{2+} is associated with the oxidative damage of the proteasome (Pena et al. 2006a, b), in a similar way as that of Cu^{2+} and Hg^{2+} (Pena et al. 2008). Cadmium decreased 20S proteasome activity, and induced the accumulation not only of the oxidized but also of the ubiquitinated proteins (Pena et al. 2006a, b). On the other hand, metals like Co^{2+} and Ni^{2+} inhibit in vitro proteasome activity (Pena et al. 2008). In this sense, Amici et al. (2002) suggested a direct effect of metals on proteasome activity that might result from the displacement of Mg^{2+} from its binding sites inside the catalytic chamber that decrease its functionality.

Interestingly, protein 20S oxidation might not be entirely detrimental to 20S proteasome function. Maize (*Zea mays*) leaf segments subjected to low cadmium treatment increased the 20S proteasome activity together with its level of protein carbonylation (Pena et al. 2007). The moderate oxidative chemical modification of 20S proteasome acts to relax the structure of the proteasome and can activate its proteolytic activity (Shringarpure et al. 2001), thus avoiding the accumulation of carbonylated proteins in Cd-treated maize segments (Pena et al. 2007). Treatments with low metal concentrations together with MG132 or higher cadmium concentrations decreased 20S proteasome activity with a concomitant accumulation of oxidized and ubiquitinated proteins. Impairment of proteasome functionality under high Cd concentration was associated with severe oxidation of 20S protein (Pena et al. 2007). Similar observations were described for proteasome isolated from maize roots submitted to a mild oxidative treatment through MCO system in vitro (Basset et al. 2002). Polge et al. (2009) demonstrated that RNAs encoding subunits of the 20S proteasome were up-regulated in response to cadmium in the leaves of *A. thaliana*. The increase in proteasome structural and catalytic subunit transcripts was followed by increases in proteasome quantity and chymotrypsin-like activity. Further, mutations of one of the subunits of the 19S regulatory particle resulted in an enhanced accumulation of the 20S versus 26S proteasome and a higher tolerance to oxidative stress in *Arabidopsis* (Kurepa et al. 2008). So far, 20S proteasome plays a major role in the degradation of moderately oxidized proteins during oxidative stress produced by metals in plants, and it is regulated at both transcriptional and post-translational level. Moreover, metal effects on proteasome functionality may be additional to other mechanisms previously demonstrated to be involved in metal toxicity in plants.

6 Conclusion

Metal stress alters the plant cell redox state causing serious reversible and irreversible oxidative changes in the proteome. Nowadays, the information about the content and distribution of metals (Salt et al. 2008) together with the cell redox proteomics (Rinalducci et al. 2008) is beginning to shed light about the impact that biological and environmental relevant metal concentrations have on plant cells. The widespread occurrence of protein modifications, the selectivity in the oxidation, the existence of regenerative mechanism of oxidative modifications and the regulated control of proteolysis are indicating a key role of the process of protein oxidation in normal plant cell physiology and in the response to stress conditions. The idea is emerging that protein oxidative modifications, together with other post-translational modifications like protein glutathionylation, nitrosylation, or even ubiquitination, could have signalling ramifications (Møller and Sweetlove 2010).

References

Ali FE, Barnham KJ, Barrow CJ, Separovic F (2004) Metal catalyzed oxidation of tyrosine residues by different oxidation systems of copper/hydrogen peroxide. J Inorg Biochem 98:173–184

Amici M, Forti K, Nobili C, Lupidi G, Angeletti M, Fioretti E, Eleuteri AM (2002) Effect of neurotoxic metal ions on the proteolytic activities of the 20S proteasome form bovine brain. J Biol Inorg Chem 7:750–756

Anand P, Kwak Y, Simha R, Donaldson RP (2009) Hydrogen peroxide induced oxidation of peroxisomal malate synthase and catalase. Arch Biochem Biophys 491:25–31
Azpilicueta CE, Benavides MP, Tomaro ML, Gallego SM (2007) Mechanism of *CATA3* induction by cadmium in sunflower leaves. Plant Physiol Biochem 45:589–595
Azpilicueta CE, Pena LB, Tomaro ML, Gallego SM (2008) Modifications in catalase activity and expression in developing sunflower seedlings under cadmium stress. Redox Rep 13:40–46
Balestrasse KB, Gardey L, Gallego SM, Tomaro ML (2001) Response of antioxidant defence system in soybean nodules and roots subjected to cadmium stress. Austr J Plant Physiol 28:23–28
Balestrasse KB, Benavides MP, Gallego SM, Tomaro ML (2003) Effect of cadmium stress on nitrogen metabolism in nodules and roots of soybean plants. Func Plant Biol 30:57–64
Balestrasse KB, Gallego SM, Tomaro ML (2006) Oxidation of the enzymes involved in nitrogen assimilation plays an important role in the cadmium-induced toxicity in soybean plants. Plant Soil 284:187–194
Basset G, Raymond P, Malek L, Brouquisse R (2002) Changes in the expression and the enzymic properties of the 20S proteasome in sugar-starved maize roots. Evidence for an in vivo oxidation of the proteasome. Plant Physiol 128:1149–1162
Belmonte Pereira L, de Melazzo A, Mazzanti C, Gonçalves JF, Cargnelutti D, Tabaldi LA, Becker AG, Spanholi Calgaroto N, Gomes Farias J, Battisti V, Bohrer D, Nicoloso FT, Morsch VM, Schetinger MRC (2010) Aluminum-induced oxidative stress in cucumber. Plant Physiol Biochem 48:683–689
Berlett BS, Stadtman ER (1997) Protein oxidation in aging, disease and oxidative stress. J Biol Chem 272:20313–20316
Bridgewater JD, Vachet RW (2005) Metal-catalyzed oxidation reactions and mass spectrometry: the roles of ascorbate and different oxidizing agents in determining Cu–protein-binding sites. Anal Biochem 341:122–130
Bridgewater JD, Lim J, Vachet RW (2006) Using metal-catalyzed oxidation reactions and mass spectrometry to identify amino acid residues within 10Å of the metal in Cu-binding proteins. J Am Soc Mass Spectrom 17:1552–1559
Buchanan BB, Gruissem W, Jones RL (2000) Biochemistry and molecular biology of plants. American Society of Plant Physiology, USA
Cargnelutti D, Tabaldi LA, Spanevello RM, de Oliveira JG, Battisti V, Redin M, Linares CE, Dressler VL, de Moraes Flores EM, Nicoloso FT, Morsch VM, Schetinger MR (2006) Mercury toxicity induces oxidative stress in growing cucumber seedlings. Chemosphere 65:999–1006
Dalling MJ (1986) Plant proteolytic enzymes. CRC Press, USA
Davies MJ (2005) The oxidative environment and protein damage. Biochim Biophys Acta 1703:93–109
Demirevska-Kepova K, Simova-Stoilova L, Stoyanova Z, Hölzer R, Feller U (2004) Biochemical changes in barley plants after excessive supply of copper and manganese. Environ Exp Bot 52:253–266
Djebali W, Gallusci P, Polge C, Boulila L, Galtier N, Raymond P, Chaibi W, Brouquisse R (2008) Modifications in endopeptidase and 20S proteasome expression and activities in cadmium treated tomato (*Solanum lycopersicum* L.) plants. Planta 227:625–639
Dudev T, Lim C (2003) Principles governing Mg, Ca, and Zn binding and selectivity in proteins. Chem Rev 103:773–787
Dudev T, Lim C (2008) Metal binding affinity and selectivity in metalloproteins: insights from computational studies. Annu Rev Biophys 37:97–116
Eising R, Trelease RN, Ni W (1990) Biogenesis of catalase in glyoxysomes and leaf-type peroxisomes of sunflower cotyledons. Arch Biochem Biophys 278:258–264
Engel N, Schmidt M, Lütz C, Feierabend J (2006) Molecular identification, heterologous expression and properties of light-insensitive plant catalases. Plant Cell Environ 29:593–607
Farrokhi N, Whitelegge JP, Brusslan JA (2008) Plant peptides and peptidomics. Plant Biotechnol J 6:105–134

Finley EL, Dillon J, Crouch RK, Schey KL (1998) Identification of tryptophan oxidation products in bovine alpha-crystallin. Protein Sci 7:2391–2397

Foyer CH, Noctor G (2003) Redox sensing and signaling associated with reactive oxygen in chloroplasts, peroxisomes and mitochondria. Physiol Plant 119:355–364

Gajewska E, Skłodowska M (2010) Differential effect of equal copper, cadmium and nickel concentration on biochemical reactions in wheat seedlings. Ecotoxicol Environ Saf 73:996–1003

Gallego SM, Benavides MP, Tomaro ML (1996) Effect of heavy metal ions excess on sunflower leaves: evidence for involvement of oxidative stress. Plant Sci 121:151–159

Gallego SM, Benavides MP, Tomaro ML (1999) Effect of cadmium ions on antioxidant defense system in sunflower cotyledons. Biol Plantarum 42:49–55

Gallego SM, Benavides MP, Tomaro ML (2002) Involvement of antioxidant defense system in the adaptive response to heavy metal ions in *Helianthus annuus* L. cells. Plant Growth Regul 36:17–21

Gallego SM, Kogan MJ, Azpilicueta CE, Peña C, Tomaro MT (2005) Glutathione-mediated antioxidative mechanisms in sunflower (*Helianthus annuus* L.) cells in response to cadmium stress. Plant Growth Regul 46:267–276

Garnier L, Simon-Plas F, Thuleau P, Agnel JP, Blein JP, Ranjeva R, Montillet JL (2006) Cadmium affects tobacco cells by a series of three waves of reactive oxygen species that contribute to cytotoxicity. Plant Cell Environ 29:1956–1969

Ghezzi P, Bonetto V (2003) Redox proteomics: identification of oxidatively, modified proteins. Proteomics 3:1145–1153

Gill SS, Tuteja N (2010) Reactive oxygen species and antioxidant machinery in abiotic stress tolerance in crop plants. Plant Physiol Biochem 48:909–930

Gómez-Baena G, Diez J, García-Fernández JM, El Alaoui S, Humanes L (2001) Regulation of glutamine synthetase by metal-catalyzed oxidative modification in the marine oxyphotobacterium *Prochlorococcus*. Biochim Biophys Acta 1568:237–244

Gonçalves JF, Becker AG, Cargnelutti D, Tabaldi LA, Pereira LB, Battisti V, Spanevello RM, Morsch VM, Nicoloso FT, Schetinger MRC (2007) Cadmium toxicity causes oxidative stress and induces response of the antioxidant system in cucumber seedlings. Braz J Plant Physiol 19:223–232

Grotjohann N, Janning A, Eising R (1997) *In vitro* photoinactivation of catalase isoforms from cotyledons of sunflower (*Helianthus annuus* L.). Arch Biochem Biophys 346:208–218

Grune T, Merker K, Sandig G, Davies KJA (2003) Selective degradation of oxidatively modified protein substrates by the proteasome. Biochem Biophys Res Commun 305:709–718

Grune T, Jung T, Merker K, Davies KJA (2004) Decreased proteolysis caused by protein aggregates, inclusion bodies, plaques, lipofuscin, ceroid, and 'aggresomes' during oxidative stress, aging, and disease. Int J Biochem Cell Biol 36:2519–2530

Guan ZQ, Chai TY, Zhang YX, Xu J, Wei W (2009) Enhancement of Cd tolerance in transgenic tobacco plants overexpressing a Cd-induced catalase cDNA. Chemosphere 76:623–630

Hardin SC, Larue CT, Oh MH, Jain V, Huber SC (2009) Coupling oxidative signals to protein phosphorylation via methionine oxidation in *Arabidopsis*. Biochem J 422:305–312

Hong J, Schöneich C (2001) The metal-catalyzed oxidation of methionine in peptides by fenton systems involves two consecutive one-electron oxidation processes. Free Radic Biol Med 31:1432–1441

Humanes L, García-Fernandez JM, López-Ruiz A, Diez J (1995) Glutamine synthetase from the green alga *Monoraphidium braunii* is regulated by oxidative modification. Plant Sci 110:269–277

Iannone MF, Rosales EP, Groppa MD, Benavides MP (2010) Reactive oxygen species formation and cell death in catalase-deficient tobacco leaf disks exposed to cadmium. Protoplasma 245:15–27

Ishida H, Makino A, Mae T (1999) Fragmentation of the large subunit of ribulose-1,5-bisphosphate carboxylase/oxygenase by reactive oxygen species. J Biol Chem 274:5222–5226

Jung T, Catalgol B, Grune T (2009) The proteasomal system. Mol Aspects Med 30:191–296

Juszczuk IM, Tybura A, Rychter AM (2008) Protein oxidation in the leaves and roots of cucumber plants (*Cucumis sativus* L.), mutant MSC16 and wild type. J Plant Physiol 165:355–365

Kato Y, Kitamoto N, Kawal Y, Osawa T (2001) The hydrogen peroxide/copper ion system, but not other metal-catalyzed oxidation systems, produces protein-bound dityrosine. Free Radic Biol Med 31:624–632

Kharenko OA, Ogawa MY (2004) Metal-induced folding of a designed metalloprotein. J Inorg Biochem 98:1971–1974

Kim K, Rhee SG, Stadtman ER (1985) Nonenzymatic cleavage of proteins by reactive oxygen species generated by dithiothreitol and iron. J Biol Chem 260:15394–15397

Kranner I, Colville L (2011) Metals and seeds: biochemical and molecular implications and their significance for seed germination. Environ Exp Bot 72:93–105

Kristensen BK, Askerlund P, Bykova NV, Egsgaard H, Møller IM (2004) Identification of oxidised proteins in the matrix of rice leaf mitochondria by immunoprecipitation and two-dimensional liquid chromatography-tandem mass spectrometry. Phytochemistry 65:1839–1851

Kurahashi T, Miyazaki A, Suwan S, Isobe M (2001) Extensive investigations on oxidized amino acid residues in H_2O_2-treated Cu, Zn-SOD protein with LC-ESI-Q-TOF-MS, MS/MS for the determination of the copper-binding site. J Am Chem Soc 123:9268–9278

Kurepa J, Toh-E A, Smalle JA (2008) 26S proteasome regulatory particle mutants have increased oxidative stress tolerance. Plant J 53:102–114

Kuriakose SV, Prasad MNV (2008) Cadmium stress affects seed germination and seedling growth in *Sorghum bicolor* (L.) Moench by changing the activities of hydrolyzing enzymes. Plant Growth Regul 54:143–156

Lancien M, Gadal P, Hodges M (2000) Enzyme redundancy and the importance of 2-oxoglutarate in higher plant ammonium assimilation. Plant Physiol 123:817–824

Levine RL, Garland D, Oliver CN, Amici A, Climent I, Lenz AG, Ahn BW, Shaltiel S, Stadtman ER (1990) Determination of carbonyl content in oxidatively modified proteins. Methods Enzymol 186:464–478

Levine RL, Willams JA, Stadtman ER, Shacter E (1994) Carbonyl assays for determination of oxidatively modified proteins. Methods Enzymol 233:346–357

Madian AG, Regnier FE (2010) Proteomic identification of carbonylated proteins and their oxidation sites. J Proteome Res 9:3766–3780

McCarthy I, Romero-Puertas MC, Palma JM, Sandalio LM, Corpas FJ, Gómez M, del Río LA (2001) Cadmium induces senescence symptoms in leaf peroxisomes of pea plants. Plant Cell Environ 24:1065–1073

Miyazaki A, Sydnes MO, Isobe M, Ohinata H, Miyazu M, Takai A (2009) Oxidatively induced Cu for Mn exchange in protein phosphatase 1γ: a new method for active site analysis. Bioorg Med Chem 17:7978–7986

Møller IM, Kristensen BK (2006) Protein oxidation in plant mitochondria detected as oxidized tryptophan. Free Radic Biol Med 40:430–435

Møller IM, Sweetlove LJ (2010) ROS signaling-specificity is required. Trends Plant Sci 15:370–374

Møller IM, Jensen PE, Hansson A (2007) Oxidative modifications to cellular components in plants. Annu Rev Plant Biol 58:459–481

Murali Achary VM, Jena S, Panda KK, Panda BB (2008) Aluminium induced oxidative stress and DNA damage in root cells of *Allium cepa* L. Ecotoxicol Environ Saf 70:300–310

Nakano R, Ishida H, Makino A, Mae T (2006) In vivo fragmentation of the large subunit of ribulose-1,5-bisphosphate carboxylase by reactive oxygen species in an intact leaf of cucumber under chilling-light conditions. Plant Cell Physiol 47:270–276

Nyström T (2005) Role of oxidative carbonylation in protein quality control and senescence. EMBO J 24:1311–1317

Ortega JL, Roche D, Sengupta-Gopalan C (1999) Oxidative turnover of soybean root glutamine synthetase. In vitro and in vivo studies. Plant Physiol 119:1483–1490

Ortega-Villasante C, Rellán-Álvarez R, Del Campo FF, Carpena-Ruiz RO, Hernández LE (2005) Cellular damage induced by cadmium and mercury in *Medicago sativa*. J Exp Bot 56:2239–2251

Palma JM, Sandalio LM, Corpas FJ, Romero-Puertas MC, McCarthy I, del Río LA (2002) Plant proteases, protein degradation, and oxidative stress: role of peroxisomes. Plant Physiol Biochem 40:521–530

Pandey N, Sharma CP (2002) Effect of heavy metals Co^{2+}, Ni^{2+} and Cd^{2+} on growth and metabolism of cabbage. Plant Sci 163:753–758

Paradiso A, Berardino R, de Pinto MC, Sanità di Toppi L, Storelli MM, Tommasi F, De Gara L (2008) Increase in ascorbate-glutathione metabolism as local and precocious systemic responses induced by cadmium in *Durum* wheat plants. Plant Cell Physiol 49:362–374

Pena LB, Pasquini LA, Tomaro ML, Gallego SM (2006a) Proteolytic system in sunflower (*Helianthus annuus* L.) leaves under cadmium stress. Plant Sci 171:531–537

Pena LB, Tomaro ML, Gallego SM (2006b) Effect of different metals on protease activity in sunflower cotyledons. Electron J Biotech 9:258–262

Pena LB, Pasquini LA, Tomaro ML, Gallego SM (2007) 20S proteasome and accumulation of oxidized and ubiquitinated proteins in maize leaves subjected to cadmium stress. Phytochemistry 68:1139–1146

Pena LB, Zawoznik MS, Tomaro ML, Gallego SM (2008) Heavy metals effects on proteolytic system in sunflower leaves. Chemosphere 72:741–746

Polge C, Jaquinod M, Holzer F, Bourguignon J, Walling L, Brouquisse R (2009) Evidence for the existence in *Arabidopsis thaliana* of the proteasome proteolytic pathway: activation in response to cadmium. J Biol Chem 284:35412–35424

Rawlings ND, Barrett AJ, Bateman A (2010) MEROPS: the peptidase database. Nucl Acids Res 38:D227–D233

Rellán-Álvarez R, Ortega-Villasante C, Álvarez-Fernández A, del Campo FF, Hernández LE (2006) Stress responses of *Zea mays* to cadmium and mercury. Plant Soil 279:41–50

Requena JR, Chao CC, Levine RL, Stadtman ER (2001) Glutamic and aminoadipic semialdehydes are the main carbonyl products of metal-catalyzed oxidation of proteins. Proc Natl Acad Sci USA 98:69–74

Requena JR, Levine RL, Stadtman ER (2003) Recent advances in the analysis of oxidized proteins. Amino Acids 25:221–226

Rey P, Bécuwe N, Barrault MB, Rumeau D, Havaux M, Biteau B, Toledano MB (2007) The *Arabidopsis thaliana* sulfiredoxin is a plastidic cysteine-sulfinic acid reductase involved in the photooxidative stress response. Plant J 49:505–514

Rinalducci S, Murgiano L, Zolla L (2008) Redox proteomics: basic principles and future perspectives for the detection of protein oxidation in plants. J Exp Bot 59:3781–3801

Romero-Puertas MC, Palma JM, Gómez M, Del Río LA, Sandalio LM (2002) Cadmium causes the oxidative modification of proteins in pea plants. Plant Cell Environ 25:677–686

Romero-Puertas MC, Rodríguez-Serrano M, Corpas FJ, Gómez M, del Río LA, Sandalio LM (2004) Cd-induced subcellular accumulation of O_2^- and H_2O_2 in pea leaves. Plant Cell Environ 27:1122–1134

Rouhier N, Vieira Dos Santos Ch, Tarrago L, Rey P (2006) Plant methionine sulfoxide reductase A and B multigenic families. Photosynth Res 89:247–262

Rouhier N, Koh CS, Gelhaye E, Corbier C, Favier F, Didierjean C, Jacquot JP (2008) Redox based anti-oxidant systems in plants: biochemical and structural analyses. Biochim Biophys Acta 1780:1249–1260

Sadineni V, Galeva NA, Schöneich C (2006) Characterization of the metal-binding site of human prolactin by site-specific metal-catalyzed oxidation. Anal Biochem 358:208–215

Salt DE, Baxter I, Lahner B (2008) Ionomics and the study of the plant ionome. Annu Rev Plant Biol 59:709–733

Schöneich C (2000) Mechanisms of metal-catalyzed oxidation of histidine to 2-oxo-histidine in peptides and proteins. J Pharm Biomed Anal 21:1093–1097
Schützendübel A, Schwanz P, Teichmann T, Gross K, Langenfeld-Heyser R, Godbold DL, Polle A (2001) Cadmium-induced changes in antioxidative systems, hydrogen peroxide content, and differentiation in Scots pine roots. Plant Physiol 127:887–898
Sharma SS, Dietz KJ (2009) The relationship between metal toxicity and cellular redox imbalance. Trends Plant Sci 14:43–50
Sharp JS, Becker JM, Hetticha RL (2003) Protein surface mapping by chemical oxidation: structural analysis by mass spectrometry. Anal Biochem 313:216–225
Shringarpure R, Grune T, Davies KJA (2001) Protein oxidation and 20S proteasome-dependent proteolysis in mammalian cells. Cell Mol Life Sci 58:1442–1450
Shringarpure R, Grune T, Mehlhase J, Davies KJA (2003) Ubiquitin conjugation is not required for the degradation of oxidized proteins by proteasome. J Biol Chem 278:311–318
Simpson DJ (2001) Proteolytic degradation of cereal prolamins: the problem with proline. Plant Sci 161:825–838
Singh S, Eapen S, D'Souza SF (2006) Cadmium accumulation and its influence on lipid peroxidation and antioxidative system in an aquatic plant, *Bacopa monnieri* L. Chemosphere 62:233–246
Smalle J, Vierstra RD (2004) The ubiquitin 26S proteasome proteolytic pathway. Annu Rev Plant Biol 55:555–590
Stadtman ER (1993) Oxidation of free amino acids and amino acid residues in proteins by radiolysis and by metal-catalyzed reactions. Annu Rev Biochem 62:797–821
Stadtman ER, Berlett BS (1991) Fenton chemistry. Amino acid oxidation. J Biol Chem 266:17201–17211
Stohs SJ, Bagchi D (1995) Oxidative mechanisms in the toxicity of metal ions. Free Radic Biol Med 18:321–336
Sweetlove LJ, Heazlewood JL, Herald V, Holtzapffel R, Day DA, Leaver CJ, Millar AH (2002) The impact of oxidative stress on Arabidopsis mitochondria. Plant J 32:891–904
Taiz L, Zeiger E (2010) Plant physiology. Sinauer Associates, USA
Tan YF, O'Toole N, Taylor NL, Harvey Millar A (2010) Divalent metal ions in plant mitochondria and their role in interactions with proteins and oxidative stress-induced damage to respiratory function. Plant Physiol 152:747–761
Temple A, Yen TY, Gronert S (2006) Identification of specific protein carbonylation sites in model oxidations of human serum albumin. J Am Soc Mass Spectrom 17:1172–1180
van der Hoorn RAL, Jones JDG (2004) The plant proteolytic machinery and its role in defence. Curr Opin Plant Biol 7:400–407
Vogt W (1995) Oxidation of methionyl residues in proteins: tools, targets, and reversal. Free Radic Biol Med 18:93–105
Williams RJP (1997) The natural selection of the chemical elements. Cell Mol Life Sci 53:816–829
Wilson CJ, Apiyo D, Wittung-Stafshede P (2004) Role of cofactors in metalloprotein folding. Q Rev Biophys 37:285–314
Xu G, Chance MR (2007) Hydroxyl radical-mediated modification of proteins as probes for structural proteomics. Chem Rev 107:3514–3543
Yan LJ, Orr WC, Sohal RJ (1998) Identification of oxidized proteins based on sodium dodecyl sulphate-polyacrylamide gel electrophoresis, immunochemical detection, isoelectric focusing, and microsequencing. Anal Biochem 263:67–71
Yang P, Fu H, Walker J, Papa CM, Smalle J, Ju YM, Vierstra RD (2004) Purification of the Arabidopsis 26S proteasome: biochemical and molecular analyses revealed the presence of multiple isoforms. J Biol Chem 279:6401–6413

Zn/Cd/Co/Pb P_{1b}-ATPases in Plants, Physiological Roles and Biological Interest

Nathalie Leonhardt, Pierre Cun, Pierre Richaud, and Alain Vavasseur

Abstract P_{1B}-ATPases are cation transporters found in all organisms. In plants, they are found in almost all membranes including plasma membrane, internal chloroplastic membrane, thylakoids, post-Golgi and tonoplast, where they manage nutrient fluxes (Cu^{+}, Zn^{2+}) as well as detoxification of toxic cations (Co^{2+}, Cd^{2+}, Pb^{2+}, high Zn^{2+}). All P_{1B}-ATPases from *Arabidopsis thaliana* have now been characterized; they are involved in plant-scale and cell-scale cations fluxes. Orthologs of these genes are being characterized in other species such as rice or soybean, and their expression domain and roles are similar to those of *Arabidopsis*. Orthologs from hyperaccumulator species, such as *Arabidopsis halleri* or *Thlaspi cearulescens*, have been found to be over-expressed due to the presence of strong promoters and gene duplications. They have recently been shown as essential in the hyperaccumulators properties. Our knowledge about plant P_{1B}-ATPases opens new fields of investigation in biotechnology on the use of these transporters in phytoremediation and biofortification strategies. In this chapter, we draw an overview of the present knowledge about plant P_{1B}-ATPases structures, their physiological roles and their phylogeny. Finally, we summarize the potential that P_{1B}-ATPases give us in crop improvement.

N. Leonhardt (✉) • P. Cun • A. Vavasseur
Laboratoire des Echanges Membranaires et Signalisation, Unité Mixte de Recherche 6191, Centre National de la Recherche Scientifique-Commissariat à l'Energie Atomique-Université Aix-Marseille II, Bat 156, St Paul-lez-Durance, France
e-mail: nathalie.leonhardt@cea.fr; pierre.cun@cea.fr; alain.vavasseur@cea.fr

P. Richaud
Laboratoire Bioenergétique et Biotechnologie des Bactéries and Microalgues, Unité Mixte de Recherche 6191, Centre National de la Recherche Scientifique-Commissariat à l'Energie Atomique-Université Aix-Marseille II, St Paul-lez-Durance, France
e-mail: pierre.richaud@cea.fr

D.K. Gupta and L.M. Sandalio (eds.), *Metal Toxicity in Plants: Perception, Signaling and Remediation*, DOI 10.1007/978-3-642-22081-4_11,

1 Introduction

ATPases are cation transporters involved in various mechanisms such as compartment acidification or ion compartmentation. They are classified into three groups depending on their function and localization: V-ATPases transport H^+ through the tonoplast, F-ATPases are ATP-synthases localized at the mitochondrial and thylakoïdal membranes and P-ATPases are cation transporters found in most organisms.

1.1 P-ATPases Subfamily

P-ATPases mainly transport metal cations (K^+, Na^+, Cu^+, Ag^+, Mg^{2+}, Co^{2+}, Cd^{2+} and Zn^{2+}) against their concentration gradient. This transport is Mg^{2+}- and ATP-dependent, and these ATPases are called P because during their catalytic cycle a phosphorylated intermediate is formed. P-ATPases are able to autophosphorylate: ATP is bound to a cytosolic loop and the P_γ is transferred to the Asp of the DKTGTLT domain (usual sequence of catalytic domains; Møller et al. 1996), responsible for the conformational change of the protein essential in the transport activity. The transport of metal by P_{1B}-ATPases follows the Albers-Post model, the protein conformation alternates between two conformations and two phosphorylation levels (Argüello et al. 2007).

Baxter et al. (2003) have classified P-ATPases in five sub-families:

P_1	P_{1A}, Bacterial K^+-ATPases
	P_{1B}, Metal transporters (Zn^{2+}, Co^{2+}, Cd^{2+}, Pb^{2+}, Cu^+, Ag^+)
P_2	P_{2A} and P_{2B}, Cu^{2+}-ATPases
	P_{2C}, Na^+/K^+- and H^+/K^+-ATPases
	P_{2D}, Unknown function
P_3	P_{3A}, Plasma membrane H^+-ATPases
	P_{3B}, Mg^{2+}-ATPases
P_4	Amino acid translocase
P_5	Homeostasis during gametogenesis

P_{1B}-ATPases are phylogenetically different from others and can be split into two clusters regarding their substrates. The first cluster contains the Cu^+/Ag^+-ATPases and is found in all organisms from prokaryotes to eukaryotes. The *Arabidopsis thaliana* genome contains four genes encoding for Cu^+/Ag^+-ATPases: AtHMA5, AtHMA6 (PAA1), AtHMA7 (RAN1) and AtHMA8 (PAA2). Orthologs of these genes have been found in other plant genomes, such as soybean (Bernal et al. 2006) and rice (Lee et al. 2007). The second cluster contains the putative Zn^{2+}-, Co^{2+}-, Cd^{2+}- and Pb^{2+}-ATPases, *At*HMA1, *At*HMA2, *At*HMA3 and *At*HMA4. *At*HMA2 and *At*HMA4 are involved in Cd^{2+} and Zn^{2+} translocation from roots to shoots (Mills et al. 2003; Hussain et al. 2004; Gravot et al. 2004; Verret et al. 2004; Wong and Cobbett 2009a). *At*HMA3 and *Os*HMA3 have been recently shown to drive Cd^{2+} influx into plant vacuoles (Morel et al. 2009; Ueno et al. 2010; Miyadate et al. 2011). *At*HMA1 which presents the atypical SPC motif is involved in copper entry into the chloroplast (Seigneurin-Berny et al. 2005) but could also participate in the transport of other cations such as Ca^{2+}(Moreno et al. 2008).

1.2 Structure of the P_{1B}-ATPases

P_{1B}-ATPases exhibit eight transmembrane domains, two cytosolic loops (one containing the actuator site and the other one the ATP binding and hydrolysing sites), and cytoplasmic N- and C- terminus domains (Fig. 1).

The transmembrane (TM) domains 6–8 contain several metal binding sites and the TM 6 has a CPX motif (mainly CPC, but also CPS, CPA, CPT and CPD). The cysteine in the CPX domain is essential for the activity of the protein, and its mutation fully impairs the transport activity of the protein (Yoshimizu et al. 1998; Fan and Rosen 2002; Mandal and Argüello 2003; Lowe et al. 2004). The presence of SPC motif constitutes an exception that can be found in the Cu^{+}-ATPases such as *At*HMA1 which is widespread in plants (Argüello et al. 2007). Since this specific domain is found in most P-ATPases, it is not sufficient to determine the enzyme specificity. Also, it has been proposed that the specificity would result from the combination of the CPX sequence and the presence of specific residues located on TM 7 and 8 (Mills et al. 2005; Dutta et al. 2007). Finally, a new classification of the P_{1B}-ATPases has been established using the nature of these residues describing

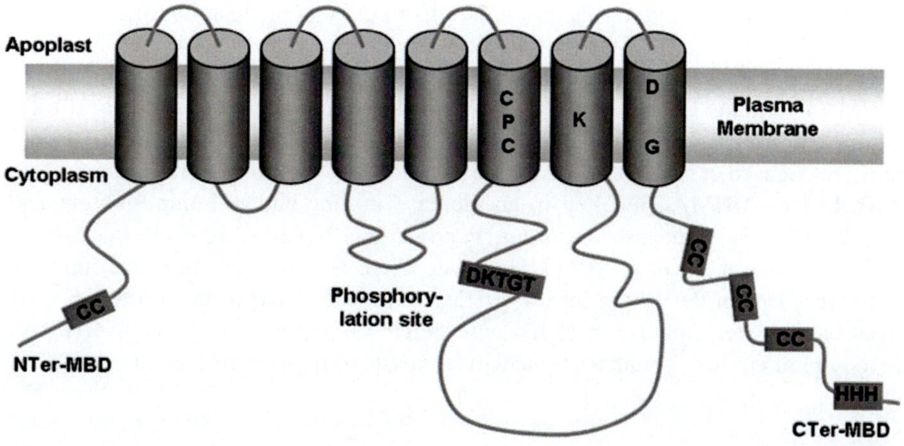

Fig. 1 *Structure of AtHMA2, given as prototypical Zn^{2+}-P_{1B}-ATPases*. Positions in transmembranes 6–8 of residues identified as essential in transport specificity (Argüello et al. 2007) are indicated, CC and His-rich metal binding domains (MBD) are indicated on N- and C-terminus

several sub-classes matching with different substrate specificity (Argüello et al. 2007). The amino acid sequence of the cytosolic loops is well conserved among the different members of the P_{1B}-ATPases. The cytosolic loop between the TM 4 and 5 constitutes the actuator domain. Structure analyses of CopA and Serca (Sazinsky et al. 2006a) have shown that a highly conserved domain exhibiting a GE sequence can interact with the ATP-binding domain at the level of the second cytosolic loop responsible for conformation changes of the protein essential for the cation transport. The second cytosolic loop, between TM 6 and 7, is divided into two domains: the P domain is the nearest to the membrane and contains the DKTGT sequence, which is the ATP-hydrolysis site, the P_{γ} being transferred to the D residue; the other part is the one called N (for nucleotide-binding), and contains the ATP-binding sequence (Sazinsky et al. 2006b). Most P_{1B}-ATPase N-termini share a βαββα folding exhibiting metal binding sites. Specific to plant Zn^{2+}-ATPases is that the sequence of the metal binding domain (MBD) contains CC in contrast to CXXC in the other P-ATPase members. This N-terminus domain is involved in the regulation of enzyme activity. Interestingly, the deletion of the entire domain of *At*HMA2 results in a loss of the activity of this protein in plants (Wong et al. 2009b). In addition, it has been shown that the CCXXE is crucial for the ATPase activity and that a mutation in this site in *At*HMA2 and *At*HMA4 impairs the transport activity (Verret et al. 2005). Finally, a truncated *At*HMA2 lacking the N-MBD showed reduced ATPase activity without significant changes in metal binding to transmembrane metal binding sites (Eren et al. 2007).

Plant Zn^{2+}-ATPases have an unusual long C-terminus with numerous His and Cys mainly localized in MBDs (CC, poly-His) as for *At*HMA2 and *At*HMA4. Surprisingly, the function of the C-terminus does not seem to be conserved from a protein to another. For example, the deletion of the whole C-terminus of *At*HMA2 does not fully impair the activity of the protein in plants (studied by complementation of *hma2, hma4* mutant plants) but seems to cause a mislocation of the protein and this shortened form of the protein did not complement the female sterility of the double mutant (Wong et al. 2009b). Interestingly, the C-terminal domain of *At*HMA4 contains 13 Cys doublets followed by a stretch of 11His which constitutes a high affinity Zn^{2+} and Cd^{2+} chelator domain that would be able to bind 10 Zn^{2+} ions (Baekgaard et al. 2010). The expression of this C-terminus domain alone from *Tc*HMA4 or *At*HMA4 resulted in an increase in zinc and cadmium content and tolerance in yeast but also *in planta* (Papoyan and Kochian 2004; Bernard et al. 2004; Siemianowski et al. 2011). However, there is still a debate concerning the regulatory role of the C-terminus of *At*HMA4. Indeed, it has been reported that the deletion of the terminal His stretch was sufficient to suppress the ability of *At*HMA4 to complement Cd^{2+} and Zn^{2+} sensitive yeast strains (Verret et al. 2005). In contrast, other work has reported an increased activity of *At*HMA4 when the C-terminus was deleted (Mills et al. 2010; Baekgaard et al. 2010). In any case, expression of a delta C-terminus in *hma2, hma4* double mutants did not rescue the sterile phenotype of these mutants demonstrating that the C-terminus of the protein plays an essential role in plants (Mills et al. 2010).

2 Physiological Roles and Expression Profiles of Plant P_{1B}-ATPases

Arabidopsis thaliana is the only plant species in which all HMAs have been functionally described. Thus, the following chapters will essentially describe the expression profiles and the functions of the eight *At*HMAs found in the genome of this model plant (Fig. 2).

2.1 *HMA1*

Although HMA1 from *A. thaliana* was initially attributed to the Zn^{2+}, Cd^{2+}, Co^{2+}, Pb^{2+}-ATPases group such as HMA2, HMA3 and HMA4 (Axelsen and Palmgren 2001), it has finally been characterized as a member of a different cluster, cluster I including P_{1B-1}-ATPases in the Baxter classification (Baxter et al. 2003). *At*HMA1 presents two original properties relative to other *At*HMAs. First, its secondary structure is not precisely determined. Instead of eight transmembrane segments found in most HMAs, only 5–7 TMs are predicted for *At*HMA1 (7 TM segments according to the topology prediction software ARAMEMNON; 6 according to Arguëllo (2003). Second, the canonical CPX motif in the sixth TM helix is changed for a SPC motif. Using heterologous expression in yeast, it was demonstrated that *At*HMA1 is able to transport copper (Seigneurin-Berny et al. 2005). This transport activity is abolished when the original SPC motif is substituted for the classical CPC (S410C) found in most other HMAs proteins, and also by a mutation of a His located in the last predicted TM α-helix (H769D). Moreover, the transport activity is partially affected by a deletion of the N-terminus His-rich domain (Seigneurin-Berny et al. 2005).

In plants, *At*HMA1 localizes at the inner membrane of the chloroplast and participates in Cu^+ loading into the stroma. Plants defective for *At*HMA1 exhibit a lower Cu^+ content into the chloroplasts leading to a reduced Cu^+/Zn^{2+} superoxide dismutase activity in this organelle (Cu^+/Zn^{2+}-SOD). In accordance, under high light, these mutants exhibit a photosensitive phenotype due to a photo-oxidative stress. The *At*HMA1 ATPase activity, measured on purified chloroplast envelopes, was specifically stimulated by Cu^+ (Seigneurin-Berny et al. 2005). *AtHMA1* expression is reduced after Cu^+ exposure and enhanced under Cu^+ deficiency (del Pozo et al. 2010). Moreover, *hma1* lines present an increased level of expression of chloroplastic ascorbate peroxidases. This observation could explain the impaired water–water cycle of photosynthesis observed in *hma1* lines, probably linked to a lower Cu^+/Zn^{2+}-SOD content (Higuchi et al. 2009). As initially predicted for *At*HMA1, Zn^{2+} and Cd^{2+} transport activities were observed through heterologous expression of *AtHMA1* in yeast. More strikingly, a Ca^{2+} influx in the lumen of the yeast Golgi apparatus was also observed with the hallmarks of a Ca^{2+} high affinity transporter ($K_m = 370$ nM, $V_{max} = 1.53$ nmol mg^{-1} min^{-1}), strongly inhibited by

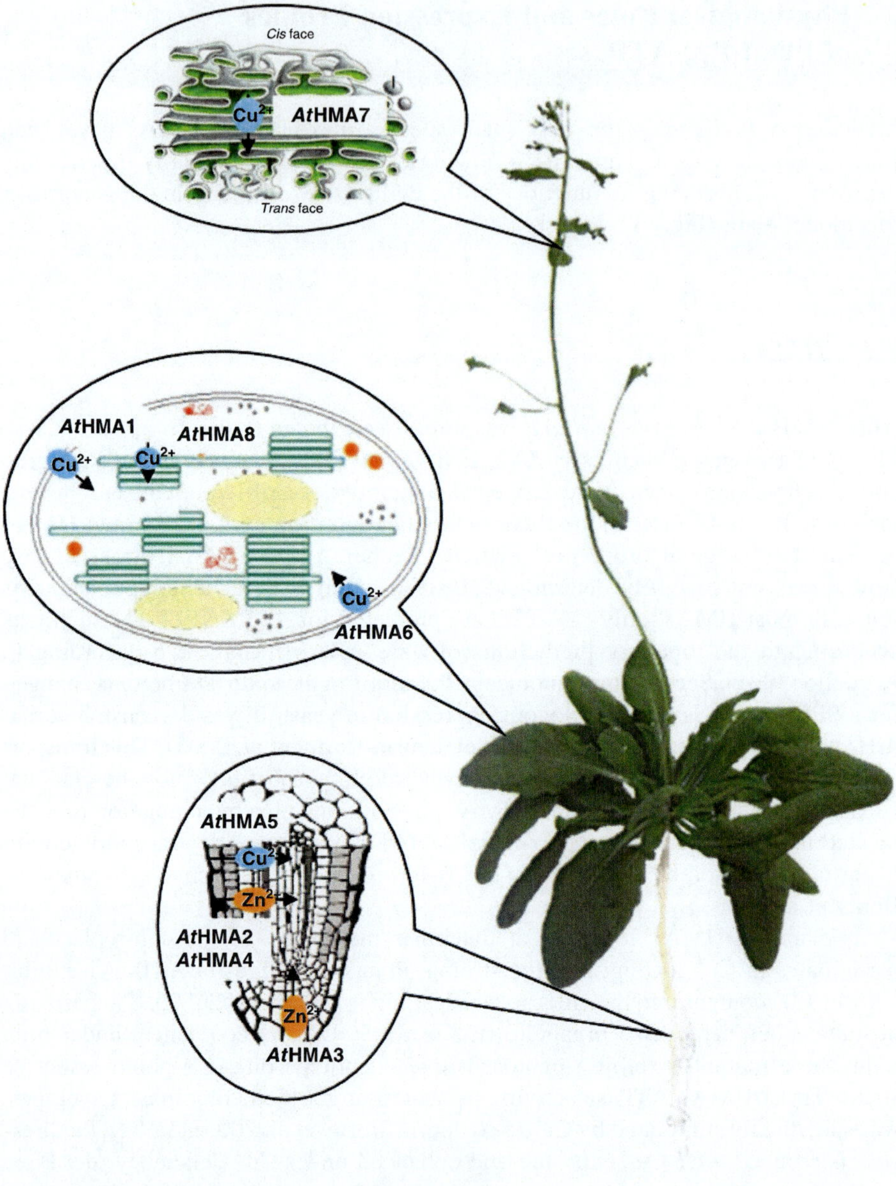

Fig. 2 *Schematic illustration of the cellular distribution of the AtHMAs in Arabidopsis. At*HMA1 localizes at the inner membrane of the chloroplast and participates in Cu^{+} loading into the stroma. *At*HMA6 localizes at the inner envelope membrane of chloroplasts and is implicated in Cu^{+} import into the stroma. *At*HMA8 is inserted in the thylakoidal membranes and delivers Cu^{+} to the lumen. *At*HMA5 is located in pericycle cells of roots and is involved in Cu^{+} compartmentalization and detoxification in root tissues. *At*HMA7 is located in the post-Golgi compartment, receiving Cu^{+} from a Cu^{+} chaperone and delivering Cu^{+} to the post-Golgi ETR1 apoprotein and plays a role in the process of Cu^{+} recycling during leaf senescence. *At*HMA2 and *At*HMA4 are located in pericycle cells and play an essential role in plant nutrition by loading Zn^{2+} into the xylem allowing the translocation of this essential micronutrient from roots to shoots. *At*HMA3 is located at the tonoplast and participates to the vacuolar storage of Zn^{2+}

thapsigargin ($IC_{(50)} = 16.74$ nm), as determined for the Serca Ca^{2+}-ATPase. Finally, ATPase activity of *AtHMA1* expressed in the yeast context was found to be stimulated by Zn^{2+}, Cd^{2+}, Ca^{2+} and Co^{2+} (Moreno et al. 2008). Thus, *At*HMA1 together with *At*HMA6 (see next chapter) has a primary role in the influx of Cu^{+} into the chloroplast, and could play an additional role in the transport of Cd^{2+}, Ca^{2+} and Co^{2+}.

2.2 HMA6 and HMA8

*At*HMA6 and *At*HMA8 are both copper transporters located at the chloroplast. *At*HMA6 mRNAs were detected in shoot and root tissues (Shikanai et al. 2003) while *At*HMA8 mRNAs were restricted to the shoot tissue (Abdel-Ghany et al. 2005). Both *hma6* and *hma8* defective mutants present a similar phenotype of abnormally high chlorophyll fluorescence. This is due to an impairment of the photosynthetic electron transfer. Both *hma6* and *hma8* lines have a reduced growth rate whereas a double mutation *hma6 hma8* leads to a lethal phenotype (Abdel-Ghany et al. 2005). In *hma6* seedlings, the Cu^{+} content is normal in the shoot but lowered in chloroplasts. In *hma8*, the Cu^{+} content in chloroplasts is also diminished but a chloroplast fractionation shows that this is only the case in thylakoids. *At*HMA6 localizes at the inner envelope membrane of chloroplasts and is implicated in Cu^{+} import into the stroma, whereas *At*HMA8 is inserted in the thylakoidal membranes and delivers Cu^{+} to the lumen. In *hma6* mutant lines, Cu^{+} deficiency leads to a limited Cu^{+}/Zn^{2+}-SOD activity while in *hma8* lines this enzymatic activity is increased, resulting in an accumulation of Cu^{+} in the stroma (Shikanai et al. 2003; Abdel-Ghany et al. 2005). Both *At*HMA6 and *At*HMA8 are necessary for a correct assembly of plastocyanin (Shikanai et al. 2003; Abdel-Ghany et al. 2005) and the strong decrease in chloroplastic holoplastocyanin in both mutant lines explains their similar phenotypes. At the moment, one ortholog gene of *AtHMA6* has been identified in rice, *OsHMA7*, and two of *AtHMA8*, *OsHMA8* in rice and *GmHMA8* in soybean. In rice, the expression pattern of these orthologs, mostly in leaves, is in agreement with a chloroplastic localization (Lee et al. 2007).

2.3 HMA5

AtHMA5 is mostly expressed in pericycle cells of roots, in flowers, and at a faintly level in above-ground tissues. *Os*HMA4 and *Os*HMA5 are related to *At*HMA5. *Os*HMA4 expression pattern is in accordance with *At*HMA5 pattern whereas *Os*HMA5 is only expressed in roots (Lee et al. 2007). In *Arabidopsis*, addition of Cu^{+} in the medium strongly and specifically induced *At*HMA5 in whole plants. Two independent T-DNA insertion alleles in *Arabidopsis*, *hma5-1* and *hma5-2* were found hypersensitive to Cu^{+}. Under Cu^{+} excess, these lines accumulated Cu^{+} in root tissues to a greater extent than wild-type plants (Andrés-Colás et al. 2006). The root growth was slowed down by Cu^+, together with a wave-like phenotype in the early development phases. Through yeast two-hybrid experiments, it has been demonstrated that *At*HMA5 interacts with *At*ATX1-like Cu^+ chaperone, and to a lesser extent with *AtCCHΔ* (AtCCH deleted of its C-ter extension). These observations suggest a regulatory role for the plant-specific domain of the CCH chaperone (MxCxxC) (Andrés-Colás et al. 2006). In conclusion, the main role of *At*HMA5 is Cu^+ compartmentalization and detoxification in root tissues.

A region in chromosome 1 was identified by QTL-search in Arabidopsis genome explaining half of the phenotypical variation in root Cu^+ tolerance of the Landsberg *erecta* (Ler) ecotype. This QTL regulates Cu^+ translocation and involves *At*HMA5 (Kobayashi et al. 2008), while the Cape Verde Islands (*Cvi*) and the Chisdra-2 (*Chi-2*) alleles together display a lesser tolerance to copper. One single substitution in each allelic HMA5 primary sequence explains this phenotype. By complementation assays using the yeast *ccc2* mutant strain, the N923T mutation (in the tightly conserved NX_6YNX_4P domain localized in predicted TM7) was found to cause a dysfunction of *At*HMA5 in the *Cvi* ecotype. In the *Chi-2* allele, a substitution was found in the $CPCX_6P$ motif, where the latter P was replaced by an L (Kobayashi et al. 2008). Among 40 Cu^+-tolerant or sensitive accessions studied, these two mutations were the only ones always inducing a Cu^+-sensitive phenotype.

2.4 HMA7

*At*HMA7 was the first plant P_{1B}-type ATPase identified and functionally characterized. *At*HMA7 plays a central role in the ethylene response pathway. A reduced *At*HMA7 activity in plants induces an altered ligand specificity of the ethylene receptors, and a severe loss-of-function mutation confers a constitutive ethylene-like response (Hirayama et al. 1999; Woeste and Kieber 2000). *At*HMA7 like *At*HMA5, is able to rescue the *ccc2* yeast mutant, suppressing the deficiency in the high affinity iron transporter Fet3 due to an abnormal regulation of cellular Cu^+. Ccc2 is a yeast Menkes-like protein implicated in the import of Cu^+ in the post-Golgi compartment. Although *At*HMA7, as Ccc2, did not present in its amino acid sequence any targeting signal for retention in this organelle, it was hypothesized that *At*HMA7 was expressed in the post-Golgi compartment, receiving Cu^+ from a Cu^+ chaperone and delivering Cu^+ to the post-Golgi ETR1 apoprotein (Hirayama et al. 1999; Woeste and Kieber 2000). As shown for *At*HMA5, yeast two-hybrid experiments established that the metal binding domains of *At*HMA7 interact with *At*ATX1 and *CCHΔ* (Puig et al. 2007), supporting the implication of *At*HMA7 in the Cu^+ delivery pathway. When the *Arabidopsis* ethylene receptor, *At*ETR1, is expressed in the *ccc2* yeast mutant background, it lacks any detectable ethylene-binding activity. However, heterologous expression of *AtHMA7* in this strain is able to rescue this function (Binder et al. 2010). In plants, the weak alleles *Atran1-1* and *Atran1-2* display the triple response phenotype in the presence of copper chelators.

Additionally, ethylene binding was absent in the severe alleles *Atran1-3* and *Atran1-4*. Altogether, these results demonstrate that *At*HMA7 is essential for the biogenesis of ethylene receptors in plants and thus plays a crucial role in plant growth and development (Binder et al. 2010). The *At*HMA7 mRNA is mostly found in roots and induced in senescent leaves (Himelblau and Amasino 2000) suggesting that *At*HMA7 plays a role in the process of Cu^{+} recycling during leaf senescence (Woeste and Kieber 2000; Himelblau and Amasino 2000). Another P_{1B}-type ATPase, *Os*HMA9, closely related to *At*HMA7, has been characterized in rice (Lee et al. 2007). Expression of this transporter in the *E. coli copA* mutant was able to partially rescue the severe sensitivity of this strain to Cu^{+}. However, *Os*HMA9 was ineffective in the *zntA* strain, defective in a Zn^{2+}-transporting ATPase. In plants, *Os*HMA9 is mostly expressed in roots and mature leaves of rice at the flowering stage and also faintly within the young panicle. The level of *Os*HMA9 transcripts responds to the rice Cu^{+} status by an increase when the plants are submitted to a high concentration of Cu^{+}, but also when Zn^{2+} and Cd^{2+} are applied, while Pb^{2+} has no effect. The protein fused to the fluorescent GFP is targeted at the plasma membrane of vascular bundles and anthers (Lee et al. 2007). Independent null alleles, *oshma9-1* and *oshma9-2*, were found highly sensitive to Cu^{+}, Cd^{2+}, Zn^{2+} and Pb^{2+}. This phenotype was accompanied by a high content of those metals in the mutant tissues. *OsHMA6* could also be related to *AtHMA7*. It is expressed in rice roots and leaves whatever the development stage, and faintly in panicle and seeds (Lee et al. 2007).

2.5 HMA2, HMA4 and HMA3

HMA2, HMA3 and HMA4 are members of the P_{1B-2}-ATPases, transporting Zn^{2+}, Co^{2+}, Cd^{2+} and Pb^{2+} ATPases according to the Baxter classification (Baxter et al. 2003). Due to their restriction to plant kingdom in Eukaryotes and their biotechnological interest, these transporters have been by far the most studied plant P_{1B}-type ATPases.

2.5.1 HMA2 and HMA4

*At*HMA2 and *At*HMA4 have a closely related expression pattern, predominantly in vascular tissues of roots, stems and leaves. They are also expressed in developing anthers in the tapetum. The only difference is that *AtHMA2* is also found in the vascular tissues of mature siliques (Hussain et al. 2004). At the cellular level, they both localize at the plasma membrane (Hussain et al. 2004; Verret et al. 2004). A *hma4* mutant line was found to be more sensitive to Zn^{2+} and Cd^{2+} than the wild-type ones (Verret et al. 2004). Such Zn^{2+} and Cd^{2+} sensitivity was not observed on mature mutant plants (Mills et al. 2005). The double defective *hma2, hma4* mutant has a severe nutritional deficiency phenotype and is sterile. The localization of those transporters in the male reproductive organs is consistent with the male-sterile phenotype observed. This phenotype can be overcome by feeding the plants in irrigated pots with a high Zn^{2+} treatment (1–3 mM), Co^{2+} or Cu^{+} being inefficient to rescue this double mutant (Hussain et al. 2004). Under normal nutritional conditions, the Zn^{2+} content in the shoot tissues was decreased by 20% for the *hma4* mutant and by 75% for the *hma2, hma4* mutant, while the other element contents were unchanged (Hussain et al. 2004). In accordance with these observations, overexpression of *AtHMA4* in *A. thaliana* increases the plant tolerance to Zn^{2+} and Cd^{2+} and their content in shoot tissues (Verret et al. 2004). It is now well recognized that *At*HMA2 and *At*HMA4 are also the main actors of Cd^{2+} translocation from the roots to the shoot. This is particularly evident in the *hma2, hma4* mutant context since in that case Cd^{2+} translocation from root to shoot is impaired by 98% (Wong et al. 2009). In conclusion, *At*HMA2 and *At*HMA4, located in pericycle cells, play an essential role in plant nutrition, by loading Zn^{2+} into the xylem allowing the translocation of this essential micronutrient from the roots to the shoot.

On a functional aspect, heterologous expression of *AtHMA4* confers a hyper-tolerance towards Zn^{2+}, Cd^{2+} and Pb^{2+} to wild-type yeast strains and rescues the hypersensitive mutant *ycf1* and *zrc1* (Verret et al. 2005). In yeast membranes, *At*HMA2 presents an ATP hydrolytic activity strongly activated at high affinity by Zn^{2+} and Cd^{2+} and at higher concentrations by Pb^{2+}, Co^{2+} and Cu^{+} (Eren and Argüello 2004). Many experiments have shown the important role of the N-terminal extension of *At*HMA2 in transport activity. Variants of *At*HMA4 with cystein or glutamic acid residues from the N-terminus substituted by an alanine in the conserved GICCTSE MBD were unable to complement the yeast mutants *ycf1* and *zrc1* (Verret et al. 2005). Still in yeast, truncation of the N-terminus in *At*HMA2 resulted in a half decrease of the V_{max} of the ATPase activity but did not affect its metal dependence (Eren et al. 2007). In plants, N-terminal mutants of *At*HMA2 and *At*HMA4 localized normally but failed to complement the *hma2, hma4* phenotype (Wong et al. 2009; Mills et al. 2010). These observations underline an essential role of this N-terminal MBD in plants. Conversely, the C-ter cytoplasmic extension of *At*HMA2 and *At*HMA4 does not appear to be essential in metal translocation, but has a role in a correct targeting of the transporter to the membrane (Wong et al. 2009b) and is necessary to fully complement the *hma2, hma4* seedlings (Mills et al. 2010).

An ortholog of *AtHMA4* has been identified in the heavy metal hyperaccumulator species *Thlaspi caerulescens*. Heterologous expression of *TcHMA4* in yeast promoted an efflux of Zn^{2+}, Cd^{2+} and Pb^{2+} and induced an increased tolerance of the wild-type yeast strain towards Cd^{2+} (Papoyan and Kochian 2004) such as the one observed through the expression of *AtHMA4*. In plants, *TcHMA4* was found almost exclusively and strongly expressed in roots where its level of expression was 20-fold that of *AtHMA4* in *Arabidopsis* (Bernard et al. 2004) and highly increased at high or low Zn^{2+} in the medium. Studies in recent years have clearly shown that the hyperaccumulator character is accompanied by a high level of expression of various metal transporters due to very strong promoters (Becher et al. 2004; Hanikenne et al. 2008). Gene duplication also participates in the hyperaccumulator character in *Arabidopsis halleri* where three orthologs of *AtHMA4* have been found. These genes are essential to this trait since silencing of *AhHMA4s* resulted in a loss of the hyperaccumulator properties (Hanikenne et al. 2008). This could be a general concept since recently Lochlainn et al. (2011) have identified four orthologs of *AtHMA4* in *Thlaspi caerulescens*.

2.5.2 HMA3

The first characterization of *At*HMA3, in yeast, identified this transporter as a Cd^{2+}/Pb^{2+} transporter able to rescue the *ycf1* yeast mutant strain (Gravot et al. 2004). In this organism, the protein-GFP fusion was detected at the vacuolar compartment (Gravot et al. 2004). In plants, *AtHMA3* is highly expressed in guard cells, hydathodes, vascular tissues and the root apex (Morel et al. 2009). Its level of expression is independent of Zn^{2+} and Cd^{2+} concentrations in the medium, in agreement with previous results (Gravot et al. 2004). In accordance with the location in yeast, *At*HMA3-GFP fusions were observed at the tonoplast (Morel et al. 2009) where the transporter participates to the vacuolar storage of Cd^{2+}. Overexpression of *AtHMA3* improves plant tolerance to Cd^{2+}, Co^{2+}, Pb^{2+}, and Zn^{2+}, and increases Cd^{2+} accumulation in the plants by two- to threefold. Thus, it was hypothesized that *At*HMA3 plays a role in the detoxification of biological (Zn^{2+}) and non-biological heavy metals (Cd^{2+}, Co^{2+} and Pb^{2+}) by participating in their vacuolar sequestration (Morel et al. 2009). Recently, the most related P_{1B}-ATPase to *At*HMA3 in rice, *Os*HMA3, was identified during a QTL search for low Cd^{2+} in grain (Ueno et al. 2010, 2011; Miyadate et al. 2011). In rice, *Os*HMA3 localizes at the tonoplast of root cells (Ueno et al. 2010) and overexpression of *OsHMA3* decreases the accumulation of Cd^{2+} in the shoot and grains, without modifying other micronutrients in the grains. *OsHMA3* plays an important role in rice since a naturally null mutant *Oryza sativa* cv Cho-Ko-Koku has an enhanced rate of root-to-shoot Cd^{2+} translocation and a high Cd^{2+} content in grain (Ueno et al. 2010, 2011; Miyadate et al. 2011). In *A. thaliana*, some ecotypes such as *Col0* present an *HMA3* pseudogene inducing the absence of the corresponding protein and a less tolerant phenotype to Cd^{2+} (Morel et al. 2009).

By hybridization on *A. thaliana* GeneChips, *AhHMA3* was found highly expressed in the Zn^{2+}/Cd^{2+} hyperaccumultor species *A. halleri* (Becher et al. 2004). This observation was confirmed by northern blot with leaf RNA and by RT-PCR on shoot RNA as well as root transcripts. Very recently, *Tc*HMA3 from a high (Ganges) and low Cd^{2+}-accumulating (Prayon) ecotypes from *T. caerulescens* have been characterized. Interestingly, the *TcHMA3* alleles from both ecotypes share 98% identity and code for a Cd^{2+} vacuolar transporter highly expressed in the roots. The main difference between these ecotypes was the expression level of *TcHMA3*, Ganges showed a sevenfold higher expression than Prayon, partly caused by a higher number of copies (Ueno et al. 2011).

3 Phylogeny

The following phylogenetic study includes P_{1B}-ATPases from 22 photosynthetic organisms whose full genomes have been published on the Phytozome v.6.0 website (http://www.phytozome.net/). Then, the recently published *HMA* sequences from the hyperaccumulator species *A. halleri* and *T. caerulescens* were added and the two sequences found in the yeast *Saccharomyces cerevisiae* (http://www.yeastgenome.org/). Finally, 205 sequences coding HMA transporters were identified and analyzed (Fig. 3). All photosynthetic land plants (moss, lycophytes, monocots and eudicots) present about the same number of HMA transporters, an average of 9 sequences, with the exception of two species, *Manihot esculenta* (genome size ~760 Mbp) with only 5 HMAs, and at the opposite end, *Glycine max* which presents the highest number of HMAs coding genes with 18 members. Unicellular organisms, photosynthetic or not, *Chlamydomonas reinhardtii, Volvox carteri* and *S. cerevisiae* possess a reduced number of sequences (four for the microalgae and two for the yeast). All HMAs analyzed in the present study respected the classification from Baxter into six clusters, although those from the unicellular organisms appear rather apart. Four of the six clusters (clusters II, IV, V and VI) present an equivalent distribution of HMAs, each being subdivided into three subclusters with generally one containing eudicot proteins, a second one the *Poaceae* or monocot proteins, and the last one including moss and lycophyte transporters. Cluster I presents a higher degree of divergence (P_{1B-1}), while cluster III is the largest including 54 proteins and contains 2 subclusters segregating the eudicot or *Poaceae* members. Sequences in each of the first subclusters showed 75–90% identities, whereas sequences belonging to the second subcluster presented only 50% identities.

Plants are the only eukaryotic organisms possessing P_{1B-2} Zn^{2+}-ATPases with the exception of PCA1 from *S. cerevisiae* which is a particular case. The P_{1B-2}-ATPase cluster II contains 48 $Zn^{2+}/Cd^{2+}/Co^{2+}/Pb^{2+}$ divalent cation transporters, in reference to the metal specificity found for those in *A. thaliana*. In this species, the three transporters of this subcluster played major roles (*At*HMA4 and *At*HMA2 in the Zn^{2+} vascular vessel loading, but also in Cd^{2+} translocation, and *At*HMA3 in the vacuolar storage of these two metals). The *Poaceae* species present in general the same number of transporters, while a larger diversity was found in eudicots, exhibiting from one to four members. Thus, due to this diversity, it is not possible to directly spread the physiological roles of HMA2–4 in *Arabidopsis* to other plant species.

The main divergences between *Arabidopsis* and other Brassicaceae occurred at around 45–20 Mya (Beilstein et al. 2010; Ueno et al. 2011). Among the Brassicaceae family, only *A. halleri* and the genus *Thlaspi* unequivocally contain characterized Zn^{2+} and Cd^{2+} hyperaccumulator species. Moreover, the occurrence of a constitutive metal tolerance phenotype in metallicolous and non-metallicolous *A. halleri* populations strongly suggests that these traits had only arisen during two recent evolutionary events (Roosens et al. 2008). Unfortunately the full genome of

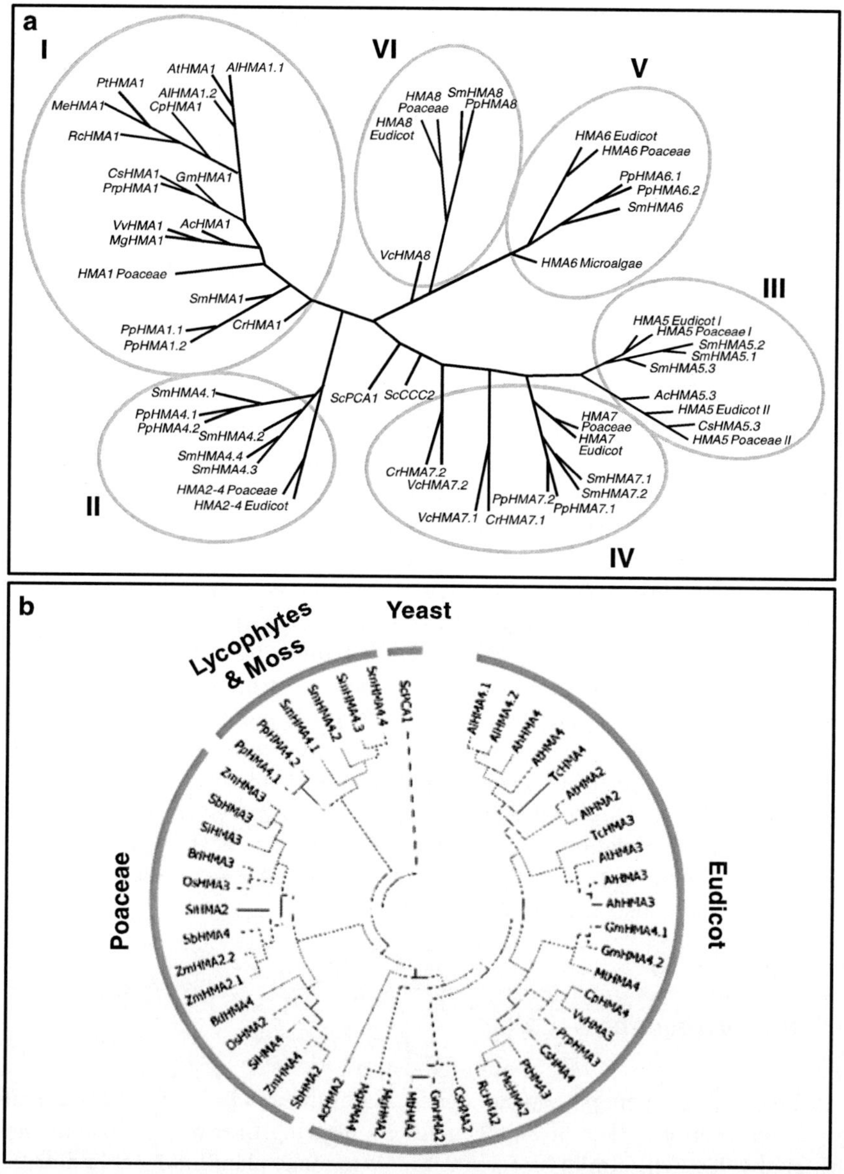

Fig. 3 *Phylogenetic analysis of HMAs from 22 complete genomes of photosynthetic species* (13 Eudicots: *Manihot esculenta* (*Me*), *Ricinus communis* (*Rc*), *Populus trichocarpa* (*Pt*), *Medicago truncatula* (*Mt*), *Glycine max* (*Gm*), *Cucumis sativa* (*Cs*), *Prunus persica* (*Prp*), *Arabidopsis thaliana* (*At*), *A. lyrata* (*Al*), *Carica papaya* (*Cp*), *Vitis vinifera* (*Vv*), *Mimulus guttatus* (*Mg*) and *Aquilegia coerulea* (*Ac*); 5 *Poaceae*: *Sorghum bicolor* (*Sb*), *Zea mays* (*Zm*), *Setaria italica* (*Si*), *Oryza sativa* (*Os*) and *Brachypodium distachyon* (*Bd*); *Selaginella moellendorffii* (*Sm*); *Physcomitrella patens* (*Pp*); 2 microalgae: *Chlamydomonas reinhardtii* (*Cr*) and *Volvox carteri* (*Vc*)), the yeast *Saccharomyces cerevisiae* (*Sc*), and published *HMA* sequences of *A. halleri* (*Ah*)

these species is not available and at present not all *HMA* coding genes from these species have been characterized. The divergence between the two non accumulator species (*A. lyrata* and *A. thaliana*) has occurred around 13–5 Mya (Beilstein et al. 2010; Yogeeswaran et al. 2005). A comparative analysis of the genomes of *A. lyrata* ($n = 8$, diploid) and *A. thaliana* ($n = 5$, around 80% of the size of *A. lyrata*) strongly suggests that three chromosomal fusion events have occurred during this period (Yogeeswaran et al. 2005). In consequence, the number of HMAs in these species is relatively close, with the exception of two HMA4 coding genes in *A. lyrata* instead of only one in *A. thaliana*, but all three proteins present a very high degree of identity.

4 Biotechnological Interest

Due to their essential role in Zn^{2+} translocation from roots to shoot (Hussain et al. 2004), *At*HMA2 and *At*HMA4 and their orthologs are considered as potential targets for biotechnological strategies in the frame of Zn^{2+} biofortification (Colangelo and Guerinot 2006; Palmgren et al. 2008; Palmer and Guerinot 2009). However, a major problem is the dual role of these transporters in the translocation of an essential micronutrient, Zn^{2+}, together with a highly toxic metal, Cd^{2+} (Krämer 2009). From another point of view, the role of these transporters in Cd^{2+} translocation from the roots to the shoot renders these proteins interesting tools in Cd^{2+} phytoremediation strategies. Last, the recently demonstrated role of *At*HMA3 in limiting Cd^{2+} transfer from roots to grain in rice (Ishikawa et al. 2010; Tezuka et al. 2010) underlines the interest of this vacuolar transporter in strategies aiming to limit Cd^{2+} contamination in edible plant tissues. The next section will try to emphasize the potential interest of *HMA2*, *HMA3* and *HMA4* in biofortification and phytoremediation (Zhao and McGrath 2009).

4.1 Biofortification

Biofortification is a general concept consisting in the amelioration of food quality through conventional breeding or genetic engineering of crops to increase their nutritional value. One aim of biofortification is to make plant foods more nutritious as the plants are growing by themselves without further treatment. One of the

---

Fig. 3 (continued) and *Thlaspi caerulescens* (*Tc*). Sequences were aligned with the MAFFT v. 6 software. The trees were obtained with the Neighbor-Joining analysis. The distances are not significant. (**a**) Phylogenetic tree including all 205 *HMA* sequences (*Cluster I* = 23 *HMA1*, *II* = 47 *HMA2-4*; *III* = 54 *HMA5*; *IV* = 36 *HMA7*; *V* = 22 *HMA6*; *VI* = 21 *HMA8*) and two *S. cerevisiae* sequences. For a simplified representation, the individual branches were collapsed when possible. (**b**) Phylogenetic tree of 48 *HMA2*, *HMA3* and *HMA4* sequences well-known examples is the Golden Rice in which two genes have been inserted by genetic engineering leading in the production and accumulation of β-carotene, a precursor of vitamin A, in the grains (Ye et al. 2000). Micronutrient deficiencies are widespread in humans, mainly concerning iron and zinc (White and Broadley 2009). Concerning Fe, it has been shown that a specific targeting of ferritin and nicotianamine synthase was able to increase the iron content in rice endosperm (Wirth et al. 2009). Zn^{2+} deficiencies are widespread, estimated to be at least 25% of the world population (Maret and Sandstead 2006). A primary reason is that many soils of calcareous or alkaline nature lack sufficient phyto-available Zn^{2+} (White and Broadley 2009) and because of the narrow food base of a large part of the world population relies on starch-rich staples that generally contain a suboptimal Zn^{2+} content. Secondary, concerning the rice, for example, the husk and the embryo containing typically 50% of the Zn^{2+} in the grain and is discarded during the milling and polishing processes. Identification of Zn^{2+} transporters able to increase the grain filling in Zn^{2+} is a first step before their manipulation in plants to optimize the Zn^{2+} grain content.

The pioneering work of Hussain et al. (2004) on *Arabidopsis* has clearly demonstrated that *At*HMA2 and *At*HMA4 are in charge of most of Zn^{2+} translocation from roots to shoot, since a double *hma2, hma4* mutant line exhibited a dwarf and sterile phenotype that was overcome by feeding the plants with 3 mM Zn^{2+} in the nutrient solution. In agreement, it was found that overexpression of *AtHMA4* in *Arabidopsis* using the strong *Cauliflower* Mosaic Virus promoter (CaMV 35S) was able to increase the shoot Zn^{2+} content by 60% at 3 μM Zn^{2+} in the nutrient solution and by 52% at 200 μM Zn^{2+} (Verret et al. 2004). These results demonstrate that manipulation of *AtHMA4* is a pertinent strategy to increase Zn^{2+} content in the edible parts of the plant. However, an adverse effect is that this increased translocation of Zn^{2+} was accompanied by a concomitant increase in Cd^{2+} content in the leaves (Verret et al. 2004). Thus, it is essential that we gain a better understanding of the molecular basis of HMA selectivity in order to select transporters with a higher selectivity for Zn^{2+} than Cd^{2+}. Molecular engineering of the transmembrane domains 6–8 exhibiting metal binding sites forming the pore and a survey of the biodiversity of *HMA* alleles in plants could allow the attainment of such a goal. A deregulation of the activity of the transporter may also be a strategy to enhance Zn^{2+} translocation in plants. Previous experiments using heterologous expression in yeast or measurements of ATP hydrolysis have shown that the C-terminal extension of these transporters has regulatory properties. Some results point to a total or partial inactivation of the transporter after a deletion of the C-terminus. Verret et al. (2004) reported an inactivation of *AtHMA4* expressed in yeast after deletion of its terminal His-stretch. In agreement, Eren et al. (2006) observed that removal of the 244 amino acid C-terminus of *At*HMA2 leads to a 43% reduction in the enzyme turnover without significant effect on the Zn^{2+} $K_{½}$ for enzyme activation. In contrast, another study reports a strong increase in Cd^{2+} tolerance conferred on the Cd^{2+}-sensitive yeast *ycf1* after deletion of its C-terminus (Mills et al., 2005). More recently, Baekgaard et al. (2010) observed that sequential removal of the His-stretch and the cystein pairs confers a gradual increase in Zn^{2+} and Cd^{2+} tolerance of sensitive yeasts through a decrease in their Zn^{2+} and Cd^{2+} content. While these observations are difficult to reconcile, expression of a deletion mutant lacking the C-terminal 244 amino acids rescued most of the *hma2, hma4* Zn^{2+}-deficiency phenotypes with the exception of embryo or seed development (Wong et al. 2009b). The GFP-tagged protein also appeared partially mis-localized in the root pericycle cells. Thus, manipulation of the C-terminus of *At*HMA4 did not at present offer an obvious way to enhance Zn^{2+} translocation in plants. Tentative expression or overexpression of heterologous genes from metallophytes did not result in an increased translocation of Zn^{2+} (Barabasz et al. 2010). This may be attributed to the fact that HMAs from metallophytes do not display a higher activity when expressed in yeast than those from non-metallophyte origin (Papoyan and Kochian 2004; Bernard et al. 2004). There are no formal data whether HMAs need to form multimeric complexes to be active but this is likely from what is known from well described P-ATPases such as the H^{+}-ATPases. Thus, another explanation of the lack of efficiency of heterologous expression in plants could be related to the formation of inactive heterologous multimers.

Increasing shoot Zn^{2+} content is interesting for some crops, but the major goal is an enhanced Zn^{2+} content in grains. In a test using five different *Arabidopsis* transgenic lines overexpressing the endogenous gene *AtHMA4*, the Zn^{2+} content measured in the seeds was on average 14% lesser than in the wild type (A Vavasseur, A. Chevalier and P. Richaud, unpublished results). This unexpected result means that the constitutive overexpression of *AtHMA4* leads to a spreading of the metal in all the shoot tissues to the detriment of the initial vectorization of metal trafficking. Thus, in order to increase the Zn^{2+} content in specific edible parts of the plant, a precise vectorisation of the transport of the micronutrient is essential. This could be achieved by (1) the introduction of multi copies of the genes of interest, mimicking what is observed in metallophytes (Hanikenne et al. 2008; Lochlainn et al. 2011), (2) the use of promoters from metallophytes that have been shown extremely active (Hanikenne et al. 2008; Lochlainn et al. 2011), (3) a modification of native promoters to enhance their activity, and (4) the use of promoters strongly expressed in the pericycle (Dembinsky et al. 2007). In any case, while *AtHMA4* and orthologs stay interesting biotechnological targets to enhance Zn^{2+} translocation, enhancing Zn^{2+} content in grain will certainly require manipulation of other genes involved in Zn^{2+} trafficking from the phloem to the different tissues of the seeds. A pioneering study from Tauris et al. (2009) using laser capture micro dissection in barley grain indicated that *Hv*HMA2 and *Hv*HMA4 were detected in transfer cells but not in the aleurone and endosperm tissues. Transporters from other families such as ZIP, YSL, MTP and CAX were detected in these specific tissues and could be in charge of Zn^{2+} trafficking in these tissues. Such studies should allow the discovery of new candidate genes for Zn^{2+} biofortification.

The biofortification concept aimed at an amelioration of food quality also includes a diminution of toxics or anti-nutrient compounds such as phytate in food (Raboy 2007). In this domain, *At*HMA3 was initially identified as a Cd^{2+}/Pb^{2+} vacuolar pump when expressed in yeast, able to rescue the metal-sensitive phenotype of *ycf1* mutant lines (Gravot et al. 2004). In plants, *At*HMA3 also localized at the tonoplast and is expressed at a low level in roots and shoot vascular tissues (Morel et al. 2009). In accordance with the function determined in yeast, a *hma3* knock-out mutant was found slightly more sensitive to Zn^{2+}, Cd^{2+}, Pb^{2+} and Co^{2+} while overexpression of the protein in *Arabidopsis* gave a strong tolerance to these metals (Morel et al. 2009). Interestingly, very recently, a major QTL explaining more than 80% of the phenotypic variance in the shoot Cd^{2+} content of rice was detected in different mapping populations (Ueno et al. 2009; Ishikawa et al. 2010; Tezuka et al. 2010). These QTL identified *Os*HMA3 as a vacuolar transporter able to sequester Cd^{2+} at the level of the roots (Ishikawa et al. 2010; Tezuka et al. 2010). Identification of *Os*HMA3 as an essential transporter for sequestration of Cd^{2+} in the roots provides an efficient way to breed rice and other crops with low Cd^{2+} accumulation.

4.2 *Interest in Phytoremediation*

Phytoremediation consists in the use of plants to remove or make safe toxins contained in soil or water (Salt et al. 1998). Essentially, it is the use of green plants to clean-up contaminated soils, sediments or water. This green technology presents many advantages such as a lower cost, a respect of the soil and of the biotope, and the possibility to remediate large areas. A major counterpart of phytoremediation is that generally decades are necessary to clean up contaminated sites to non-toxic concentrations of pollutant (Rascio and Navari-Izzo 2011). A second problem is that polluted sites generally do not contain a single pollutant but a mix of metals and eventually organic toxins. The first large-scale experiments have revealed that each case is specific, depending on multiple parameters such as the nature of pollutants, the depth at which they have to be extracted, the nature of the soil and the climate.

*At*HMA4 and its orthologs are interesting since they allow a higher translocation of Zn^{2+} and Cd^{2+} and presumably Pb^{2+} and Co^{2+} from the roots to the shoot (Verret et al. 2004), allowing a detoxification of the roots and thus a better tolerance of the plant and a higher yield of phytoextraction. While the increase in Zn^{2+} and Cd^{2+} transfer in *35S::AtHMA4* transgenic plants was less than a doubling, the first experiments on actual polluted soils have shown that the transgenic lines were able to extract 3.1-fold more Cd^{2+} and 2.8-fold more Zn^{2+} than the wild-type plants (C. Sarrobert, Institut de Biologie Environnementale et Biotechnologies, France, personal communication). Thus, a major interest of HMA4 is to give a selective advantage in case of a multi metallic pollution and this may be the reason why this gene has been under a high selection pressure in hyperaccumulators which have to face such types of environment. Genes encoding *At*HMA4 orthologs have been found highly expressed in Zn^{2+}/Cd^{2+} hyperaccumulators, triplicated in *Arabidopsis halleri* (Hanikenne et al. 2008) and even quadruplicated in *Noccaea (T.) caerulescens* (Lochlainn et al. 2011). Silencing of *Ah*HMA4 encoding genes resulted in a total loss of metal hyperaccumulation (Hanikenne et al. 2008), demonstrating the crucial role of *AhHMA4* genes in this property. Thus, introducing

HMA4 under the control of strong promoters, as discussed above for Zn^{2+}, in plants producing high biomass such as tobacco or poplar could result in lines specifically dedicated to phytoextraction of heavy metals, also able to decrease the length of the decontamination treatment which is the major drawback. Additionally, a coexpression of *HMA3* could result in a better tolerance of these plants (Morel et al. 2009). Cys and His residues in the soluble C-termini of *At*HMA2 and *At*HMA4 have been identified as metal-binding residues (Wong et al. 2009b; Baekgaard et al. 2010). Interestingly, expression of the unique soluble C-terminus of *At*HMA4 or *Tc*HMA4 was sufficient to rescue Cd^{2+}/Zn^{2+}-sensitive yeast mutants (Bernard et al. 2004; Papoyan and Kochian 2004). In plants, overexpression of the *AtHMA4* C-terminus in tobacco increased cadmium and zinc concentrations in roots and shoots up to fourfold (Siemianowski et al. 2011). This last result shows that expression of chelating peptides in plants is an interesting strategic way. These peptides could be optimized by DNA-shuffling or molecular engineering according to the pollutant of interest, and be previously tested in simple models such as yeast mutants.

5 Conclusion

Plant P_{1B}-ATPases have been the target of many studies in recent years that have enlightened their crucial role in plant Zn and Cu nutrition and homeostasis. Their major contribution in the Zn and Cd hyperaccumulator trait is also now well defined. Despite these findings many questions are still open. How are these proteins regulated at the cellular level? What is the precise role of the N- and C-termini in the regulation of the transport? Which molecular domains contribute to the allocation of these proteins to various intracellular membranes? Which counter ions are involved in the transport? Do these proteins work as monomer, homo or heteromers? The structure of P_{1B}-ATPases will certainly be solved in the next few years. This will be a determinant in understanding the metal specificity of these transporters and will open new gates for their use in biotechnological strategies.

Acknowledgments The authors want to thank Emmanuelle Billon for her help in the phylogenetic analysis, the Nuclear Toxicology program of the CEA and the Region Provence-Alpes-Côte d'Azur (PACA).

References

Abdel-Ghany SE, Müller-Moulé P, Niyogi KK, Pilon M, Shikanai T (2005) Two P-type ATPases are required for copper delivery in *Arabidopsis thaliana* chloroplasts. Plant Cell 17:1233–1251

Andrés-Colás N, Sancenón V, Rodríguez-Navarro S, Mayo S, Thiele DJ, Ecker JR, Puig S, Peñarrubia L (2006) The *Arabidopsis* heavy metal P-type ATPase HMA5 interacts with metallochaperones and functions in copper detoxification of roots. Plant J 45:225–236

Argüello JM (2003) Identification of ion-selectivity determinants in heavy-metal transport P_{1B}-type ATPases. J Membr Biol 195:93–108

Argüello JM, Eren E, González-Guerrero M (2007) The structure and function of heavy metal transport P_{1B}-ATPases. Biometals 20:233–248

Axelsen KB, Palmgren MG (2001) Inventory of the superfamily of P-type ion pumps in *Arabidopsis*. Plant Physiol 126:696–706

Baekgaard L, Mikkelsen MD, Sørensen DM, Hegelund JN, Persson DP, Mills RF, Yang Z, Husted S, Andersen JP, Buch-Pedersen MJ, Schjoerring JK, Williams LE, Palmgren MG (2010) A combined zinc/cadmium sensor and zinc/cadmium export regulator in a heavy metal pump. J Biol Chem 285:31243–31252

Barabasz A, Krämer U, Hanikenne M, Rudzka J, Antosiewicz DM (2010) Metal accumulation in tobacco expressing *Arabidopsis halleri* metal hyperaccumulation gene depends on external supply. J Exp Bot 61:3057–3067

Baxter I, Tchieu J, Sussman MR, Boutry M, Palmgren MG, Gribskov M, Harper JF, Axelsen KB (2003) Genomic comparison of P-type ATPase ion pumps in *Arabidopsis* and rice. Plant Physiol 132:618–628

Becher M, Talke IN, Krall L, Krämer U (2004) Cross-species microarray transcript profiling reveals high constitutive expression of metal homeostasis genes in shoots of the zinc hyperaccumulator *Arabidopsis halleri*. Plant J 37:251–268

Beilstein MA, Nagalingum NS, Clements MD, Manchester SR, Mathews S (2010) Dated molecular phylogenies indicate a Miocene origin for *Arabidopsis thaliana*. Proc Natl Acad Sci USA 107:18724–18728

Bernal M, Testillano PS, Alfonso M, del Carmen RM, Picorel R, Yruela I (2006) Identification and subcellular localization of the soybean copper P_{1B}-ATPase *Gm*HMA8 transporter. J Struct Biol 158:46–58

Bernard C, Roosens N, Czernic P, Lebrun M, Verbruggen N (2004) A novel CPx-ATPase from the cadmium hyperaccumulator *Thlaspi caerulescens*. FEBS Lett 569:140–148

Binder BM, Rodríguez FI, Bleecker AB (2010) The copper transporter RAN1 is essential for biogenesis of ethylene receptors in *Arabidopsis*. J Biol Chem 285:37263–37270

Brooks RR, Lee J, Reeves RD, Jaffre T (1977) Detection of nickeliferous rocks by analysis of herbarium species of indicator plants. J Geochem Explorat 7:49–57

Colangelo EP, Guerinot ML (2006) Put the metal to the petal: metal uptake and transport throughout plants. Curr Opin Plant Biol 9:322–330

Courbot M, Willems G, Motte P, Arvidsson S, Roosens N, Saumitou-Laprade P, Verbruggen N (2007) A major quantitative trait locus for cadmium tolerance in *Arabidopsis halleri* colocalizes with HMA4, a gene encoding a heavy metal ATPase. Plant Physiol 144:1052–1065

del Pozo T, Cambiazo V, González M (2010) Gene expression profiling analysis of copper homeostasis in *Arabidopsis thaliana*. Biochem Biophys Res Commun 393:248–252

Dembinsky D, Woll K, Saleem M, Liu Y, Fu Y, Borsuk LA, Lamkemeyer T, Fladerer C, Madlung J, Barbazuk B, Nordheim A, Nettleton D, Schnable PS, Hochholdinger F (2007) Transcriptomic and proteomic analyses of pericycle cells of the maize primary root. Plant Physiol 145:575–588

Dutta SJ, Liu J, Stemmler AJ, Mitra B (2007) Conservative and nonconservative mutations of the transmembrane CPC motif in ZntA: effect on metal selectivity and activity. Biochemistry 46:3692–3703

Eren E, Argüello JM (2004) *Arabidopsis* HMA2, a divalent heavy metal-transporting P(IB)-type ATPase, is involved in cytoplasmic Zn^{2+} homeostasis. Plant Physiol 136:3712–3723

Eren E, Kennedy DC, Maroney MJ, Argüello JM (2006) A novel regulatory metal binding domain is present in the C terminus of *Arabidopsis* Zn^{2+}-ATPase HMA2. J Biol Chem 281:33881–33891

Eren E, González-Guerrero M, Kaufman BM, Argüello JM (2007) Novel Zn^{2+} coordination by the regulatory N-terminus metal binding domain of *Arabidopsis thaliana* Zn(2+)-ATPase HMA2. Biochemistry 46:7754–7764

Fan B, Rosen BP (2002) Biochemical characterization of CopA, the *Escherichia coli* Cu(I)-translocating P-type ATPase. J Biol Chem 277:46987–46992

Gómez-Galera S, Rojas E, Sudhakar D, Zhu C, Pelacho AM, Capell T, Christou P (2010) Critical evaluation of strategies for mineral fortification of staple food crops. Transgenic Res 19:165–180

Gravot A, Lieutaud A, Verret F, Auroy P, Vavasseur A, Richaud P (2004) *At*HMA3, a plant P_{1B}-ATPase, functions as a Cd/Pb transporter in yeast. FEBS Lett 561:22–28

Hanikenne M, Talke IN, Haydon MJ, Lanz C, Nolte A, Motte P, Kroymann J, Weigel D, Krämer U (2008) Evolution of metal hyperaccumulation required cis-regulatory changes and triplication of HMA4. Nature 453:391–395

Higuchi M, Ozaki H, Matsui M, Sonoike K (2009) A T-DNA insertion mutant of AtHMA1 gene encoding a Cu transporting ATPase in *Arabidopsis thaliana* has a defect in the water-water cycle of photosynthesis. J Photochem Photobiol B 94:205–213

Himelblau E, Amasino RM (2000) Delivering copper within plant cells. Curr Opin Plant Biol 3:205–210

Hirayama T, Kieber JJ, Hirayama N, Kogan M, Guzman P, Nourizadeh S, Alonso JM, Dailey WP, Dancis A, Ecker JR (1999) RESPONSIVE-TO-ANTAGONIST1, a Menkes/Wilson disease-related copper transporter, is required for ethylene signaling in *Arabidopsis*. Cell 97:383–393

Hussain D, Haydon MJ, Wang Y, Wong E, Sherson SM, Young J, Camakaris J, Harper JF, Cobbett CS (2004) P-type ATPase heavy metal transporters with roles in essential zinc homeostasis in *Arabidopsis*. Plant Cell 16:1327–1339

Ishikawa S, Abe T, Kuramata M, Yamaguchi M, Ando T, Yamamoto T, Yano M (2010) A major quantitative trait locus for increasing cadmium-specific concentration in rice grain is located on the short arm of chromosome 7. J Exp Bot 61(3):923–934

Kobayashi Y, Kuroda K, Kimura K, Southron-Francis JL, Furuzawa A, Kimura K, Iuchi S, Kobayashi M, Taylor GJ, Koyama H (2008) Amino acid polymorphisms in strictly conserved domains of a P-type ATPase HMA5 are involved in the mechanism of copper tolerance variation in *Arabidopsis*. Plant Physiol 148(2):969–980

Krämer U (2009) The dilemma of controlling heavy metal accumulation in plants. New Phytol 181:3–5

Lee SKim, Y-Y LY, An G (2007) Rice P1B-type heavy-metal ATPase, *Os*HMA9, is a metal efflux protein. Plant Physiol 145:831–842

Liang HM, Lin TH, Chiou JM, Yeh KC (2009) Model evaluation of the phytoextraction potential of heavy metal hyperaccumulators and non-hyperaccumulators. Environ Pollut 157:1945–1952

Lochlainn SP, Bowen HC, Fray RG, Hammond JP, King GJ, White PJ, Graham NS, Broadley MR (2011) Tandem quadruplication of HMA4 in the Zinc (Zn) and Cadmium (Cd) hyperaccumulator *Noccaea caerulescens*. PLoS One 6:e17814

Lowe J, Vieyra A, Catty P, Guillain F, Mintz E, Cuillel M (2004) A mutational study in the transmembrane domain of Ccc2p, the yeast Cu(I)-ATPase, shows different roles for each Cys-Pro-Cys cysteine. J Biol Chem 279:25986–25994

Mandal AK, Argüello JM (2003) Functional roles of metal binding domains of the *Archaeglobus fulgidus* Cu(+)-ATPase CopA. Biochemistry 42:11040–11047

Maret W, Sandstead HH (2006) Zinc requirements and the risks and benefits of zinc supplementation. J Trace Elem Med Biol 20:3–18

Migeon A, Blaudez D, Wilkins O, Montanini B, Campbell MM, Richaud P, Thomine S, Chalot M (2010) Genome-wide analysis of plant metal transporters, with an emphasis on poplar. Cell Mol Life Sci 67:3763–3784

Mills RF, Krijger GC, Baccarini PJ, Hall JL, Williams LE (2003) Functional expression of AtHMA4, a P1B-type ATPase of the Zn/Co/Cd/Pb subclass. Plant J 35:164–176

Mills RF, Francini A, Ferreira da Rocha PS, Baccarini PJ, Aylett M, Krijger GC, Williams LE (2005) The plant P_{1B}-type ATPase *At*HMA4 transports Zn and Cd and plays a role in detoxification of transition metals supplied at elevated levels. FEBS Lett 579:783–791

Mills RF, Valdes B, Duke M, Peaston KA, Lahner B, Salt DE, Williams LE (2010) Functional significance of *At*HMA4 C-terminal domain in plants. PLoS One 5:e13388

Miyadate H, Adachi S, Hiraizumi A, Tezuka K, Nakazawa N, Kawamoto T, Katou K, Kodama I, Sakurai K, Takahashi H, Satoh-Nagasawa N, Watanabe A, Fujimura T, Akagi H (2011) *Os*HMA3, a P_{1B}-type of ATPase affects root-to-shoot cadmium translocation in rice by mediating efflux into vacuoles. New Phytol 189:190–199

Møller JV, Juul B, le Maire M (1996) Structural organization, ion transport, and energy transduction of P-type ATPases. Biochim Biophys Acta 1286:1–51

Morel M, Crouzet J, Gravot A, Auroy P, Leonhardt N, Vavasseur A, Richaud P (2009) *At*HMA3, a P_{1B}-ATPase allowing Cd/Zn/Co/Pb vacuolar storage in *Arabidopsis*. Plant Physiol 149:894–904

Moreno I, Norambuena L, Maturana D, Toro M, Vergara C, Orellana A, Zurita-Silva A, Ordenes VR (2008) AtHMA1 Is a thapsigargin-sensitive Ca^{2+}/heavy metal pump. J Biol Chem 283:9633–9641

Palmer CM, Guerinot ML (2009) Facing the challenges of Cu, Fe and Zn homeostasis in plants. Nat Chem Biol 5:333–340

Palmgren MG, Clemens S, Williams LE, Krämer U, Borg S, Schjørring JK, Sanders D (2008) Zinc biofortification of cereals: problems and solutions. Trends Plant Sci 13:464–473

Papoyan A, Kochian LV (2004) Identification of *Thlaspi caerulescens* genes that may be involved in heavy metal hyperaccumulation and tolerance: characterization of a novel heavy metal transporting ATPase. Plant Physiol 136:3814–3823

Puig S, Mira H, Dorcey E, Sancenón V, Andrés-Colás N, Garcia-Molina A, Burkhead JL, Gogolin KA, Abdel-Ghany SE, Thiele DJ, Ecker JR, Pilon M, Peñarrubia L (2007) Higher plants possess two different types of ATX1-like copper chaperones. Biochem Biophys Res Commun 354:385–390

Raboy V (2007) The ABCs of low-phytate crops. Nat Biotechnol 25:874–875

Rascio N, Navari-Izzo F (2011) Heavy metal hyperaccumulating plants: how and why do they do it? and what makes them so interesting? Plant Sci 180:169–181

Roosens NHCJ, Willems G, Saumitoi-Laprade P (2008) Using *Arabidopsis* to explore zinc tolerance and hyperaccumulation. Trends Plant Sci 13:208–215

Salt DE, Smith RD, Raskin I (1998) Phytoremediation. Annu Rev Plant Physiol Plant Mol Biol 49:643–668

Sazinsky MH, Agarwal S, Argüello JM, Rosenzweig AC (2006a) Structure of the actuator domain from the *Archaeoglobus fulgidus* Cu(+)-ATPase. Biochemistry 45:9949–9955

Sazinsky MH, Mandal AK, Argüello JM, Rosenzweig AC (2006b) Structure of the ATP binding domain from the *Archaeoglobus fulgidus* Cu + −ATPase. J Biol Chem 281:11161–11166

Seigneurin-Berny D, Gravot A, Auroy P, Mazard C, Kraut A, Finazzi G, Grunwald D, Rappaport F, Vavasseur A, Joyard J, Richaud P, Rolland N (2005) HMA1, a new Cu-ATPase of the chloroplast envelope, is essential for growth under adverse light conditions. J Biol Chem 281:2882–2892

Shikanai T, Müller-Moulé P, Munekage Y, Niyogi KK, Pilon M (2003) PAA1, a P-type ATPase of *Arabidopsis*, functions in copper transport in chloroplasts. Plant Cell 15(6):1333–1346

Siemianowski O, Mills RF, Williams LE, Antosiewicz DM (2011) Expression of the $P(_1B)$ -type ATPase *At*HMA4 in tobacco modifies Zn and Cd root to shoot partitioning and metal tolerance. Plant Biotechnol J 9:64–74

Talke IN, Hanikenne M, Krämer U (2006) Zinc-dependent global transcriptional control, transcriptional deregulation, and higher gene copy number for genes in metal homeostasis of the hyperaccumulator *Arabidopsis halleri*. Plant Physiol 142:148–167

Tauris B, Borg S, Gregersen PL, Holm PB (2009) A roadmap for zinc trafficking in the developing barley grainbased on laser capture microdissection and gene expression profiling. J Exp Bot 60:1333–1347

Tezuka K, Miyadate H, Katou K, Kodama I, Matsumoto S, Kawamoto T, Masaki S, Satoh H, Yamaguchi M, Sakurai K, Takahashi H, Satoh-Nagasawa N, Watanabe A, Fujimura T,

Akagi H (2010) A single recessive gene controls cadmium translocation in the cadmium hyperaccumulating rice cultivar Cho-Ko-Koku. Theor Appl Genet 120:1175–1182

Ueno D, Koyama E, Kono I, Ando T, Yano M, Ma JF (2009) Identification of a novel major quantitative trait locus controlling distribution of Cd between roots and shoots in rice. Plant Cell Physiol 50:2223–2233

Ueno D, Yamaji N, Kono I, Huang CF, Ando T, Yano M, Ma JF (2010) Gene limiting cadmium accumulation in rice. Proc Natl Acad Sci USA 107:16500–16505

Ueno D, Milner MJ, Yamaji N, Yokosho K, Koyama E, Zambrano MC, Kaskie M, Ebbs S, Kochian LV, Ma JF (2011) Elevated expression of *TcHMA3* plays a role in the extreme Cd tolerance in a Cd-hyperaccumulating ecotype of *Thlaspi cearulescens*. Plant J. doi: doi:10.1111/j.1365-313X.2011.04548.x

Van Dillewijn P, Couselo JL, Corredoira E, Delgado A, Wittich RM, Ballester A, Ramos JL (2008) Bioremediation of 2,4,6-trinitrotoluene by bacterial nitroreductase expressing transgenic aspen. Environ Sci Technol 42:7405–7410

Verret F, Gravot A, Auroy P, Leonhardt N, David P, Nussaume L, Vavasseur A, Richaud P (2004) Overexpression of *At*HMA4 enhances root-to-shoot translocation of zinc and cadmium and plant metal tolerance. FEBS Lett 576:306–312

Verret F, Gravot A, Auroy P, Preveral S, Forestier C, Vavasseur A, Richaud P (2005) Heavy metal transport by *At*HMA4 involves the N-terminal degenerated metal binding domain and the C-terminal His11 stretch. FEBS Lett 579:1515–1522

White PJ, Broadley MR (2009) Biofortification of crops with seven mineral elements often lacking in human diets–iron, zinc, copper, calcium, magnesium, selenium and iodine. New Phytol 182:49–84

Willems G, Dräger DB, Courbot M, Godé C, Verbruggen N, Saumitou-Laprade P (2007) The genetic basis of zinc tolerance in the metallophyte *Arabidopsis halleri* ssp. *Halleri* (*Brassicaceae*): an analysis of quantitative trait loci. Genetics 176:659–674

Wirth J, Poletti S, Aeschlimann B, Yakandawala N, Drosse B, Osorio S, Tohge T, Fernie AR, Günther D, Gruissem W, Sautter C (2009) Rice endosperm iron biofortification by targeted and synergistic action of nicotianamine synthase and ferritin. Plant Biotechnol J 7:631–644

Woeste KE, Kieber JJ (2000) A strong loss-of-function mutation in RAN1 results in constitutive activation of the ethylene response pathway as well as a rosette-lethal phenotype. Plant Cell 12:443–455

Wong CK, Cobbett CS (2009) HMA P-type ATPases are the major mechanism for root-to-shoot Cd translocation in *Arabidopsis thaliana*. New Phytol 181:71–78

Wong CK, Jarvis RS, Sherson SM, Cobbett CS (2009) Functional analysis of the heavy metal binding domains of the Zn/Cd-transporting ATPase, HMA2, in *Arabidopsis thaliana*. New Phytol 181:79–88

Ye X, Al-Babili S, Klöti A, Zhang J, Lucca P, Beyer P, Potrykus I (2000) Engineering the provitamin a (beta-carotene) biosynthetic pathway into (carotenoid-free) rice endosperm. Science 287:303–305

Yogeeswaran K, Frary A, York TL, Amenta A, Lesser AH, Nasrallah JB, Tanksley SD, Nasrallah ME (2005) Comparative genome analyses of *Arabidopsis* spp.: inferring chromosomal rearrangement events in the evolutionary history of *A. thaliana*. Genome Res 15:505–515

Yoshimizu T, Omote H, Wakabayashi T, Sambongi Y, Futai M (1998) Essential Cys-Pro-Cys motif of *Caenorhabditis elegans* copper transport ATPase. Biosci Biotechnol Biochem 62:1258–1260

Zhao FJ, McGrath SP (2009) Biofortification and phytoremediation. Curr Opin Plant Biol 12:373–380

Zimmermann M, Clarke O, Gulbis JM, Keizer DW, Jarvis RS, Cobbett CS, Hinds MG, Xiao Z, Wedd AG (2009) Metal binding affinities of *Arabidopsis* zinc and copper transporters: selectivity's match the relative, but not the absolute, affinities of their amino-terminal domains. Biochemistry 48:11640–11654

Interference of Heavy Metal Toxicity with Auxin Physiology

Mudawi Elobeid and Andrea Polle

Abstract Auxins are important phytohormones involved in the coordination of plant growth and defence. In this chapter, we summarize auxin functions in plant biology and identify interactions with heavy metal toxicity. Cadmium induces the formation of reactive oxygen species, which in turn activate auxin oxidases. Auxin oxidases lead to an increased degradation of auxin and thereby are likely to decrease the activities of many genes involved in growth processes. Evidence supporting this hypothetic signalling cascade from heavy metals to eventual growth reductions comes from auxin feeding experiments that ameliorate Cd toxicity, positive effects of auxin producing endophytes on Cd tolerance and by monitoring endogenous auxin physiology with auxin reporter lines of *Arabidopsis thaliana* and *Populus* × *canescens*. Available data for essential micronutrients suggest that they interact with auxin physiology in a manner similar to Cd when present in excess.

1 Introduction: Auxin as a Growth Regulator

Auxins represent an important class of plant hormones playing critical roles in the coordination of plant growth and development (Vanneste and Friml 2009). The most abundant naturally occurring auxin is indole-3-acetic acid (IAA) that controls

---

M. Elobeid
Department of Tree Physiology and Forest Botany, Büsgen-Institut, Georg-August University, Göttingen, Germany

Faculty of Forestry, Department of Silviculture, University of Khartoum, Shambat, Khartoum North, Sudan
e-mail: melobei@gwdg.de

A. Polle (✉)
Department of Tree Physiology and Forest Botany, Büsgen-Institut, Georg-August University, Göttingen, Germany
e-mail: apolle@gwdg.de

D.K. Gupta and L.M. Sandalio (eds.), *Metal Toxicity in Plants: Perception, Signaling and Remediation*, DOI 10.1007/978-3-642-22081-4_12,

a vast array of physiological processes in plants including apical dominance, tropism, shoot elongation and root initiation (Vanneste and Friml 2009). In plants, its biosynthesis is complex and can occur via different pathways, usually starting with tryptophane (Zhao 2010). The main tissues for its biosynthesis are the apical meristems and young leaves (Bartel et al. 2001). Auxin is transported basi-petally through the plant body, i.e., from the tip to the base and the maintenance of the gradient is important for plant architecture (Leyser 2009). In young poplar trees, the auxin gradient through the plant body has been visualized by employing an auxin-sensitive reporter gene construct (*GH3::GUS*, Teichmann et al. 2008).

Polar auxin transport involves (1) diffusion-controlled and pH-dependent uptake of the protonated form IAAH into the cell or uptake via H^+ co-transport though AUX1 carrier proteins (Pickett et al. 1990) and (2) directed auxin efflux via PIN proteins (Vernoux et al. 2010). Differential activities of PIN and AUX1 proteins are creating local auxin maxima and minima and, thus, these proteins are crucial regulators of auxin distribution. Polar transport of auxin can be inhibited by chemical compounds such as NPA (naphthylphthalamic acid) and TIBA (2,3,5-triiodobenzoic acid) (Lomax et al. 1995).

Inside the cell, auxin acts by promoting the degradation of transcriptional repressors (AUX/IAA proteins) that are bound to auxin response factors (ARFs) and thereby inhibit the transcription of auxin response genes (Vanneste and Friml 2009). The degradation involves binding of an auxin receptor (TIR1) to form an Aux/IAA-SCF^{TIR1} complex, which is decomposed by the ubiqitin/S26proteasome pathway (Chapman and Estelle 2009). Thereby auxin eventually alters the expression of many genes, thus controlling plant growth and development (Jain et al. 2006; Jain and Khurana 2009). To fulfil these actions, endogenous IAA concentrations are regulated by three main processes: biosynthesis, degradation, and conjugation (Normanly et al. 1995). The mechanisms by which plants modulate levels of the phytohormone IAA are complex (Clere et al. 2002; Rampey et al. 2004). Auxin can induce the expression of three functional groups of genes: *AUX/IAA*, *GH3*, and *SAUR* (Woodward and Bartel 2005). Members of the *GH3* gene family are IAA-amido synthetases that conjugate auxin with amino acids and thereby diminish the pool of active, i.e., unconjugated auxin (Staswick et al. 2005). The latter process is important since most auxin is present in conjugated forms such as glycosides or amides.

2 Auxin: A Mediator Between Growth and Stress Adaptation

Under varying environmental conditions, growth and development of plants has to be counterbalanced with the utilization of plant resources for defences and stress adaption. The crucial role of auxin and GH3 proteins for modulating signalling cascades during pathogen attack has recently been reviewed (Bari and Jones 2009). Auxin is also an essential component in pathways mediating plant responses to abiotic stresses such as drought, salt, cold, etc. (Popko et al. 2010). Junghans et al.

(2006) found that a strong decline of radial growth in the xylem of salt-stressed poplars corresponds to a strong decrease in auxin levels, whereas in salt-tolerant poplars the concentrations of auxin were maintained. They identified an auxin isoleucine amido hydrolase (ILL3) that was able to release auxin from its amino acid conjugate, thereby maintaining homeostasis under stress and decreasing salt-induced growth reductions. Application of exogenous auxin rescued biomass production in stressed salt-sensitive poplars (Popko et al. 2010). While these examples show that the phytohormone auxin is a central regulator of growth and adaptation to stress, knowledge on how auxin physiology is affected by metal toxicity is scattered.

Plant defence reactions against metal toxicity and their implications for plant growth have been extensively investigated. Heavy metals are defined as metals with a density above 5 g cm^{-3} and encompass compounds such as Fe, Mn, Zn, Ni, and Cu, which are required by plants as essential micronutrients whereas others such as Hg, Pb, and Cd have no known functions and act merely as toxic compounds (Polle and Schützendübel 2003). Plants, like all other organisms, possess homeostatic cellular mechanisms to maintain the correct concentrations of essential and other metal ions inside the different cellular compartments to minimize the potential damage that could result from the exposure to excess metal ions (Clemens 2001; Benavides et al. 2005). Many studies showed that micronutrients may cause oxidative stress when present in excess in the environment (Schützendübel and Polle 2002; Ducic and Polle 2005; Gratao et al. 2005; Sharma and Dietz 2009). It is known for a long time that ROS (reactive oxygen species) may result in direct oxidation of IAA (Beffa et al. 1990). It is therefore possible that the observed growth reductions in response to metal toxicity are mediated by effects on auxin physiology. The goal of this chapter is to give an overview on the current knowledge regarding the interference of excess metals with auxin physiology and to discuss the involvement of possible metabolic pathways resulting in the observed growth reductions.

3 Auxin and Heavy Metal Stress

Among the nonessential heavy metals, cadmium (Cd) is the most well studied compound. It is mainly released into the environment by anthropogenic activities and its emissions have been estimated to be in the range of 30,000 t per annum (Sanita di Toppi and Gabrielli 1999). As Cd is highly toxic its over-accumulation in plants results in growth reductions and death. Environmental pollution with Cd is of great concern since its accumulation in the food chain may eventually affect human health. Although other heavy metals such as Hg and Pb have similar sources and consequences, we will mainly focus on Cd because of the wealth of information available on plant responses to the latter element.

In a polluted environment plant roots are the first to get in contact with Cd. Many studies show that Cd is intensively retained in the roots and that only smaller amounts are directly transported to the shoot (Cataldo et al. 1983; Kastori et al. 1992; Dixit et al. 2001; Liu et al. 2007). In *Zea mays* most of the Cd accumulated in roots inside the cells of the endodermis, pericycle and central cylinder parenchyma and the transport to the aboveground compartment correlated with the concentration applied (Wojcik and Tukiendorf 2005). In herbaceous plants, uptake of Cd appears to take part mainly in the elongation zone of roots (Pineros et al. 1998; Farrell et al. 2005). In contrast, the apex of poplar roots was the major entry site for Cd (He et al. 2011). A well-developed root cap provides protection against metal uptake in the root-tip zone of tree species (Arduini et al. 1996). In trees, besides Cd retention in roots, accumulation in bark played a major role for its deposition (He et al. 2011). At low levels of Cd exposure, the carbohydrate pools of poplar trees were not depleted and, thus, the plants were not energy-limited (He et al. 2011). It is therefore conceivable that moderate Cd levels in aerial parts compared to those in roots lead to strong growth reductions of the shoot because of a disturbance of the internal phytohormone balance and not because of carbohydrate starvation.

Direct measurements of endogenous auxin levels in Cd-exposed plants are scarce. In *Glycine max* exposure to growth-affecting Cd concentrations caused inhibition of auxin biosynthesis and decreased auxin levels, whereas low Cd concentrations increased auxin biosynthesis (Huang et al. 2006). Analysis of Cd-exposed *Arabidopsis* seedling expressing β-glucuronidase (GUS) under the auxin-sensitive *DR5* promoter suggested significant reduction in auxin in the cotyledons (Xu et al. 2010a). In fact, the *GH3* sequence employed for the auxin GUS reporter was initially identified as a Cd and auxin inducible mRNA (Hagen et al. 1988). Elobeid (2008) studied the effects of Cd on GH3::GUS and found evidence for diminished auxin in the apex of poplars stems. In other studies it was found that the activity of IAA oxidase, an enzyme involved in the degradation of auxin, was increased in response Cd (Chaoui and El Ferjani 2005). Therefore, it was suggested that the stimulation of IAA oxidase by Cd may eventually lead to depletion in the level of auxin with consequences for growth.

Further support for the hypothesis that growth reductions caused by Cd may at least partially be caused by depletion of active auxin comes from several auxin feeding experiments. For example, in *Zea mays* uptake of Cd and Ni increased concurrently with growth reduction; however, exogenous application of the synthetic auxin NAA rescued growth (Guo et al. 1999). Similarly, Zhou et al. (2007) found that in *Zea mays* subjected to metal stress (Cu, Zn, Cd and Pb) IAA caused growth promotion, increased biomass, and even resulted in increased accumulation of heavy metals in the shoot. In *Helianthus annuus* positive effects of externally added auxin on growth and heavy metal uptake (Cd, Zn) were also found (Faessler et al. 2010). Thus, addition of IAA served alleviation of the phytotoxicity of metals, perhaps acting as a chelating agent. Notably, many growth promoting bacteria can also generate auxins and may thereby potentially modulate plant responses to heavy metals (Dimkpa et al. 2009). Furthermore, measures to prevent the depletion of endogenous auxin can also improve plant tolerance against Cd (Xu et al. 2010b). Exposure to NO inhibited Cd-induced activation of IAA oxidase, thereby prevented auxin break-down, stimulated the production of proline and glutathione, and lowered oxidative damage (Xu et al. 2010b). It was therefore suggested that NO improves cadmium tolerance in plants by reducing oxidative damage and maintaining the auxin homeostasis.

Auxin is not only a signalling molecule for gene regulation but is also involved in cell wall loosening, a pre-requisite for elongation growth (Schopfer and Liszkay 2006). The cleavage of cell wall polysaccharides can be accomplished by hydroxyl radicals produced from hydrogen peroxide and superoxide in a reaction catalyzed by cell wall peroxidase (Schopfer 2001). These ROS are specifically produced in the apoplast of actively growing tissues such as the elongation zone of the root. Auxin promotes the formation of hydroxyl radicals, thereby increasing wall extensibility (Liszkay et al. 2003). In Cd-exposed plants, the redox balance is disturbed and increased lignification, counteracting wall extensibility, has been observed (Schützendübel et al. 2001, 2002). It is therefore possible that Cd shifts the redox balance by stimulation IAA-oxidase mediated auxin breakdown. In line with this suggestion, Haluskova et al. (2010) reported that application of exogenous H_2O_2 stimulated IAA-oxidase activity, which in turn causes increased auxin degradation.

In conclusion, the literature provides evidence that auxin metabolism is crucially involved in protecting plants from Cd-induced damage. The molecular mechanisms inside the cell and the suspected interference with auxin-mediated growth in the apoplast require further studies.

4 Auxin and Essential Metals

Among the essential micronutrients, transition metals such as copper, iron, and manganese (Cu, Fe, Mn) are important cofactors in metalloproteins, for example in peroxidases, oxidases, and superoxide dismutases. Free Cu^{2+} or Fe^{3+} catalyse the formation of hydroxyl radicals (OH*) via the Fenton reaction. $OH^{\cdot}$ is highly reactive and leads to unspecific oxidative degradation. Therefore, the intracellular levels of free Cu or Fe have to be kept low to avoid triggering the Fenton reaction inside the cell, which would cause unspecific injury (Polle and Schützendübel 2003). Exposure to excess metals leads to over-accumulation and toxic effects when tissue concentrations exceed the optimal demand (Ducic and Polle 2005; Yruela 2005).

Different tissues exhibit different sensitivity to metal stress. For example, in poplar stem radial growth declined already when the plants were grown in the presence of 1 μM Cu in hydroponic nutrient solutions, whereas elongation growth and photosynthesis declined only when the plants were exposed to 50-fold higher Cu concentrations (Elobeid and Polle 2010). It was suspected that sub-toxic concentrations might already affect the phytohormone balance and thereby influence growth. In *Helianthus annuus* elongation growth of roots was more strongly reduced in the presence of excess Cu (80 μM) than that of shoots (Ouzounidou and Ilias 2005). Application of auxin diminished the negative influence of excess Cu and improved the water use efficiency of the plants. Thus, auxin application alleviated metabolic and physiological disturbances of Cu stress. The negative influence of excess Cu on the auxin physiology of roots was also documented in GUS reporter lines of *Arabidopsis thaliana* (Lequeux et al. 2010). Under these conditions, lignification was stimulated and growth of the main root inhibited (Lequeux et al. 2010). Overall, the results of these studies suggest that excess Cu activates stress pathways involving auxin degradation, perhaps in a similar way as Cd.

The involvement of Mn with auxin physiology is known for a long time. Already in the 1960s in vitro studies showed that Mn stimulated auxin degradation involving auxin oxidase activities (e.g., Stonier et al. 1968). An early investigation of Morgan and others (1966) with *Gossypium hirsutum* revealed that excess Mn caused toxicity symptoms, which resembled those of auxin deficiency. Their study showed that Mn stimulated IAA-oxidase activity and suggested that the enzyme can function in vivo to regulate auxin levels and down-stream processes. Similar hydroponic experiments with Mn-stressed *G. hirsutum* demonstrated that growth, IAA oxidase, leaf abscission, internode length, and similar symptoms were consistently manipulated by raising or lowering Mn levels in the plant culture medium (Morgan et al. 1976). It was concluded that there is a causal relationship between IAA oxidase and the responses that involve destruction of auxin (Morgan et al. 1976). Beffa et al. (1990) showed that auxin oxidase could be separated from peroxidase activities and that the product of the oxidation was indo-3yl-methanol. They suggested the existence of a specific enzymatic system, which catalyzes the oxidative degradation of IAA. Therefore, IAA oxidase may have an important role in regard to the fine regulation of the cellular level of auxin and, thus, growth. To investigate the influence of Mn stress on auxin physiology in poplar, *GH3::GUS* reporter lines have been employed (Elobeid 2008). Using this technique it was shown that the strong auxin signal in the vascular strands of the elongation zone of the stem was faded under the influence of Mn stress (Fig. 1a, b). Similarly the auxin signal in the lower parts of the stem, where radial growth occurs (Fig. 1c), and in roots, where side root emerge (Fig. 1e), disappeared under the influence of Mn (Fig. 1d, f). Since these effects occur before irreversible leaf injury and biomass loss, they support that Mn interferes with auxin physiology. Links between auxin and Mn have also been inferred from the analysis of an *Arabidopsis thaliana* mutant (ilr2) that had lost its responsiveness to application of conjugated auxin (auxin-isoleucine), but showed normal behavior to auxin (Magidin et al. 2003). This mutant was apparently unable to activate the required amido hydrolase and at the same time was insensitive to excess Mn suggesting modulation of Mn transport in the root by auxin (Magidin et al. 2003). When a vacuolar Ca^{2+}/H^{+} antiporter (CAX4) was suppressed, application of Cd^{2+}, Mn^{2+} and auxin altered root growth (Mei et al. 2009). *DR5::GUS* auxin reporter detected decreased auxin responses in the *cax4* lines. Mei et al. (2009) concluded that the cation/H^{+} antiporter CAX4 plays an important function in root growth under heavy metal stress conditions.

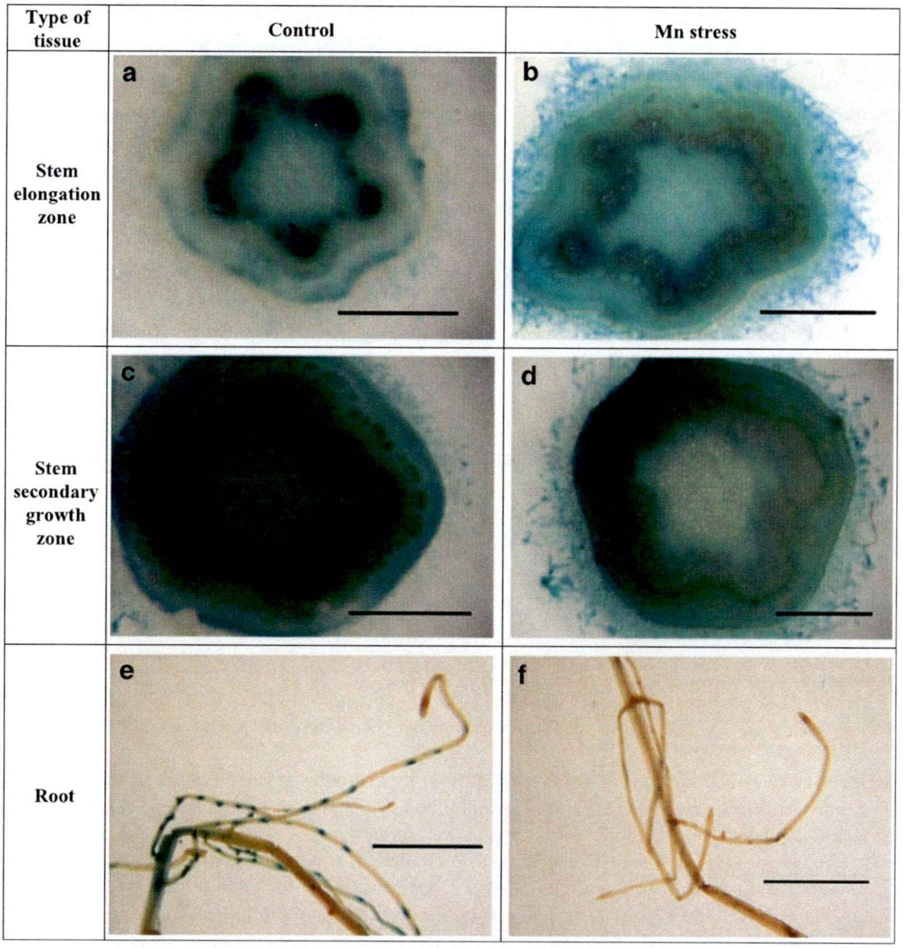

Fig. 1 Cross-sections of stem and root tissues of *GH3::GUS*-transformed grey poplar (*Populus* × *canescens*). GUS staining was performed following exposure of *P.* × *canescens* to manganese treatment (10 mM supplied as $MnCl_2$ for 14 days) in a hydroponic medium. Sections (**a** and **b**) were taken from the elongation zone (2.5 cm beneath the stem apex) and sections of secondary growth zone (**c** and **d**) were taken from the middle of stem. Root staining was performed using fine roots. The scale bar is equivalent to 1 mm (**a–d**) and 3 mm (**e** and **f**) (After Elobeid 2008, modified)

5 Conclusions

Plants monitor their environment to balance their metabolic needs between growth and defence. The importance of auxin homeostasis to mediate between abiotic stresses such as cold, drought, and heat and growth has been demonstrated (Park et al. 2007). This chapter highlights that auxin and heavy metal homeostasis are also intimately linked. The signalling sequence is still speculative. However, some studies provide evidence that heavy metals induce oxidative stress, which in turn might activate auxin oxidases. These enzymes inactivate auxin and thereby might play roles in growth regulation. Therefore, we propose that reductions in plant productivity under the influence of (moderate) metal stress may initially not be caused by photosynthetic reduction and energy limitation but may be the result of interference of heavy metals with auxin metabolism. This would explain why exogenous addition of auxin or stimulation of endogenous auxin levels prevents growth inhibition and increase heavy metal tolerance. Metal transporters and GH3-regulated genes are involved in these processes but the mechanisms are still elusive. A better understanding of these basic processes will not only increase knowledge on the regulation of plant metal homeostasis but may also be useful to improve plants for phytoremediation.

Acknowledgment We are grateful to the Ulrich Foundation (Göttingen, Germany) for financial support to M. Elobeid for the preparation of this chapter and the German Science Foundation for continuous support.

References

Arduini I, Godbold DL, Onnis A (1996) Cadmium and copper uptake and distribution in Mediterranean tree seedlings. Physiol Plant 97:111–117

Bari R, Jones JDG (2009) Role of plant hormones in plant defence responses. Plant Mol Biol 69:473–488

Bartel B, LeClere S, Magidin M, Zolman BK (2001) Inputs to the active indole-3-acetic acid pool: de novo synthesis, conjugate hydrolysis, and indole-3-butyric acid β-oxidation. J Plant Growth Regul 20:198–216

Beffa R, Martin HV, Pilet PE (1990) In vitro oxidation of indole acetic acid by soluble auxin-oxidases and peroxidases from maize roots. Plant Physiol 94:485–491

Benavides MP, Gallego SM, Tomaro ML (2005) Cadmium toxicity in plants. Braz J Plant Physiol 17:21–34

Cataldo DA, Garland TR, Wildung RE (1983) Cadmium uptake kinetics in intact soybean plants. Plant Physiol 73:844–848

Chaoui A, El Ferjani E (2005) Effects of cadmium and copper on antioxidant capacities, lignification and auxin degradation in leaves of pea (*Pisum sativum* L.) seedlings. Plant Biol Pathol 328:23–31

Chapman EJ, Estelle M (2009) Mechanisms of auxin-regulated gene expression in plants. Annu Rev Genet 43:265–285

Clemens S (2001) Molecular mechanisms of plant metal tolerance and homeostasis. Planta 212:475–486

Clere SL, Tellez R, Rampey RA, Matsuda SPT, Bartel B (2002) Characterization of a family of IAA-amino acid conjugate hydrolases from *Arabidopsis*. J Biol Chem 277:20446–20452

Dimkpa CO, Merten D, Svatos A, Buechel G, Kothe E (2009) Metal-induced oxidative stress impacting plant growth in contaminated soil is alleviated by microbial siderophores. Soil Biol Biochem 41:154–162

Dixit V, Pandey V, Shyam R (2001) Differential antioxidative responses to cadmium in roots and leaves of pea (*Pisum sativum* L. cv. Azad). J Exp Bot 52:1101–1109

Ducic T, Polle A (2005) Manganese and copper toxicity and detoxification in plants. Braz J Plant Physiol 172:115–122

Elobeid M (2008) Physiology of auxin in response to environmental stress and heavy metal pollution. Thesis University of Göttingen, Optimus Verlag, Göttingen, Germany
Elobeid M, Polle A (2010) Response of grey poplar (*Populus* × *canescens*) to copper stress. Global Science Books, Plant Stress (Special Issue 1):82–86
Faessler E, Evangelou MW, Robinson BH, Schulin R (2010) Effects of indole-3-acetic acid (IAA) on sunflower growth and heavy metal uptake in combination with ethylene diamine disuccinic acid (EDDS). Chemosphere 80:901–907
Farrell RE, McArthur DFE, Van Rees KCJ (2005) Net Cd^{2+} flux at the root surface of durum wheat (*Triticum turgidum* L. var. *durum*) cultivars in relation to cultivar differences in Cd accumulation. Can J Plant Sci 85:103–107
Gratao PL, Polle A, Lea PJ, Azevedo RA (2005) Making the life of heavy metal-stressed plants a little easier. Funct Plant Biol 32:481–494
Guo DS, Xi YY, Wang AY, Zhang J, Yuan XY (1999) Contribution of auxin to the uptake of nickel and cadmium in maize plants. Biomed Environ Sci 12:170–176
Hagen G, Uhrhammers N, Guilfoyle TJ (1988) Regulation of expression of an auxin-induced soybean sequence by cadmium. J Biol Chem 263:6442–6446
Haluskova L, Valentovicova K, Huttova J, Mistrik I, Tamas L (2010) Elevated indole-3-acetic acid peroxidase activity is involved in the cadmium-induced hydrogen peroxide production in barley root tip. Plant Growth Regul 62:59–64
He J, Qin J, Long LY, Ma YL, Li H, Li K, Jiang XN, Liu TX, Polle A, Luo ZB (2011) Net cadmium flux and accumulation reveal tissue-specific oxidative stress and detoxification in *Populus* × *canescens*. Physiol Plant. doi:10.1111/j.1399-3054.2011.01487.x
Huang YX, Liao BH, Xiao LT, Liu SC, Wang ZK (2006) Effects of Cd^{2+} on seedling growth and phytohormone contents of *Glycine max*. J Environ Sci 27:1398–1401 (In Chinese)
Jain M, Khurana JP (2009) Transcript profiling reveals diverse roles of auxin-responsive genes during reproductive development and abiotic stress in rice. FEBS Lett 276:3148–3162
Jain M, Kaur N, Tyagi AK, Khurana JP (2006) The auxin-responsive GH3 family in rice (*Oryza sativa*). Funct Integr Genomics 6:36–46
Junghans U, Polle A, Düchting P, Weiler E, Kuhlmann B, Gruber F, Teichmann T (2006) Adaptation to high salinity in poplar involves changes in the xylem anatomy and auxin physiology. Plant Cell Environ 29:1519–1531
Kastori R, Petrovic M, Petrovic N (1992) Effect of excess lead, cadmium, copper and zinc on water relations in sunflower. J Plant Nutr 15:2427–2439
Lequeux H, Hermans C, Lutts S, Verbruggen N (2010) Response to copper excess in *Arabidopsis thaliana*: Impact on the root system architecture, hormone distribution, lignin accumulation and mineral profile. Plant Physiol Biochem 48:673–682
Leyser O (2009) The control of shoot branching: an example of plant information processing. Plant Cell Environ 32:694–703
Liszkay A, Kenk B, Schopfer P (2003) Evidence for the involvement of cell wall peroxidase in the generation of hydroxyl radicals mediating extension growth. Planta 217:658–667
Liu CP, Shen ZG, Li XD (2007) Accumulation and detoxification of cadmium in *Brassica pekinesis* and *B. chinesis*. Plant Biol 51:116–120
Lomax TL, Muday GK, Rubery PH (1995) Auxin transport. In: Davies PJ (ed) Plant hormones: physiology, biochemistry and molecular biology, 2nd edn. Kluwer, Dordrecht, pp 509–530
Magidin M, Pittman JK, Hirschi KD, Bartel B (2003) ILR2, a novel gene regulating IAA conjugate sensitivity and metal transport in *Arabidopsis thaliana*. Plant J 35:523–534
Mei H, Cheng NH, Zhao J, Park SH, Escareno RA, Pittman JK, Hirsch KD (2009) Root development under metal stress in *Arabidopsis thaliana* requires the H^+/cation antiporter CAX4. New Phytol 183:95–105
Morgan PW, Joham HE, Amin JV (1966) Effect of manganese toxicity on the indolacetic acid oxidase system of cotton. Plant Physiol 41:718–724

Morgan PW, Taylor DM, Joham HE (1976) Manipulation of IAA-oxidase activity and auxin-deficiency symptoms in intact cotton plants with manganese nutrition. Physiol Plant 37:149–156

Normanly J, Slovin JP, Cohen JD (1995) Rethinking auxin biosynthesis and metabolism. Plant Physiol 107:323–329

Ouzounidou G, Ilias I (2005) Hormone-induced protection of sunflower photosynthetic apparatus against copper stress. Plant Biol 49:223–228

Park JE, Park JY, Kim YS, Staswick PE, Jeon J, Yun J, Kim SY, Kim JM, Lee YH, Park CM (2007) GH3-mediated auxin homeostasis links growth regulation with stress adaptation response in *Arabidopsis*. J Biol Chem 282:10036–10046

Pickett FB, Wilson AK, Estelle M (1990) The *aux1* mutation of *Arabidopsis* confers both auxin and ethylene resistance. Plant Physiol 94:1462–1466

Pineros MA, Shaff JE, Kochian V (1998) Development, characterization, and application of a cadmium-selective microelectrode for the measurement of cadmium fluxes in roots of Thlaspi species and wheat. Plant Physiol 116:1393–1401

Polle A, Schützendübel A (2003) Heavy metal signalling in plants: linking cellular and organismic responses. In: Hirt H, Shinozaki K (eds) Topics in current genetics. Plant stress responses, vol 4. Springer, Berlin, pp 187–215

Popko J, Hänsch R, Mendel RR, Polle A, Teichmann T (2010) The role of abscisic acid and auxin in the response of poplar to abiotic stress. Plant Biol 12:242–258

Rampey RA, Clere SL, Kowalczyk M, Ljung K, Sandberg G, Bartel B (2004) A family of auxin-conjugate hydrolases that contribute to free indole-3-acetic acid levels during *Arabidopsis* germination. Plant Physiol 135:978–988

Sanita di Toppi L, Gabrielli R (1999) Response to cadmium in higher plants. Environ Exp Bot 41:105–130

Schopfer P (2001) Hydroxyl radical-induced cell-wall loosening in vitro and in vivo: implications for the control of elongation growth. Plant J 28:679–688

Schopfer P, Liszkay A (2006) Plasma membrane-generated reactive oxygen intermediates and their role in cell growth of plants. Biofactors 28:73–81

Schützendübel A, Polle A (2002) Plant responses to abiotic stresses: heavy metal-induced oxidative stress and protection by mycorrhization. J Exp Bot 53:1351–1365

Schützendübel A, Schwanz P, Teichmann T, Gross K, Langenfeld-Heyser R, Godbold DL, Polle A (2001) Cadmium-induced changes in antioxidative systems, hydrogen peroxide content, and differentiation in pine roots. Plant Physiol 127:887–898

Schützendübel A, Nikolova P, Rudolf C, Polle A (2002) Cadmium and H_2O_2-induced oxidative stress in *Populus* × *canescens* roots. Plant Physiol Biochem 40:577–584

Sharma SS, Dietz KJ (2009) The relationship between metal toxicity and cellular redox imbalance. Trends Plant Sci 14:43–50

Staswick PE, Serban B, Rowe M, Tiryaki I, Maldonado MT, Maldonado MC, Suza W (2005) Characterisation of an *Arabidopsis* enzyme family that conjugates amino acids to indole-3-acetic acid. Plant Cell 17:616–627

Stonier T, Rodriguez-Tormes F, Yoneda Y (1968) Studies on auxin protectors. IV. The effect of manganese on auxin protector-I of the Japanese morning glory. Plant Physiol 43:69–72

Teichmann T, Bolu-Arianto WH, Olbrich A, Langenfeld-Heyser R, Göbel C, Grzeganek P, Feussner I, Hänsch R, Polle A (2008) GH3::GUS depicts cell-specific developmental patterns and stress-induced changes in wood anatomy in the poplar stem. Tree Physiol 28:1305–1315

Vanneste S, Friml J (2009) Auxin: a trigger for change in plant development. Cell 136:1005–1016

Vernoux T, Besnard F, Traas J (2010) Auxin at the shoot apical meristem. Cold Spring Harb Perspect Biol 2:1–14

Wojcik M, Tukiendorf A (2005) Cadmium uptake, localization and detoxification in *Zea mays*. Plant Biol 49:237–245

Woodward WW, Bartel B (2005) Auxin: regulation, action and interaction. Ann Bot 95:707–735

Xu J, Yin H, Liu XJ, Li X (2010a) Salt affects plant Cd-stress responses by modulating growth and Cd accumulation. Planta 231:449–459

Xu J, Wang W, Yin H, Liu X, Sun H, Mi Q (2010b) Exogenous nitric oxide improves antioxidative capacity and reduces auxin degradation in roots of *Medicago truncatula* seedlings under cadmium stress. Plant Soil 326:321–330

Yruela I (2005) Copper in plants. Braz J Plant Physiol 17:145–156

Zhao Y (2010) Auxin biosynthesis and its role in plant development. Annu Rev Plant Biol 61:49–64

Zhou JM, Dang Z, Chen NC, Xu SG, Xie ZY (2007) Enhanced phytoextraction of heavy metal contaminated soil by chelating agents and auxin indole-3-acetic acid. J Environ Sci 28:2085–2088 (In Chinese)

Index

D.K. Gupta and L.M. Sandalio (eds.), *Metal Toxicity in Plants: Perception, Signaling and Remediation*, DOI 10.1007/978-3-642-22081-4,